Birds of a Feather

Osteological and archaeological papers from the South Pacific in honour of R. J. Scarlett

edited by

Atholl Anderson

New Zealand Archaeological Association
Monograph 11

B A R International Series 62

1979

B.A.R.

B.A.R., 122 Banbury Road, Oxford OX2 7BP, England

B.A.R. -S62, 1979: "Birds of a Feather."

ISBN 9780860540625 paperback
ISBN 9781407349367 e-book
DOI https://doi.org/10.30861/9780860540625
A catalogue record for this book is available from the British Library

This book is available at www.barpublishing.com

CONTENTS

PREFACE AND ACKNOWLEDGEMENTS

In 1975 a suggestion that friends and colleagues of Ron Scarlett might contribute to a volume in his honour was received with an enthusiasm quite unreflected in the time it has taken to bring this work to fruition. Initially only archaeologists were party to the preparations, but soon Ron's colleagues in other fields found out and were delighted to participate. We might have expected that this publication would mark Ron's retirement; that it does not is regarded with pleasure by all of us.

I would like to take this opportunity to thank Ann Harrison and Michael Trotter for assistance in various aspects of the preparations and the Council of the New Zealand Archaeological Association for its continued interest in the project as well as the inclusion of this volume in the Association's monograph series. To the authors I can only express my admiration of their patience and courtesy in waiting so long to see their contributions published.

Editor

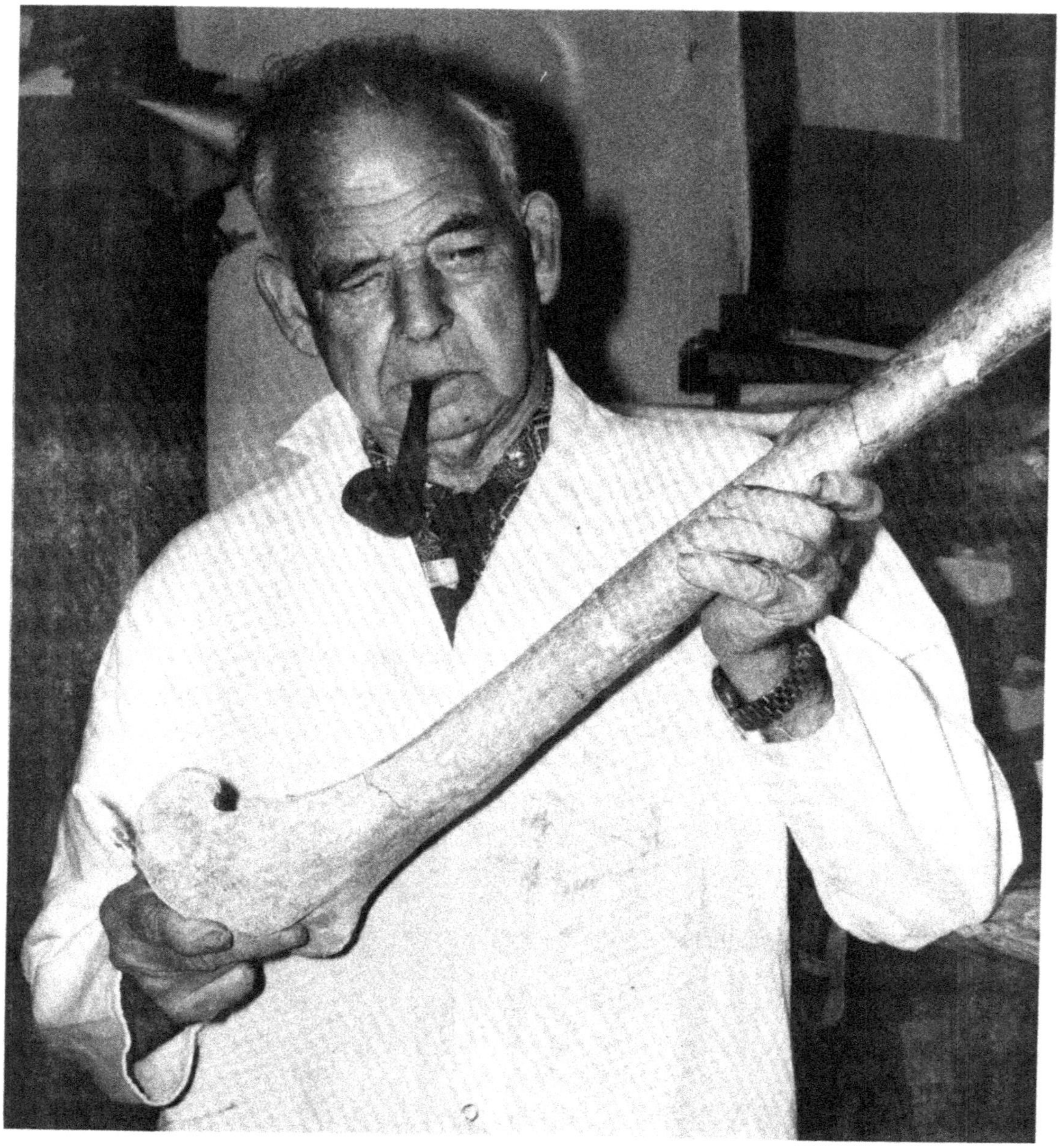

RON SCARLETT

Canterbury Museum

1979

RONALD JACK SCARLETT

Ron was born in 1911 at Stoke, a village near Nelson and, following an almost itinerant primary education, spent the uncertain years of the 1920s and 1930s at farming, labouring, gardening and mining. At the age of 26 he enrolled for a B.A. degree at Canterbury University College and then, upon graduation, studied Anthropology under H. D. Skinner at Otago. His interest in this field, coupled with a boyhood enthusiasm for natural history and an invitation to join the Pyramid Valley excavations in 1949, formed the foundation of his subsequent career. In 1950 he was appointed to the Canterbury Museum as the Recorder of Collections and later became Osteologist. Apart from short periods abroad, most recently at the Royal Ontario Museum as a Research Associate, he has made Canterbury Museum his professional home.

The study of avifaunal osteology has always been Ron Scarlett's main field, and this has proved a very fortunate circumstance indeed for archaeologists in the South Pacific; most particularly in New Zealand where birds were an unusually important component of prehistoric economies. For thirty years Ron has provided a prompt, painstaking bone identification service and, what is more, assisted at numerous excavations throughout New Zealand and Oceania. Nor does his contribution to archaeology stop there, for he was a foundation member of the New Zealand Archaeological Association and the first editor of the Association's Newsletter. Amongst his contributions in this latter capacity, his admonitions to archaeologists on the subject of retaining faunal remains are still worth recalling. More widely, his active interest in ornithological and anthropological fieldwork has demonstrated a concern not just for bones, but for the birds themselves and their relationships to people of this and other ages.

The contributors to this volume wish to express their respect for Ron's talents and devotion to his profession, their gratitude for his ever-willing assistance and their appreciation of his company in the field.

Scientific Publications

1953 A Sub-fossil Hawk from New Zealand Records of the Canterbury Museum 6(3):245-252

1955 Further Report on Bird Remains from Pyramid Valley Records of the Canterbury Museum 6(4):261-264

1955 A New Rail from South Island Swamps in New Zealand Records of the Canterbury Museum 6(4):265-266

1955 Falcon in Chatham Islands Notornis 6(3):57

1957 Former Faunal Areas: some sub-fossil evidence Proceedings of the New Zealand Ecological Society 4:17-18

1958 A Dentalium Shell Necklace Workshop *Journal of the Polynesian Society* 67(1):75

1960 Excavations at Pari Whakatau, Claverley *New Zealand Archaeological Association Newsletter* 4(1):2-7

1962 Sub-fossil Records of the Little Grey Kiwi in the North Island *Notornis* 10(2):84-85

1962 Interim List of Moa Species from North Island Archaeological Sites *New Zealand Archaeological Association Newsletter* 5(4):245-246.

1963 Comments on Excavations at South Bay *New Zealand Archaeological Association Newsletter* 6(3):159-160

1964 Australian Stragglers in New Zealand, with First Record of the Fan-Tailed Cuckoo *Notornis* 11(2):107-109 (with E. G. Turbott)

1964 Further Heaphy River Excavations *New Zealand Archaeological Association Newsletter* 7(3):128-132 (with O. R. Wilkes)

1966 A Pelican in New Zealand *Notornis* 13(4):204-217

1967 Excavation of a Moa-hunter site at the Mouth of the Heaphy River *Records of the Canterbury Museum* 8(3):181-212 (with O. R. Wilkes)

1967 Adzes and Chisels from the Heaphy River *Records of the Canterbury Museum* 8(3):213-234

1967 Birds of the Heaphy River Area *Notornis* 14(1):31-33

1968 An Owlet-nightjar from New Zealand *Notornis* 15(4):254-266

1968 A Second North Island Locality for *Pachyornis elephantopus* (Owen) *Notornis* 15(1):36

1969 The Occurrence of the Musk Duck, *Biziura lobata* (Shaw) in New Zealand *Notornis* 16(1):57-59

1969 Moas and Other Extinct Birds *In* G. A. Knox (Ed) *The Natural History of Canterbury*:565-568 Reed, Wellington

1969 On the Alleged Queensland Moa, *Dinornis queenslandiae* de vis *Memoirs of the Queensland Museum* 15(3):207-212

1970 The Genus Capellirallus *Notornis* 17(4):303-319

1970 The Auckland Island Merganser Wildfols 21:78-86 (with HJanet Kear)

1970 The Auckland Island Merganser *Wildfowl* 21:78-86 (with Janet Kear)

1970 A Small Woodhen from New Zealand *Notornis* 17(1):68-74

1972 Bone of a Presumed Odontopterygian Bird from the Miocene of New Zealand *New Zealand Journal of Geology and Geophysics* 15(2):269-274

1972 *Bones for the New Zealand Archaeologist* Canterbury Museum Bulletin 4

1974 Moa and Man in New Zealand *Notornis* 21(1):1-12

1975 Thyroid Bones of Moas *Records of the Canterbury Museum* 9(3):247-258

1976 King Penguins at Chatham Island *Notornis* 23(4):355

This volume: A Deceptive Terrestrial Vertebrate Fossil Site on the Waipara River, North Canterbury, New Zealand (with T. H. V. Rich, P. V. Rich, R. E. Fordyce and P. Gatehouse)

This volume: Avifauna and Man

LIST OF CONTRIBUTORS

J. Allo Bay-Petersen,
Aarhus, Denmark.

A. J. Anderson,
Anthropology Department, University of Otago.

R. Bulmer,
Anthropology Department, University of Auckland.

J. M. Davidson,
Auckland Institute and Museum.

E. W. Dawson,
Eastbourne, Wellington.

R. E. Fordyce,
Zoology Department, University of Canterbury.

P. Gatehouse,
Christchurch Training College.

P. Gathercole,
University Museum of Archaeology and Anthropology, Cambridge.

G. E. Hamel,
Anthropology Department, University of Otago.

G. Law,
Gowing Drive, Auckland.

B. F. Leach,
Anthropology Department, University of Otago.

H. M. Leach,
Anthropology Department, University of Otago.

P. V. Rich,
National Museum of Victoria, Australia.

T. H. V. Rich,
National Museum of Victoria, Australia.

R. J. Scarlett,
Canterbury Museum, Christchurch.

J. Specht,
The Australian Museum, Sydney.

S. Specht,
Roseville, New South Wales.

D. G. Sutton,
Anthropology Department, University of Auckland.

M. M. Trotter,
Canterbury Museum, Christchurch.

G. F. van Tets,
Division of Wildlife Research, CSIRO.

J. C. Yaldwyn,
National Museum, Wellington.

THE TYPES OF W. R. B. OLIVER'S MOAS AND NOTES ON OLIVER'S METHODS OF MEASURING MOA BONES

J. C. Yaldwyn

Introduction

W. R. B. Oliver (b. 1883, d. 1957) was the last of a long line of New Zealanders publishing on the systematics of moas to describe new genera, subgenera and species of these extinct birds. His New Zealand predecessors and contemporaries in this field, Julius von Haast, James Hector, F. W. Hutton, Augustus Hamilton, T. J. Parker, H. O. Forbes and Gilbert Archey, publishing between 1863 and 1941, built up and refined a systematic structure for the moas from Richard Owen's basic scheme of two genera and 19 species (as summarized in Owen, 1879, 1883), to the six genera and 20 species (several different from Owen's 19 spp.), of Archey's Auckland Museum monograph of 1941. Owen's fellow countrymen R. Lydekker and W. Rothschild, described (in 1891 and 1907 respectively) an additional new genus and 12 new species of moas from material in the British and other United Kingdom Museums, but none of those named "species" are recognized unequivocally today (see Scarlett, 1972; Cracraft, 1976).

Oliver in his 1949 Dominion Museum bulletin on moas "in the main adopted Archey's scheme of classification" (*ibid.*, 56) but added a new genus, two new subgenera and six new species to the existing systematic structure. He had already described a new species of *Euryapteryx* in his 1930 book, *New Zealand Birds*. His new moa taxa can be listed as follows:

Zelornis Oliver, 1949 (type species *Euryapteryx exilis* Hutton) - a genus close to *Euryapteryx* Haast, 1874

Mauiornis Oliver, 1949 (type species *Pachyornis septentrionalis* Oliver) - a subgenus of *Pachyornis* Lydekker, 1891

Pounamua Oliver, 1949 (type species *Pachyornis murihiku* Oliver) - a subgenus of *Pachyornis* Lydekker, 1891

Pachyornis (Mauiornis) septentrionalis Oliver, 1949

Pachyornis (Pounamua) murihiku Oliver, 1949

Pachyornis (Pachyornis) australis Oliver, 1949

Euryapteryx tane Oliver, 1949

Euryapteryx kuranui Oliver, 1930 - regarded by Archey (1941) and Oliver (1949) as a synonym of *E. gravis* (Owen, 1870)

Dinornis gazella Oliver, 1949

Dinornis hercules Oliver, 1949

In the second edition of his New Zealand Birds in 1955, Oliver made no change to the moa systematic structure (now 7 genera and 28 species) but did not use the Pachyornis subgenera. R. J. Scarlett in his 1972 Canterbury Museum bulletin Bones for the New Zealand Archaeologist was the next New Zealander to review moa systematics. He accepted, with some doubts, all of Oliver's 1949 species except E. tane (which he considered a synonym of E. geranoides Owen), but did not use Zelornis (he synonymized it with Euryapteryx), or the Pachyornis subgenera. Scarlett also, it should be noted, listed (ibid., 21) but did not name a new species of Dinornis ("known ...from one unpublished tibio-tarsus") and a new species of Euryapteryx (" a variable small form...from Stewart Island and the Murihiku area of the South Island"). In the latest review of moa systematics known to me, Cracraft (1976) did not recognize any of Oliver's taxa (specific or generic), and reduced the systematic structure to six genera and 13 species - the lowest number of species recognized since before 1879.

As the status of some (Scarlett, 1972) or all (Cracraft, 1976) of Oliver's new species is now questioned it is important to understand clearly just what type material his new names are based on. Oliver clearly states in his 1949 monograph what the "type" of each of his new species is, but in some cases these types can not be recognized from his text without additional information. In some species, photographic figures published in 1949, which a reader might justifiably assume to be of type material, are not of type material at all and in fact are not recognizable as to locality, collection or registered skeleton from Oliver's text alone. It is important then to go through each of Oliver's new moa species and firmly identify the actual type material and then to identify the actual figured material of each of the new forms. This task is attempted in the first part of this paper. It should be pointed out here that Recommendation 73C of the International Code of Zoological Nomenclature (ICZN, 1964) states that the "full locality, date, and other data on the labels accompanying the holotype" as well as "the collection in which it is situated and any collection - or register - number assigned to it" should be published where possible.

Anyone using Oliver's works on moas is aware of the great importance he gave to the measurement of moa bones, especially leg bones. In 1930 he provided (ibid., 35) an "artificial key to the genera of moas, based on the proportions of the leg bones... To use the key, it is necessary first to take the following measurements of each bone...". By 1949, Oliver, had abandoned (ibid., 56) the "artificial key for the identification of leg bones", but provided "tables of measurements...under each genus, in which any bone can be identified by inspection". He did not explain anywhere in his 1949 monograph just what the four leg bone measurements L, P, M and D used in these tables (e.g. ibid., 86) represent, though it is clear from his earlier work (1930: 35) and from Archey (1941:13) that they must be "length", "proximal width", "middle width" and "distal width" respecitvely. In the 1955 edition of New Zealand Birds, Oliver did not use proportions or width measurements for moa leg bones but gave maximum and minimum lengths for the three main leg bones of each species (ibid., 574 et seq.). In 1930 his dimensions were in centimetres (ibid., viii) for "birds" (including his moa bone measurements, see ibid., 37), but in millimetres for eggs! In 1949 he used millimetres in his moa bone and egg measurement tables though this is not clearly stated anywhere but merely inferred from text reference (e.g. ibid., 46) and from footnotes to tables (e.g. ibid., 85). By 1955 Oliver was using millimetres for "birds and their eggs" (ibid., 15) including moa eggs, but had reverted to centimetres for moa bone measurements (ibid., 574). Both Archey (1941: 13) and Scarlett (1972) use centimetres consistently throughout their main works on moas and bird bones.

As the LPMD system of moa leg bone measurements is used throughout Oliver's 1949 monograph and was applied by him to the large unpublished collections of moa remains then available in the "Dominion" (now National) Museum, it is important that the methods Oliver used to obtain these measurement should be recorded. The second part of this paper gives an outline of Oliver's methods based on the information given in his 1930 book and on some details provided by him in a letter to W. B. Benham in 1932. Benham (1934), Archey (1941) and Scarlett (1972) each provide additional details useful in understanding Oliver's methods.

"Dominion Museum" register numbers for moa bones were prefixed by the letters "DM". With the change of name to National Museum of New Zealand the prefix has been changed to "NMNZ S"., with "S." referring to the Subfossil Bird Register.

THE TYPE SPECIMENS OF OLIVER'S MOA SPECIES

Family EMEIDAE

(Emeidae Bonaparte, 1854, predates Anomalopterygidae Oliver, 1930 - see Brodkorb, 1963:208)

Genus Pachyornis Lydekker

Pachyornis septentrionalis Oliver, 1949 (Figs. 1.1, 1.2)

Pachyornis (Mauiornis) septentrionalis Oliver, 1949:61, figs. 29 (left), 30-34, 35 (left), 36, 37 (left bone of each pair).

Pachyornis septentrionalis Oliver, 1955:574, unnumbered fig. (repeat of fig. 34 of 1949).

Type material

According to Oliver (1949) the holotype is a "portion of skeleton of individual bird" from a limestone cave at Pohue, Hawke's Bay, collected by H. Hill, in the "Dominion" Museum with registration number DM 129 (now to be referred to as NMNZ S. 129). "Pohue" is presumably Te Pohue on the Napier-Taupo road, at a height of about 1465 ft about 29 miles north-west of Napier. H. Hill (1914:345) in a paper on moa remains from the Gisborne-Hawke's Bay area refers to "many specimens of the moa" as having been found "at Pohui, on the Napier-Taupo Road...among the limestone rock-masses and caves that occur along the east valley and deep gorges at the foot of the Te Waka Range". The spelling "Pohue" is used on his map (ibid., 331) and table (ibid., 350) where "several skeletons" are listed as having been collected at Pohue by Mr. Crawford, Mr. King and Hill, in "1890 and onward". A note by Augustus Hamilton among the Hamilton papers on moas in the National Museum records this incomplete holotype skeleton as "some bones of a small Pachyornis found with several incomplete skeletons of Anomalopteryx when cutting a road in the Pohui Bush, 1909. The bones were in a cave exposed during the cutting of a road round a spur to get timber out. When Mr. Hill went to see the place the cave had been filled in.

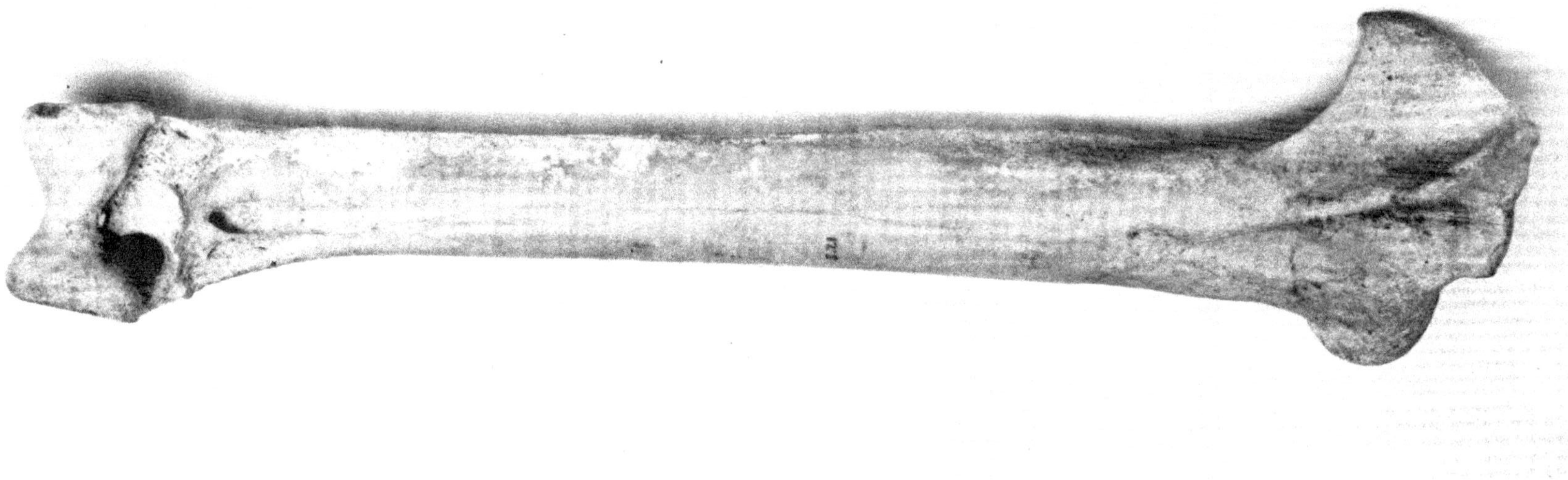

Fig. 1.1 Pachyornis septentrionalis Oliver. Right tibia of holotype from Te Pohue, Hawke's Bay, NMNZ S. 129.

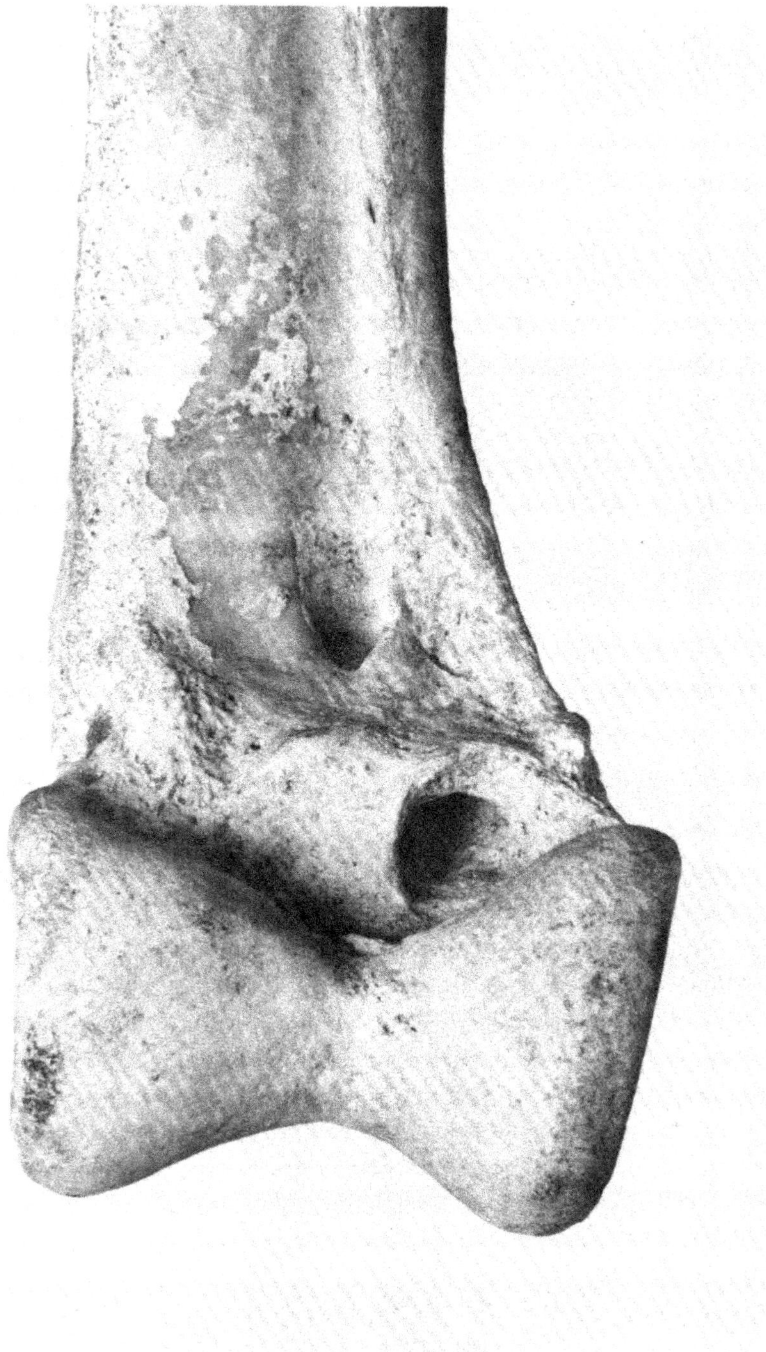

Fig. 1.2 Pachyornis septentrionalis. Distal part of right tibia of holotype.

The holotype has been catalogued as "9 bones", though the tag labels attached have the words "7 bones" inexplicably written on them, and these are as follows: sternum (with left lateral process missing but xiphoid process complete), pelvis (damaged posteriorly, with both pubes and left ischium missing, but with right ischium present and detached), right and left femora (left with trochanter crest alongside head sheared off in a recent break), right and left tibiae (left mostly covered with a thin coating of lime), a fibula, right and left metatarsi (i.e. 9 bones in 10 pieces). The bones are somewhat immature but, except for parts of the pelvis, in a very solid condition. Their colour is offwhite to pale yellow brown. The sternum, both femora and the right tibia have an old and faded label written in script on the surface of each reading "Hill Pohue HB". The principal leg bones have a number written on the shaft (see fig. 1.1) which appears to be the total length in inches and decimals of an inch. All 10 pieces have the following modern label printed on their surface in indian ink "DM 129 Pachyornis septentrionalis Oliver HOLOTYPE". Most have in addition a brown "Dominion Museum" tag label attached with the words "No. 129 Pachy. septentrionalis TYPE H. Hill 7 bones" in Oliver's hand on its printed front face and, in the case of the leg bones other than the fibula, measurements on the back face. The right tibia is figured here entire (fig. 1.1) and with the distal portion shown in greater detail (fig. 1.2; cf. Archey, 1941: fig. 9a). The measurements of this bone as given by Oliver on the label in the system explained in the second part of this paper, are as follows: total length 320 mm (cf. "12.5" presumably in inches, written on shaft", proximal width 85 mm, middle width 28 mm, distal width 48 mm. There are some irregular and thin lime patches on the shaft of this bone as shown in fig. 1.2.

As Oliver included only the Pohue skeleton under his heading of "type" material, none of the other specimens he referred to P. septentrionalis on pp. 61 and 63 can be considered as paratypes.

Original figures

The figures of this species published by Oliver in 1949 are as follows: fig. 29 (left) lower mandible dorsal view photograph; fig. 30 skull dorsal view photograph; fig. 31 skull lateral view photograph; fig. 32 palate drawing; fig. 33 preorbital plate drawing; fig. 34 sternum ventral view photograph; fig. 35 (left) pelvis dorsal view photograph; fig. 36 pelvis ventral view photograph; fig. 37 (left bone of each pair) right femur, left tibia, right metatarsus, anterior views photographs.

The origins of these figures are as follows:

Figs. 30-34 of skull and sternum are not attributed to any specimens in their captions and associated text, but the original photographs and drawings of these figures are labelled AM. 84 (a skeleton from Waikaremoana in the Auckland Museum) in Oliver's manuscript notes on moas in the National Museum. Details of this skeleton are given on page 61 of Oliver (1949), and fig. 34 of the sternum is captioned as from "Waikaremoana Auckland Museum" when republished in Oliver (1955) unnumbered on page 574.

Fig. 29 of the lower mandible is probably from the same Auckland Museum specimen but is not attributed in the caption and is not identified in Oliver's manuscript notes.

Fig. 35 of a pelvis labelled with the letter "R" is not attributed in the caption but is labelled Wanganui Museum on the original photograph. Oliver's manuscript notes imply that the letter "R" may stand for the locality "Riverlands".

Fig. 36 is definiately the holotype pelvis and is labelled as such in the caption.

In fig. 37 the femur and tibia can be recognized as holotype bones from characteristic marks on the bones themselves, and the metatarsus is probably holotypic as well but this can not be demonstrated with certainty.

Present status

Oliver distinguished P. septentrionalis from P. mappini Archey by its smaller size (e.g. femur length 161-192 mm compared with 181-224 mm) and by the "leg bones usually being proportionately more slender" (1949:61, 59). Scarlett (1972:21) lists P. septentrionalis as a separate species but suggests that it may represent merely the smaller members of the earlier described Pachyornis mappini. Cracraft (1976:197) regards P. septentrionalis as a synonym of P. mappini. He shows statistically that the leg bones of "P. septentrionalis" are smaller, but are not relatively more slender, than those of P. mappini. Cracraft considers that the two skeletal groupings covered by these two names "probably represent different sexes" and considers the separation of P. septentrionalis from P. mappini Archey as "unwarranted ... until firm evidence can be offered that they are distinct" (ibid., 198).

Pachyornis murihiku Oliver, 1949

Pachyornis sorenseni Oliver, 1944, Dominion Newspaper, Wellington, 17 May 1944 (invalid publication of nomen nudum).

Pachyornis (Pounamua) murihiku Oliver, 1949:67, fig. 41-46.

Pachyornis murihiku Oliver, 1955:574.

Type material

The holotype is described by Oliver as a "full-grown but not quite mature ... skeleton from Greenhills, near Bluff Hill, Southland" (1949:67) found by "Mr. Black in 1939 among sand-dunes" (1949:14) and now in the Southland Museum, Invercargill. The skeleton could not be found in the Southland Museum in the recent past (fide Scarlett, 1972:21, repeated in Cracraft, 1976: 196) but is now confirmed (August 1977) as present in the collections of the Museum with the registered number E 73210. Greenhills is a locality on the Invercargill-Bluff road and railway at the head of Bluff harbour, 12 miles from Invercargill, 5 miles from Bluff, with an altitude of 20 ft above sea level. No other bones were referred to this species by Oliver in 1949 or 1955.

The name Pachyornis sorenseni used for this skeleton by Oliver in some of his manuscript notes, and on some photographs, in the National Museum is deliberately published here so that its status can be discussed. It was used in a newspaper article on the Greenhills skeleton published in the Dominion on 17 May 1944, so it might be regarded as having priority over the name

P. murihiku used for the same skeleton but published first in 1949. Under Article 8 (2) of the International Code of Zoological Nomenclature (ICZN, 1964), however, it is clear that the 1944 use of the name P. sorenseni was not a valid publication as the newspaper was clearly not "issued for the purpose of scientific, public, permanent record". Regardless of whether the 1944 publication is valid or not, the name was a nomen nudum (under Article 13 (a) of the Code) when it appeared in 1944 as it was not accompanied by an indication of characters differentiating it from other species and as such was not available for systematic use.

Details of original discovery

No details of the original finding of the Greenhills skeleton have been published. Oliver was so concerned about the combination of characters in this skeleton that he wrote to the Director of the Southland Museum (Mr. J. H. Sorensen) on 28 June 1940 (National Museum file 9/1/8) to say that it "contains features quite different from those of any other described species. It agrees mostly with Emeus; but the sternum is quite different from those of any other species of Moas. The pelvis, too, is different, having a narrow escutcheon sloping outwards on either side. The upper mandible, too, is very steep; but I do not regard this as an important character... Is there any doubt about the sternum belonging to the rest of the skeleton? As far as I can judge from other species of Moas, the broad sterna are associated with broad bills". Sorensen replied on 2 July 1940 saying "I have interviewed the finder [of the Greenhills skeleton] and he is quite definite in his statement that the bones are from one bird only. He came across the pelvis showing out of the sand after a high wind, carefully removed the larger bones, noted that the vertebrae were coiled back over the rest of the skeleton, and wishing to save all the smaller bones, constructed a frame covered with fine netting and so sieved the balance out. As already mentioned, the bones of a seal were lying very close to the moa, and this accounts for the presence of a few foreign bones with the rest. I have no reason to doubt any part of the story at all and now know the finder fairly well... From my knowledge of the finder, locality, and the story of the finding I am of the opinion that the sternum belongs to the bird in question".

Original figures

The figures of this species published by Oliver in 1949 are as follows: figs. 41-42 skull dorsal and lateral views; fig. 43 sternum ventral view: figs 44-46 pelvis dorsal, lateral and ventral views. All are photographs and all are attributed in their captions to the holotype skeleton from Greenhills.

Present status

Oliver considered that "the skeleton possesses such distinctive features in the skull, sternum and pelvis, that I have no hesitation in giving it both a subgeneric and specific name" (1949:67). In size range of leg bones P. murihiku fitted between the smaller North Island P. mappini Archey and the larger South Island P. elephantopus (Owen), being distinctly closer to the former species (1949: 59). Scarlett lists P. murihiku without systematic comment (1972:21), while Cracraft (1976:196) places it firmly as a synonym of P. elephantopus regarding it as only "slightly smaller" than the known range of that species. As the species was based on a single "immature individual", Cracraft doubted whether the described morphological differences between the two named forms could be accepted as valid without additional specimens.

Pachyornis australis Oliver, 1949

Mesopteryx species b Parker, 1893:4, pl. 1.

Mesopteryx species B Parker, 1895:378, pl. 60 figs. 20, 21.

Emeus (not specifically identified) Oliver, 1930: two unnumbered figs. on p. 47.

Pachyornis pygmaeus (Hutton). Archey, 1941:41.

Pachyornis (Pachyornis) australis Oliver, 1949:70, figs. 47-52, ?fig. 53.

Pachyornis australis Oliver, 1955:575.

Type material

The holotype is a "skull and 20 vertebrae found by A. McKay in 1879 in a cave on the Salisbury Tableland, head of Takaka River, Nelson" in the National Museum, registered number DM 26, now to be known as NMNZ S. 26 (Oliver, 1949:70). The "Salisbury Tableland" is commonly referred to as the "Mt. Arthur Tableland", though neither name appears on published Lands and Survey maps of the area. The locality is a limestone mesa at the eastern end of the Peel Range, north-west of Mt. Arthur on the Arthur Range, lying between Mt. Arthur and the Cobb Reservoir. Tableland is a trig station at 4086 ft and the Salisbury Hut is on one of the north-flowing tributaries of the upper Takaka River. McKay (1879:131) describes this specimen as "part of a skeleton, comprising the head, neck and fore part of the trunk, in a fine state of preservation" but does not give any further details of its discovery.

The holotype skull and lower mandible are labelled "HOLOTYPE DM. 26" in several places but no vertebrae are now associated with this specimen. There are numerous unlabelled and unlocalized moa vertebrae in the National Museum collections but none are obviously the missing set from the holotype of P. australis. The skull is in good condition but has a distinctively-shaped dark scar (visible in all photographs) repaired with some waxy material in the centre of the dorsal surface of the cranium. The general colour of the bone varies between offwhite and yellowish-brown.

Two crania in the Southland Museum (one from Greenhills), are referred to P. australis by Oliver and some of the smaller leg bones listed under P. elephantopus in the tables of measurements are regarded as "possibly" belonging to the new species (Oliver, 1949:74). A sternum from the Southland Museum and a pelvis from the Nelson Museum are listed as P. australis with a question mark in Table 2 (ibid., 85). As Oliver regarded only the Salisbury Tableland skull and vertebrae as "type" material none of the other specimens can be regarded as paratypic.

Published figures

The holotype skull, described by Oliver (1949:72) as "one of the only two perfect skulls of moas that I have ever seen", was figured in three different publications before Oliver based his species *P. australis* on it. Parker (1893:pl. 1) gave a dorsal view photograph to show the pattern of pits on the cranium which he interpreted as indicating the original presence of a frontal feather crest, a possible male-only feature. He later (1895:pl. 60 figs. 20, 21) published very fine, natural size, dorsal and ventral view engravings (the best figures available of this specimen) in a general work on moa cranial morphology. Parker considered it "one of the most perfect" skulls "ever discovered" (1895:378) and thought that it must "belong to a species the skull of which has not hitherto been described" as it did not correspond with any of the figures or descriptions he had met with. He referred to it as *Mesopteryx* species b in 1893 and species *B* in 1895. Oliver (1930:47) used a dorsal and a lateral view photograph of this same specimen to illustrate the skull form of the genus *Emeus* but he did not specifically identify the skull in caption or text.

The figures published by Oliver in 1949 are as follows: figs. 47-49 photographs of skull dorsal, lateral and ventral views; figs. 50-52 drawings of palate, preorbital plate and junction of antorbital and lacrymal (lateral and front views of junction on one fig.); fig. 53 a photograph of the dorsal view of a pelvis captioned as "*Pachyornis australis* (?)". Figs. 47-49 are captioned "Salisbury Tableland" and the original photographs of figs. 47-49 and the original drawings of figs. 50-52 in Oliver's manuscript notes are all labelled as from the holotype skull. In the original photograph of fig. 53 the pelvis is labelled "Nelson Museum" in Oliver's manuscript notes and is clearly not type material. It is presumably the pelvis whose measurements are listed as P. australis with a question mark in Table 2 on page 85 of Oliver (1949).

Present status

Scarlett (1972:21) regarded *P. australis* as a possible variant of the South Island *P. elephantopus* (Owen) the "most variable of all Moas". It is firmly synonymized with *P. elephantopus* "one of the more common species of moas" by Cracraft (1976:196). Cracraft compared the holotype skull of *P. australis* with "a large series of skulls of *P. elephantopus* and can find no significant differences that can be regarded as being of specific value". He could match the distinctive characters attributed by Oliver to the skull of "*P australis*" within a series of skulls of *P. elephantopus* in the National Museum collections and several skulls in the collections referred by Oliver to *P. elephantopus* were of comparable size (i.e. at the lower end of the size range of *P. elephantopus*). Cracraft found that crania of this common South Island moa species exhibited considerable variability in shape and in the development of processes and muscle scars.

Genus *Euryapteryx* Haast

Euryapteryx tane Oliver, 1949

Euryapteryx exilis Hutton. Archey, 1941:57 (part).

Euryapteryx tane Oliver, 1949:105, figs. 73-78.

Euryapteryx tane Oliver, 1955:577.

Type material

The holotype is a skeleton from Doubtless Bay in the Auckland Museum, registered number AM. 3. Doubtless Bay is in the north of the North Auckland Peninsula, north-east of Kaitaia. The bone deposits in the consolidated sands of the sandhills are described in some detail by Archey (1941:93) and to a lesser extent by Oliver (1949:16).

The holotype skeleton includes at least the following parts: skull (but no lower mandible), sternum, pelvis; femur, tibia and metatarsus (both right and left). Descriptions and measurements of these elements are given in Archey (1941:49, table H) and Oliver (1949:105, tables 8-10). The bones are labelled "3" as can be seen in Oliver's photographic figures.

As Oliver included only the one skeleton under his heading of "type" material none of the other specimens he referred to E. tane on p. 105 can be considered as paratypes, especially as he thought some could possibly belong to the species he considered to be "Zelornis exilis".

Original figures

The figures of this species published by Oliver in 1949 are as follows: figs. 73-75 skull dorsal, lateral and ventral view photographs; fig. 76 pelvis ventral view photograph; fig. 77 sternum dorsal view photograph; fig. 78 preorbital plate drawing. All the photographs are attributed to the Doubtless Bay holotype in their captions and the original of the drawing in Oliver's manuscript notes in the National Museum is labelled as being made from the holotype specimen.

Present status

Oliver considered that "the characters of the type specimen " of E. tane are just those of E. curtus, the only difference being in size, E. tane being larger, though the size difference is less marked in the skull than in the remainder of the skeleton" (1949:105). He placed four species in the genus Euryapteryx. In order of increasing size these were E. curtus (Owen) and E. tane (both known from the North Island only), E. geranoides (Owen) and E. gravis (Owen) both known from both North and South Islands. In his closely similar genus Zelornis, Oliver placed two species - the smaller North Island Z. exilis (Hutton) and the larger South Island Z. haasti (Rothschild). He saw "a kind of parallel evolution in the two genera" (ibid., 103) with the only constant differences between them being in the form of the premaxilla and mandible of the skull (ibid., 101), each genus consisting of a series of species of different sizes forming an "evolutionary line" (ibid., 118) with the largest species of each genus (E. gravis and Z. haasti) showing close resemblances from "what appears to be covergence" (ibid., 103). He was "quite unable to give a definition either by form or measurements to separate the leg bones of E. gravis from those of Z. haasti" (ibid., 110).

Scarlett (1972:21) regarded Zelornis as a synonym of Euryapteryx and listed five of the six species recognized by Oliver - E. curtus, E. exilis, E. geranoides (including E. tane as a synonym), E. gravis and E. haasti (which might be a broad-billed subspecies of E. gravis). He thought the first three "species" would be united eventually as the gaps in the size ranges between them were closed, but added a sixth "variable small form" of Euryapteryx as an unnamed new species from Stewart Island and the "Murihiku area" of the South Island.

Cracraft (1976:198-199) reduced this systematic grouping of six so-called "species" to two sexually dimorphic species — Euryapteryx curtus (with Z. exilis and E. tane as synonyms) from the North Island and E. geranoides (with E. gravis and Z. haasti as synonyms) present on both North and South Islands. He could not distinguish cranial material referred by Oliver to Zelornis from Euryapteryx at the generic level, and as E. tane only differed from Z. elilis in cranial features the two could be considered conspecific. Multivariate analysis of leg bone measurements substantiated the separation of an E. exilis (including E. tane) population from an E. curtus population but the statistics of the combined sample give coefficients of variation for bone lengths which indicate to Cracraft that (as in the case of Pachyornis mappini/septentrionalis and E. geranoides/gravis) the difference is one of sexual size dimorphism.

Euryapteryx kuranui Oliver, 1930

Emeus crassus (Owen). Hutton, 1906:66.

Euryapteryx kuranui Oliver, 1930:52.

Euryapteryx gravis (Owen). Archey, 1941:54, 56.

Euryapteryx gravis (Owen). Oliver, 1949:108, fig. 85.

Euryapteryx gravis (Owen). Oliver, 1955:578.

Type material

The holotype is "a complete skeleton discovered at Castle Point by Mr. Merideth-Kaye in 1905, and now in the Canterbury Museum" (Oliver, 1930: 52). The finding of the skeleton and a description of the main elements present is given in a paper by Hutton (1906:66). He describes it as "an imperfect skeleton of a moa...received from Mr. C. K. Meredith-Kaye...found in sand by Mr. Meredith-Kaye's son, on his run, about eighteen miles south of Castle Point, on the east coast of Wellington Province. With the exception of the legs and feet, the bones are brittle and much broken...it was a full-grown bird, but the twenty-eighth vertebra was not anchylosed to the pelvis ...The sternum is much broken...The pelvis is very imcomplete...The third and ungual phalanges of the right outer toe are diseased and there is an osseous growth at the distal end of the right femur".

R. J. Scarlett informs me (in litt., 4 August 1977) that the skeleton is registered in the Canterbury Museum as AV 9285 and now consists of the following bones: cranium, premaxilla, 1 quadrate, 19 vertebrae (nos. 10-28, fide Hutton, 1906), 7 caudal vertebrae, pygostyle, 76 trachial rings, sternum,

18 (plus 4 broken) ribs, pelvis; right and left femora, tibiae, fibulae and metatarsi; front toes complete and a right hallux. A few gizzard stones and 3 pieces of egg shell were found with the skeleton and are in the Canterbury Museum. Hutton regarded the presence of the latter as "showing that the bird was a female, for the place in which the skeleton was found precludes us from supposing that it might have been a male sitting on an egg" (1906:66). Scarlett and the present writer regard the sex of the skeleton, in the absence of additional evidence, as in doubt.

Castlepoint (now written as one word) is on the east coast of the Wairarapa, 45 miles north-east by road from Masterton. Moa bones and egg shell have often been found in coastal sand dunes along this part of the east coast of the North Island (e.g. Brodie, 1950).

As Oliver states in 1930 that E. kuranui is "founded" on the Castlepoint skeleton and refers no other individually identifiable bones to this species, there is no paratypic material. The leg bone measurements given under the heading 'B' are described as Hutton's "average measurements" of leg bones from Te Aute (ibid., 54) rather than measurements of individual bones from that swamp site. In 1949 Oliver refers to the Castlepoint skeleton as the "type skeleton" (ibid., 108) and calls the skull from this skeleton the "type of E. kuranui" in the caption to fig. 85. He gives measurements of the skull, sternum, pelvis and leg bones of the holotype on pages 113, 114 and 116.

Published figure

The only published figure of any part of the holotype skeleton of E. kuranui is Oliver's fig. 85 of 1949 which is a photograph of the skull in lateral view.

Present status

Both Archey (1941) and Oliver (1949, 1955) listed E. kuranui as a synonym of the largest species of Euryapteryx they recognized, E. gravis (Owen). Archey placed all South Island specimens of this genus in E. gravis but considered that "an occasional example has been obtained in the south-eastern portion of the North Island" (ibid., 54), while Oliver recorded E. gravis from the southern North Island and from "throughout" the South Island (1949:112). Both authors were conscious of the relatively small size of the Castlepoint skeleton within the range they accepted for E. gravis. Archey thought that "E. kuranui Oliver might perhaps have been regarded as a more slender North Island form" of E. gravis "but it is matched in slenderness by the Stewart Island specimen" of his table H (1941:56). Oliver (1949:110) saw E. gravis as consisting of specimens showing "two distinct sizes of skulls". To the smaller size range "belong the types of E. gravis, E. boothi and E. kuranui", to the larger belong Pyramid Valley Euryapteryx material such as skeleton XXD (now in the National Museum) studied and measured by Oliver.

Neither Scarlett (1972) nor Cracraft (1976) mention Oliver's "E. kuranui". As explained in the "Present status" section of Euryapteryx tane, Cracraft recognizes only two species in the genus Euryapteryx, the larger being E. geranoides (Owen), which includes both E. gravis and Zelornis haasti as synonyms.

Family DINORNITHIDAE

Genus Dinornis Owen

Dinornis gazella Oliver, 1949

Dinornis gazella Oliver, 1949:166, fig. 136 (left).

Dinornis gazella Oliver, 1955:585.

Type material

The holotype is a "pelvis from Te Aute" in the National Museum, registered number DM 107 now to be known as NMNZ S. 107, "much smaller than that of novaezealandiae" (Oliver, 1949:166, 164; note, here and elsewhere in this paper the specific name novaezealandiae has been corrected under the International Code of Zoological Nomenclature to read as one word). The locality "Te Aute" refers to the swamp deposit in Hawke's Bay, investigated by Hamilton (1889) and described by Oliver (1949:8). This deposit was somewhere to the east of Te Aute College on the Woodville-Napier road about 18 miles south of Hastings. The exact position of the deposit is unknown to the writer, but Park (1888, a reference not included in the standard moa bibliographies) and Buick (1931) locate the site in a drain cut through the Patangata swamp near the west bank of the Tukituki (or Waipawa) River.

The holotype pelvis is shown on the left of Oliver's fig. 136. The bone is strong and firm, dark brown in colour with traces of grey mud in all broken surfaces. In the hand the bone appears "heavy and semi-mineralised" as Oliver describes bones from the Te Aute swamp in his manuscript notes. Both pubes are broken off, the median iliac crest is broken away and the posterior part of the escutcheon is irregularly broken on both sides. On the right dorsal surface of the escutcheon (visible in Oliver's fig. 136) there is an old damaged paper label attached reading "Dinornis sp. Te Aute, Napier" in faded script. On the ventral surface there is a modern label "Dinornis gazella HOLOTYPE DM. 107 TE AUTE" printed on the bone in indian ink. Finally there is a brown "Dominion Museum" tag label attached with the words "No. 107 Dinornis gazella Oliver HOLOTYPE Te Aute. A. Hamilton 1888".

As Oliver included only the Te Aute pelvis under his heading of "type" material none of the other specimens he "associated" with D. gazella on pages 166-167 can be considered as paratypes.

Original figure

The only published figure of the holotype pelvis is in Oliver's 1949 fig. 136. Although not referred to as the type, the left hand pelvis in this photograph is captioned "Dinornis gazella, Te Aute" and is clearly a dorsal view of S. 107 from the obvious and distinctive damage and the old label on the right side of the escutcheon.

Present status

Oliver (1949:164) recognized eight species in the genus Dinornis with the North Island D. gazella distinctly smaller than all the others. Its pelvis was described by Oliver as "a miniature of the typical pelvis of the larger species of Dinornis" and the leg bones referred to it (especially femur S. 108 from Te Aute) as being "in form...exactly like that of Dinornis novaezealandiae but much smaller" (ibid., 166). In spite of this complete reliance on size range for species limits, Oliver firmly believed that "not until good series of individual skeletons, including the skulls, are available, will we be able satisfactorily to define the species of Dinornis and their distribution" (ibid., 164). We still move towards this "satisfactory" stage.

Scarlett (1972:21) listed Oliver's eight species of Dinornis and added an unnamed new species from the South Island "known as yet from only one unpublished tibio-tarsus". In contrast Cracraft (1976:201) reduced this complex of species to four, one restricted to each island and two found on both islands. These were (in order of increasing size), the North Island D. struthoides Owen (with D. gazella as a synonym) the South Island D. torosus Hutton, D. novaezealandiae Owen (with D. ingens Owen, D. robustus Owen and D. hercules Oliver as synonyms), from both North and South Islands, and D. giganteus Owen (with D. maximus as a synonym) also from both islands. Cracraft (1976:202) found that the size difference between the few bones Oliver placed in D. gazella and material of D. struthoides (referred to by Oliver as "D. novaezealandiae") were "slight...and the variation exhibited by the combined sample of struthoides and gazella is easily within the range of a single spcies". The small series of D. gazella leg bones from Te Aute were "nearly equal in length" to a sample of D. struthoides but had "thinner shafts". This Cracraft believed to be "entirely related to age" as the D. gazella bones were of immature individuals (ibid., 202).

Dinornis hercules Oliver, 1949 (fig. 1.3)

Dinornis hercules Oliver, 1949:174, fig. 140 (2nd from left), 141 (2nd and 3rd from left), 142 (2nd from left).

Dinornis hercules Oliver, 1955:588, unnumbered fig. (repeat of part of fig. 141 of 1949) tibia on right.

Type material

The holotype is described by Oliver (1949:174) as being "a tibia from Coonoor (DM. 217) in the Dominion Museum. Length 747, breadth proximal 179, mid 59, distal 110 mm". In fig. 141 the third tibia from the left is described in the caption as "D. hercules, Coonoor" but not labelled as "type". It is a right tibia with writing showing across the shaft which can not be read in the published photographic figure. Prints of this photograph in Oliver's manuscript notes in the National Museum, and the figure of this tibia in Oliver's 1955 volume (1955:588), show that the label reads "COONOOR CAVE NI NZ 1914 DM. 217". This right tibia is thus the holotype. When examined in the Museum's collection by the writer in 1959 the holotype was found broken completely in two about one third of the way down the shaft and somewhat more worn at both ends than shown in the original photograph. The break was carefully repaired with plaster and as the whole bone was rather chalky it was given a coating of synthetic resin. In the process of cleaning and repair the original written label was removed and a new label written along the shaft in indian ink reading "HOLOTYPE Dinornis hercules Oliver 1949 COONOOR

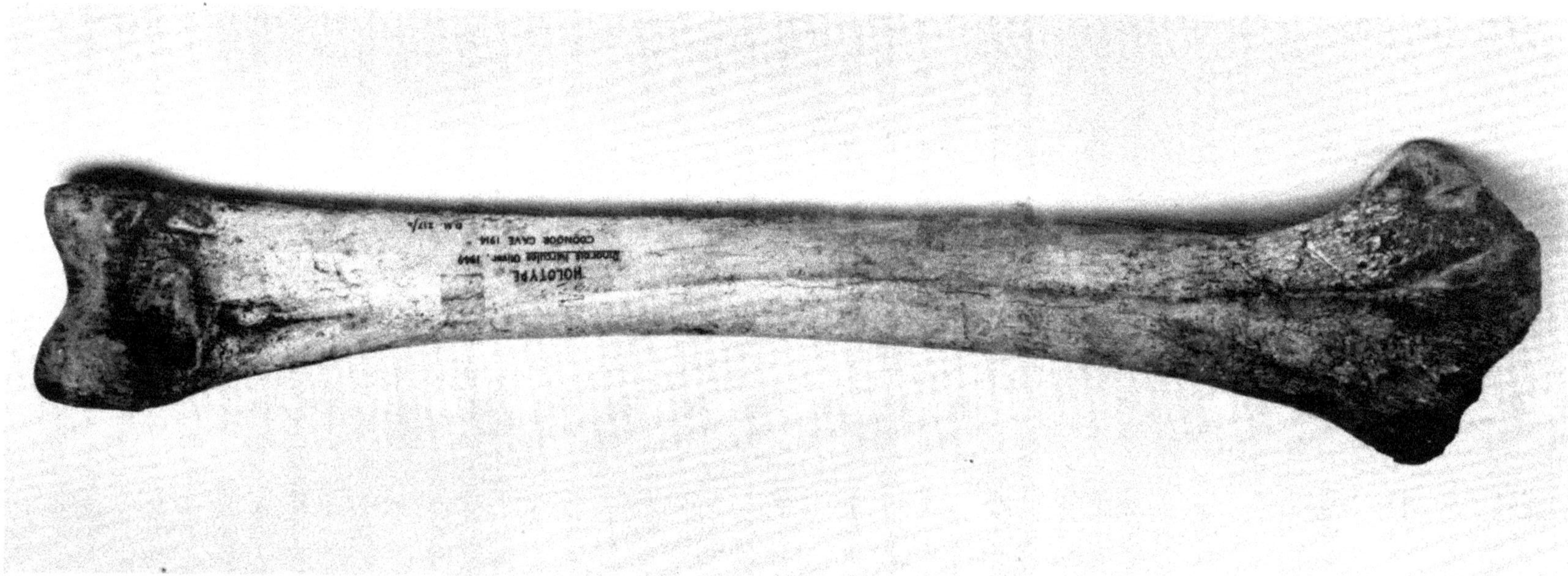

Fig. 1.3 Dinornis hercules Oliver. Right tibia, holotype, from Coonoor Cave, Wairarapa, NMNZ S. 217/a.

CAVE 1914 D.M. 217/a" (now to be known as MNNZ S. 217/a). A red square was painted on the shaft just below this label. The bone was originally pale yellow but is now light yellowish-brown in colour. Fig. 3 in the present paper shows the holotype as it now appears. Its measurements (taken by the writer) are now length 740, breadth proximal 172 +, mid 58, distal 108 mm.

There is also a left tibia in the National Museum collections labelled DM 217. It is identified as D. hercules and comes from the Coonoor Cave. It is now labelled DM 217/b and has Oliver's measurements on the tag label 741, 176, 57 and 105 + mm. As Oliver included only the right Coonoor tibia under his heading of "type" material" this left tibia and the other bones Oliver "associated" with this species in 1949 (p. 174) can not be considered as paratypes.

Some of the bones in the Museum collections listed by Oliver in 1949 as D. hercules have the manuscript name "coonoorensis" written in pencil in Oliver's hand on their tag labels. This name is deliberately published here, in a non-binomial fashion and without "indication" so that it is not an available name under the International Code of Zoological Nomenclature. This allows its relationship to the specific name hercules Oliver to be firmly established. It is clear that "coonoorensis" was considered, but not adopted, by Oliver as a possible specific name for the taxon he validly named Dinornis hercules in 1949.

Coonoor Cave is on property held by Mr. M. Conway in 1914 at Coonoor on the western side of the Puketoi Range in the northern Wairarapa, about 15 miles by road south of Dannevirke. For a description of the site and an account of the species found in the 1914 excavations see Oliver (1949:14, figs. 11, 16).

Original figures

The figures of Dinornis hercules published by Oliver in 1949 are as follows: fig. 140 (second femur from left) right femur posterior view; fig. 141 (second and third tibia from left) left tibia (with proximal end missing) and right tibia, both anterior view; fig. 142 (second metatarsus from left) right metatarsus anterior view. All are photographs. The femur in fig. 140 has the number DM 216 written on it (clearly visible in a print of this photograph in Oliver's meanuscript notes) and comes from Coonoor Cave; its measurements are given in Table 31 (ibid., 180). The broken left tibia in fig. 141 can be recognized as DM 114 in the National Museum collections and comes from Te Aute as stated in the caption. There are two holes drilled through this figured specimen, one proximally and one distally (the latter clearly visible as a black spot in fig. 141). There is an obvious open longitudinal crack on both anterior and posterior faces along the central half of the shaft. The measurements of this tibia are given in Oliver's table 31. The right tibia in fig. 141 is the holotype as established above. The metatarsus in fig. 142 can be recognized as a bone labelled DM 216 in the National Museum collections (there are six Dinornis bones from Coonoor registered under the same aumber DM. 216). The figured metatarsus is clearly labelled as from Coonoor on its shaft and in the figure caption. Its measurements are given in Oliver's table 31. The three "associated" leg bones discussed here are now labelled as "figured" specimens of D. hercules. All register numbers are now to be prefixed by MNNZ S. instead of DM if used in publication.

Present status

As outlined above in the Dinornis gazella section, Oliver recognized eight species in the genus Dinornis. He gives his reasons for describing D. hercules as a new species as follows (1949:164): "A study of the series of bones in the Dominion Museum shows that both giganteus [the biggest North Island species in Oliver's system] and maximus [the biggest South Island species] as generally understood are probably compound. Each consist of tall birds with straight-shafted tibiae and of shorter birds with curved-shafted tibiae. In maximus this distinction is not well-marked; but in giganteus, as shown in fig. 141, the difference is so great that I have thought it advisable to found a new name for the shorterbird and trust that individual skeletons someday will turn up to support my conclusions. Hutton's name excelsus was applied to a tall bird [and is thus a synonym of giganteus sensu strictu] so I have founded a new specific name, hercules, with a tibia from Coonoor as type". He explains his term "curved-shafted" tibia as saying on page 174 that the shaft in D. hercules "is straight on the outside but conspicuously curved on the inside" (see Oliver's fig. 141 and fig. 3 in the present paper).

In Oliver's view D. hercules was the second to biggest of the five North Island Dinornis species (ibid., 164). It was larger on average than the relatively common North Island D. novaezealandiae Owen (N.B. Oliver used the name D. ingens for this species but as explained in Scarlett, 1972:20, and Cracraft, 1976:202, the name D. novaezealandiae must be used for birds referred by Archey and Oliver to D. ingens, and the name D. struthoides for birds referred by these authors to D. novaezealandiae). Large specimens of that species (referred to as "long type" in table 30, ibid.:179) overlap the D. hercules range but were "very much more slender than the same bones of D. hercules" (ibid., 171). D. hercules was smaller (by definition) than the largest North Island species, D. giganteus Owen.

Scarlett (1972:21) thought D. hercules might "prive to be only a bow-legged variant of Dinornis giganteus", while Cracraft (1976) firmly synonymized it with his D. novaezealandiae (see discussion under "Present status" section of D. gazella). He found it closer in size to that species than to the bigger D. giganteus. He was able to compare the type of D. hercules to other species of Dinornis and in his opinion (1976:203) "the differences in stoutness and the curvature of the shaft of the type are attributable to individual variation".

Summary of present status of the moa species
described by W. R. B. Oliver (following Cracraft, 1976)

Pachyornis septentrionalis is a synonym of P. mappini Archey.

Pachyornis murihiku and P. australis are synonyms of P. elephantopus (Owen).

Euryapteryx tane is a synonym of E. curtus (Owen).

Euryapteryx kuranui is a synonym of E. geranoides (Owen).

Dinornis gazella is a synonym of D. struthoides Owen.

Dinornis hercules is a synonym of D. novaezealandiae Owen.

NOTES ON OLIVER 'S METHODS OF MEASURING MOA BONES

The Origin of the LPMD System used by Oliver (1949)

The first extensive paper published on moa measurements in New Zealand appeared in the first volume of the Transactions of the New Zealand Institute in 1869. In it, Haast used the following four measurements (in inches and decimals of an inch) when discussing leg bones: "Length of bone/ Grith of proximal end/Girth of shaft, thinnest part/Girth of distal end" (ibid. 82). This is the local forerunner of Oliver's LPMD system, though it uses girths rather than widths.

Owen in his 1879 memoir re-presented the moa measurements he had published since the 1840s. Altering the order somewhat, the first system he used (ibid. 78) was as follows "Length/Breadth of middle of shaft/Thickness or antero-posterior diameter of ditto/Circumference [of middle]/Breadth/of distal end", while the latest ibid., 253) was "Length/Breadth, transverse, of proximal end/Breadth of middle of shaft/Circumference of middle of shaft/ Breadth, transverse, of distal end". The latter is a LPMD system with the girth (or "circumference") at the middle added. In 1875 Hutton presented measurements of "two or three hundred" moas from the Hamilton swamp in Otago arranging them "according to the dimensions given by Professor Owen" (ibid. 274) in his papers. Hutton used the second system quoted here from Owen's memoir adding "Thickness middle" to give a six measurement system.

In the 1880s A. Hamilton, working in Napier, had special labels printed which he glued on to moa leg bones from Te Aute and other sites in the southern North Island. They were individually printed for femur, tibia and metatarsus with spaces left for a series of measurements. Labels for the latter element read "METATARSUS OF [blank for name]/Length/Circumf./Breadth, distal/ Breadth, middle/Thickness, middle/Breadth, prox." within an oblong border. These are still present on several bones in the main National Museum collections and on other bones transferred to the National collections from the Napier Museum in the 1950s. Hamilton wrote the locality along the bottom of the label outside the border but the ink used has faded badly and these labels are now fragile and difficult to read.

Hutton in 1892 used a leg bone ratio consisting of "the length (L) divided by the girth (G) at the middle of ths shaft" (ibid. 106), and later (1897 :545) gave leg bone measurements under the headings "Length/Prox. Width/Mid. Width/Distal Width". These are clearly the LPMD headings of Oliver's 1949 moa bulletin.

In the first edition of New Zealand Birds (1930:35), Oliver used the following measurements for moa leg bones: "total length between parallels... greatest width of proximal end...greatest width at middle of shaft...greatest width at distal end" and these descriptions can be taken as indicating his methods of obtaining his figures. For the key based on leg bones proportions, P, M and D were "expressed in terms of the length taken as 100" (ibid. 35). In a letter to Professor W. B. Benham dated 21 January 1932 (actually dated "1931" but it is clear that this is a typing error — the letter is in answer to two from Benham dated 4 and 18 January 1932), in the National Museum files, Oliver gives further details on his methods of taking leg bone measurements "In the measurement of femora of moas...I take the measurement along the axis of the neck, that is the greatest measurement that can be taken at the proximal end of the bone. I have been able to take this measurement in mounted specimens by means of a craniometer...I am recording both the transverse and antero-posterior measurements of all the leg bones." Benham (1934:91) followed Oliver and used the LPMD system ("Length/ Proximal Breadth/Middle Breadth/Distal Breadth") adding "Girth at Middle" (referred to below as "G"). He comments on these measurements by saying "The breadth at the proximal end of femur is taken along the axis of the head, as Owen measured it. The dimensions of all the bones were taken between two vertical uprights. Hutton does not mention how he took his measurements; in earlier days, at any rate, he seems to have used a tape, and given his figures in inches. The girth is liable to vary with the personal equation, i.e. with the precise point at which it is measured". Benham used "percentage of breadth to length" for the proportional dimensions of leg bones as did Oliver.

Oliver's Figures on Tag Labels

In his 1941 monograph on moas, Archey used the LPMDG system of Oliver and Benham but introduced on page 13 a new way of presenting these figures and the percentage of width to length figures. This consisted of placing the figures in two lines with the upper line giving the measured dimensions and the second line giving the widths as percentages of the length. For example the figures for one of the tibiae of the holotype skeleton of *Euryapteryx tane* Oliver (AM. 3) as given in Archey's table H (under the name *E. exilis*) would be presented as follows (in mm):

L	P	M	D	G
337	98	28	56	74
	29.1	8.4	16.6	21.9

It is important to know just what Archey meant in his presentation because Oliver used a similar but different method of presenting measurements on the back of the brown, "Dominion Museum", tie-on tag lebels used throughout the National Museum moa collection. He did not use the LPMD heading and did not take the measurement G.

An example of Oliver's different presentation would be the left tibia from Nuhaka, Hawke's Bay, identified by Oliver as *Anomalopteryx didiformis* (Owen), from an incomplete skeleton collected by A. Hamilton in 1913 (NMNZ S. 209). On the back of the tag label are the following figures in Oliver's hand:

350	99	31	60
	59	21	58
			51

The top line is LPMD with P, M and D all being Oliver's "greatest widths"; P is taken more or less diagonally to the transverse plane of the bone, while M and D are both "transverse" measurements. In the second line P is "transverse", M is the "antero-posterior" measurement (the least "width"; i.e. depth) and D is the "antero-posterior" depth of the inner (and larger) condyle. D in the third line is the "antero-posterior" measurement of the outer (and smaller) condyle. This three-line method of presenting measurements can be found on a great many of the bones examined by Oliver in the National Museum.

Sometimes Oliver added a fourth line to this presentation to give the percentage of width to length. Thus the right tibia of the holotype skeleton of Pachyornis septentrionlis (shown in fig. 1.1) shows Oliver's figures on the back of the tag label at their most complicated. Ignoring an alteration, and expressing the measurements in millimetres rather than in centimeters as they are written, the label has the following figures in Oliver's hand:

320	85	28	48
	52	19	48
			43
	26.4	8.7	14.3

The first three lines are as explained for the Nuhaka tibia while the fourth line gives percentages of width to length for the top line (figures surprisingly inaccurate!).

Archey's and Scarlett's Methods of measuring Moa Bones

Archey (1941:13) states that his "measurement have all been made between uprights and, to measure width, the bones have been placed at right angles to the direction of the measuring slide". His fig. 1 (ibid. 14) shows the distal end of a moa leg bone with its width being measured between the uprights of the measuring "apparatus".

In Bones for the New Zealand Archaeologist, Scarlett (1972:2) explains his methods of measuring bird bones as follows: "All bird limb bones should be given four measurements. Length, width at proximal (top) and distal (bottom) ends, and width of the shaft. These are expressed as L., P., M., D. M. is taken by some workers at the exact centre (Middle) of the shaft. Others, myself included, measure at the narrowest point. This usually differs little from the width obtained at the centre and facilitates comparison with broken bones, where the exact length is unknown, and is much quicker. It does not matter which method is used, provided that one states where the measurement is taken. ...In the field, an approximate length for larger bones, such as those of Moa or Aptornis can be obtained with a steel tape. Much greater accuracy can be obtained with a measuring device consisting of a base with rule, a fixed upright and a sliding bar and upright (see illustration). This is useful for measuring adzes, as well as bones". The illustration referred to shows a moa bone being measured on a sliding scale between uprights as described. The apparatus Scarlett uses is the same one used by Dr. (later Sir Gilbert) Archey in his work on moas at the Auckland Museum.

To test Oliver's method of taking measurements against Scarlett's, the writer asked Scarlett to remeasure two moa leg bones on Archey's sliding scale with the following results:

MNNZ S. 209, left tibia Anomalopteryx didiformis, Nuhaka

	L	P	M	D
Oliver	350	99	31	60
Scarlett	350	97	29	60

NMNZ S. 78, left femur Pachyornis mappini from Eketahuna, Wairarapa, coll. John Golder, 11 May 1894.

Oliver	211	79	33	92
Scarlett	212	78	34	90

The main difference would appear to be in M which Oliver took across the "middle of the shaft" and Ron Scarlett takes across the "narrowest point". The P and D discrepancies in the femur are probably due to abrasion since Oliver measured the bone.

When these bones were being remeasured, Scarlett commented on shrinkage and warping in moa bones as follows (in litt. 16 February 1962): "bones can shrink a few mm. Aptornis bones and moa bones that I have measured soon after they came out of Pyramid Valley, and re-measured a few years later, after they had well dried, did shrink longitudinally a little, not very much... something like 3-4 mm for some of the Dinornis tibia, but less for Aptornis. Transversely the shrinkage was negligible. Most bones don't warp in drying unless there is too much heat. Have had warped bones occasionally from Caves and swamps where pressure seems to be the cause".

Summary of W. R. B. Oliver's Measurement Method

When recording dimensions of moa leg bones Oliver used the following measurements:

L - total length between uprights;
P - greatest width at proximal end (e.g. along axis of neck in femur);
M - greatest width at middle of shaft, and
D - greatest width at distal end.

When writing measurement on tag labels in the National Museum collections, Oliver recorded LPMD in one ling adding on a second line a transverse width at P, a least "width" (i.e. depth) at M, and a greatest depth at D. In a third line he gave the depth of the lesser condyle at D (i.e. not the least depth possible at D, which would be through the intercondylar groove). The percentages of widths to length for the top line measurements, if given, appeared as a fourth line.

SUMMARY

The holotype of Pachyornis septentrionalis (synonymized by Cracraft (1976) with P. mappini), is an incomplete skeleton from Te Pohue, Hawke's Bay. Oliver figures the pelvis and leg bones of this type. The holotype of P. australis (syn.with P. elephantopus), is a skull and 20 vertebrae from the Salisbury Tableland, Nelson. Only Oliver's figures of the skull of this species are from the type. The holotype of Dinornis gazella (syn. with D. struthoides), is a pelvis figured by Oliver from Te Aute, Hawke's Bay. The holotype of D. hercules (syn. with D. novaezealandiae), is a tibia from Coonoor, Wairarapa, figured by Oliver but since broken and relabelled. These four species were described by Oliver in 1949 and the type material, except for the P. australis vertebrae, are in the National Museum. The holotype of P. murihiku Oliver, 1949 (syn. with P. elephantopus) is a skeleton from Bluff in the Southland Museum, and that of Euryapteryx tane Oliver, 1949 (syn. with E. curtus) is a skeleton from Doubtless Bay, North Auckland, in the Auckland Museum. Oliver's figures of both these species are from their type skeletons. The holotype of E. kuranui Oliver, 1930 (syn. with E. geranoides) is a skeleton from Castlepoint, Wairarapa, in the Canterbury Museum. The type skull was figured by Oliver in 1949. No paratypes were designated by Oliver for any of his species.

Oliver's LPMD headings for moa leg bone measurements in 1949 refer to length between uprights, greatest width at proximal end, transverse width at middle of shaft, and greatest width at distal end.

Literature Cited

Archey, G., 1941. The moa a study of the Dinornithiformes. Bull. Auckland Mus. 1:1-119.

Benham, W. B., 1934. The skeleton of a small moa, Emeus huttoni Owen. Trans. Roy. Soc. N.Z. 64:87-104.

Brodie, J. W., 1950. Moa remains at Castlepoint. New Zealand Science Review 8:87-88.

Brodkorb, P., 1963. Catalogue of fossil birds Part 1 (Archaeopterygiformes through Ardeiformes), Bulletin Florida State Museum Biological Sciences 7:179-293.

Buick, T. L., 1931. The mystery of the Moa: New Zealand's avian giant. Thomas Avery and Sons, New Plymouth.

Cracraft, J., 1976. The species of moas (Aves: Dinornithidae). IN: Olson, S. L., (ed.), Collected papers in avian paleontology honoring the 90 th birthday of Alexander Wetmore. Smithsonian Contributions to Paleobiology 27:189-205. (Reviewed in Fleming, 1977).

Fleming, C. A., 1977. Wishbones for Wetmore. Notornis 24:144-146. (Review of Cracraft, 1976).

Haast, J., 1869, On the measurements of Dinornis bones, obtained from excavations in a swamp, situated at Glenmark, on the property of Messrs. Kermode and Co., up to February 15, 1868. Trans. N.Z. Inst. 1:80-89 (2nd Ed. 1875:21-30).

Hamilton, A., 1889. Notes on a deposit of moa-bones in the Te Aute Swamp, Hawke's Bay. Trans. N.Z. Inst. 21:311-318.

Hill, H., 1914. The moa — lengendary, historical, and geological: why and when the moa disappeared. Trans. N.Z. Inst. 46:330-351.

Hutton, F. W., 1875. On the dimensions of Dinornis bones. Trans. N.Z. Inst. 7:274-278.

Hutton, F. W., 1892. The moas of New Zealand. Trans. N.Z. Inst. 24:93-172.

Hutton, F. W., 1897. The moas of the North Island of New Zealand. Trans. N.Z. Inst. 29:541-557.

Hutton, F. W., 1906. On a skeleton of Emeus crassus from the North Island. Trans. N.Z. Inst. 38:66-67.

ICZN, 1964. International Code of Zoological Nomenclature adopted by the XV International Congress of Zoology. Interna ional Trust for Zoological Nomenclature, London.

Lydekker, R., 1891. Catalogue of the fossil birds in the British Museum (Natural History), Cromwell Road, S.W. British Museum, London.

McKay, A., 1879. The Baton River and Wangapeka Districts, and Mount Arthur Range. Reports of Geological Explorations during 1878-79, with Maps and Sections. Colonial Museum and Geological Survey of New Zealand, Wellington.

Oliver, W. R. B., 1930. New Zealand Birds. Fine Arts, Wellington.

Oliver, W. R. B., 1949. The moas of New Zealand and Australia. Dom. Mus. Bull. 15:1-206.

Oliver, W. R. B., 1955. New Zealand Birds. A. H. & A. W. Reed, Wellington. 2nd Edition.

Owen, R., 1879. Memoirs on the extinct wingless birds of New Zealand; with an appendix on those of England, Australia, Newfoundland, Mauritius, and Rodriguez. John Van Voorst, London. 2 Vols.

Owen, R., 1883. On Dinornis (Part XXIV): containing a description of the head and feet, with their dried integuments, of an individual of the species Dinornis didinus, Owen. Trans. Zool. Soc. Lond. 11:257-261.

Park, J., 1888. On the exploration of moa-bone deposit at Patangata swamp, Te Aute. Colonial Museum and Geological Survey of New Zealand Reports of Geological Explorations 19:88-90.

Parker, T. J., 1893. On the presence of a crest of feathers in certain species of moa. Trans. N.Z. Inst. 25:3-6.

Parker, T. J., 1895. On the cranial osteology, classification and phylogeny of the Dinornithidae. Trans. Zool. Soc. Lond. 13:373-431.

Rothschild, W., 1907. Extinct birds. An attempt to unite in one volume a short account of those birds which have become extinct in historical times — that is, within the last six or seven hundred years. To which are added a few which still exist, but are on the verge of extinction. Hutchinson, London.

Scarlett, R. J., 1972. Bones for the New Zealand archaeologist. Canterbury Mus. Bull. 4:1-69.

A DECEPTIVE TERRESTRIAL VERTEBRATE FOSSIL SITE ON THE WAIPARA RIVER, NORTH CANTERBURY, NEW ZEALAND.

T. H. V. Rich, P. V. Rich, R. E. Fordyce,
P. Gatehouse, and R. J. Scarlett*

Introduction

As students from the University of Canterbury, Christchurch, two of us (PG in 1972, RF in 1973) unbeknown to each other, found a few fossil bones near the Horseshoe (Fig. 2.1) in the lower Waipara Gorge, North Canterbury, New Zealand. In April 1974, the bone-producing deposit, a lens, was located. The nature of the lithology of the lens was such that, even after it had been located by use of the underwater sieving technique described by McKenna (1965), examination of the excavated surface showed no evidence of bone. Only after the matrix was sieved was bone noticeable.

Mimiomoko Pocket, the name given to the fossil deposit, was located on the south or right bank of the Waipara River near the Horseshoe (Fig. 2.1) at 43° 06' 30" S., 172° 46' 27" E., grid reference S68/137074, New Zealand Mapping Service (N.Z.M.S.) 1, Sheet S68 and S69, Amberley and Motunau, 3rd ed., 1 Oct. 1964, reprint 1970; metric grid reference N34:917887, N.Z.M.S. 260 (1975). It was about 12 metres above river level. The word "Mimiomoko", the Maori name for a nearby hill, means "Sleeping Lizard" (Hawkins, 1957). The appellation "Pocket" is appropriate, as the productive unit was a lens of limited dimensions 1.0 x 0.3 x 0.1 metres.

Mimiomoko Pocket occurs in an exposure of the Double Corner Shell-beds (Gregg, 1959) 18 metres below the contact with the overlying Greenwood Formation (Fig. 2.2). Gregg regarded the Double Corner Shell-beds as Waiauan in age in the framework of New Zealand Tertiary Stages (*ibid.*) (late middle Miocene, 10.5-14 million years before present [Berggren, 1972]). Owing to a fault, the upper Double Corner Shell-beds and lower part of the Greenwood Formation are repeated in the vicinity of Mimiomoko Pocket, the fossil site being on the downthrown block immediately east of the fault, which is clearly indicated in Gregg's fig. 2 (*ibid.*).

Two features in particular distinguish the productive lens from the surrounding rock: the presence of greywacke granules and bones of terrestrial animals. Only with difficulty could these traits be detected in the field. Both the productive lens and the surrounding rock are shell-rich calcareous quartzarenites with the venerid bivalve, *Turia waiauensis* Marwick, forming the bulk of the shells.

The New Zealand fossil record file numbers for collections from Mimiomoko Pocket are S68/f1258, S68/f1258A, and S68/f1258B.

* Published without Scarlett knowing where it would appear. Ed.

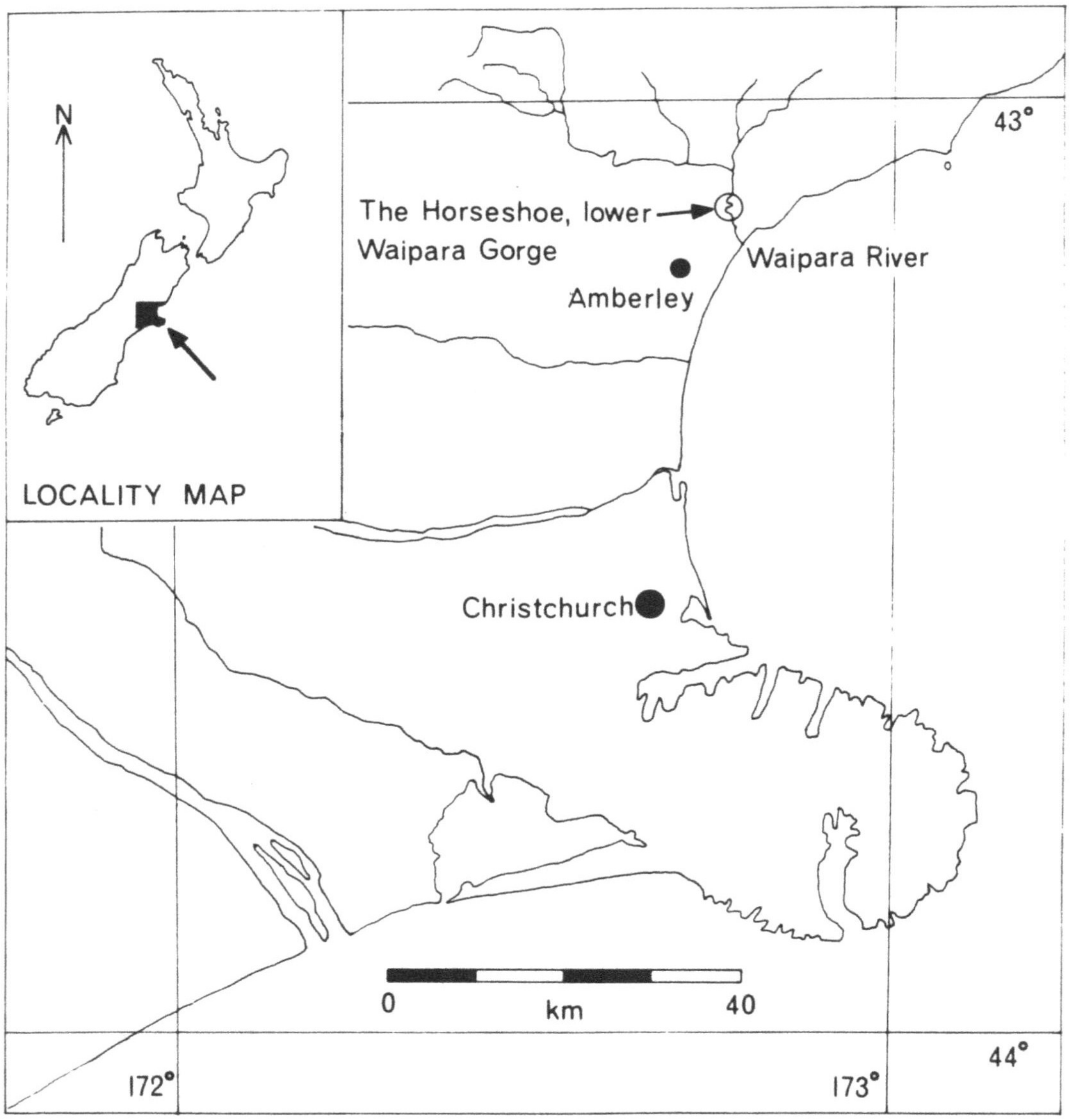

Fig. 2.1 General location of late Quaternary tetrapod side, Mimiomoko Pocket (43° 06' 30" S 172° 46' 27" E).

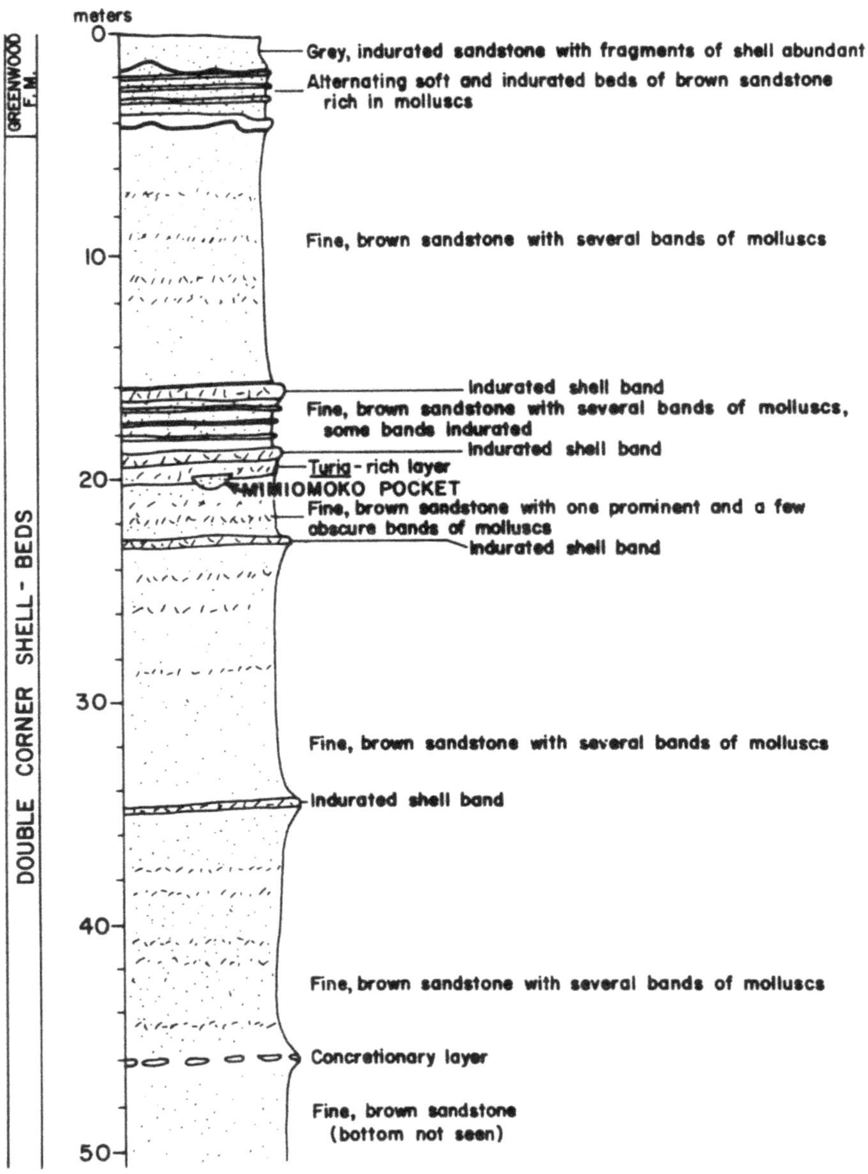

Fig. 2.2 Stratigraphic section showing position of Mimiomoko Pocket. Drawn by R. Suddarth.

Abbreviations

AV	Canterbury Museum, Christchurch, Bird Catalogue
CSIRO BS	Commonwealth Scientific and Industrial Research Organisation, Wildlife Research, Canberra
DM	National Museum of New Zealand, Wellington
NMV	Department of Ornithology, National Museum of Victoria, Melbourne
Rep	Canterbury Museum, Christchurch, Reptile Catalogue
>	greater than
⋗	slightly greater than
@	approximately

Comparative Material Available

Reptilia: Osteological specimens of the following taxa were available for study. The number in parentheses after each taxon indicates the number of individuals available: *Leiopelma* sp. (1), *Sphenodon punctatus* (2), *Naultinus elegans* (1), *Hoplodactylus duvauceli* (1), and *Leiolopisma infrapunctatum* (1).

Aves: The general avian osteological collection at the National Museum of Victoria was available, which contains at least one representative of every Australasian non-passeriform avian family except Rhynochetidae as well as most passeriform families. More specifically, the following appropriate taxa were studied. Numbers of individuals in each taxon available are indicated in parentheses: Apterygidae: *Apteryx australis* (1); Procellariidae: *Macronectes giganteus* (3), *Fulmarus glacialoides* (7), *Thalassoica antarctica* (1), *Daption capense* (17), *Pterodroma macroptera* (46), *P. lessonii* (30), *P. brevirostris* (1), *P. inexpectata* (2), *Halobaena caerulea* (32), *Pachyptila vittata* (2), *P. salvini* (26), *P. desolata* (51), *P. belcheri* (31), *P. turtur* (42), *Procellaria cinerea* (2), *P. aequinoctialis* (1), *Puffinus carneipes* (21), *P. pacificus* (1), *P. griseus* (20), *P. tenuirostris* (17), *P. gavia* (32), *P. huttoni* (1), *P. assimilis* (3), Hydrobatidae: *Oceanites oceanicus* (5), *Pelagodroma marina* (28); Pelecanoididae: *Pelecanoides urinatrix* (18), *P. georgicus* (3); Rallidae: *Capellirallus karamu* (2), *Diaphorapteryx hawkinsi* (2), *Nesolimnas dieffenbachi* (1), *Nesophalaris chathamensis* (1), *Tribonyx ventralis* (6), *T. mortierii* (4), *Porphyrio porphyrio* (11), *Notornis mantelli* (1), *Rallus philippensis* (18), *Porzana pusilla* (1), *P. fluminea* (1), *P. tabuensis* (1), *Gallirallus australis* (8), *Fulica atra* (12); Rhinochetidae: *Rhinochetos jubatus* (1).

Classification for the reptiles follows Romer (1956), while that for the New Zealand birds follows Kinsky (1970) with modifications suggested by Olson (1975).

Age of the Assemblage

The following evidence has a bearing on the age of Mimiomoko Pocket.

1. The lens appears to be comformable within the sequence. Its dip (35° west) and strike (15°) parallels that of the immediately surrounding Double Corner Shell-beds.

2. None of the plant and invertebrate fossils found in the Pocket is exclusively a post-Waiauan form, and some, e.g. Turia waiauensis, are known to be no younger than Waiauan.

3. The present position of the Pocket relative to the nearby Waipara River, and the direction and extent of river erosion appear incompatible with emplacement of the Pocket by scouring, undercutting, or cave infilling by the river in recent times. If it is supposed that a shallow cave was formed during Quaternary time by action of the river when it was 12 metres higher than at present, it is necessary to infer that the bank has not receded a significant distance horizontally during the time of down cutting to the present level. Otherwise, the cave deposit would have been totally destroyed as the bank receded unless the original cave was quite deep. Today, no such deep caves are developed in the outcrop of the Double Corner Shell-beds where Mimiomoko Pocket was found.

4. Dr. R. J. Weston of the Chemistry Division, Department of Scientific and Industrial Research, Petone, New Zealand, carried out an analysis of three samples of tetrapod bones from Mimiomoko Pocket and one on a fragment of a vertebra of a marine mammal collected about sixteen metres on strike from the lens. The bones from Mimiomoko Pocket were rich in collagen indicating an age, " less than 10,000 years old and probably only about 1000 years old;" while in contrast, the marine mammal vertebral fragment contained no detectable collagen, a result consistent with a Waiauan age for that specimen (Weston, pers. comm.). Weston, Repenning, and Fleming (1973) described the technique of collagen analysis used.

5. The presence of small, subangular graywacke pebbles in Mimiomoko Pocket but not in the surrounding Double Corner Shell-beds suggests a sediment source for the lens not active when the Double Corner Shell-beds were deposited.

6. Tuatara, one of the most abundant tetrapods in the fauna, had disappeared from the mainland by two hundred years ago (Crook, 1975). This coupled with the absence of any tetrapods introduced by man supports a probable minimum age of two centuries for the site.

Evaluating the above evidence, unless some mechanism can be invoked to explain the preservation of collagen in the bones from Mimiomoko Pocket since the Miocene, the most likely interpretation is that the lens represents a late Quaternary deposit accumulated in a shallow cave. If this was the case, the cave was scoured out or excavated as a burrow site between resistant beds when well above river level, which resulted in the lens being inclined parallel to the surrounding Double Corner Shell-beds. This coupled with the cave being subsequently filled primarily by a matrix reworked from the immediate surroundings resulted in the apparent conformity between the lens and the Double Corner Shell-beds. In this reworked material were the marine invertebrates not known from rocks younger than the Waiauan Stage. During the time the cave was being filled tetrapods were present and greywacke pebbles entered the area of deposition; events that took place at least two centuries and possibly more than a thousand years ago.

This site was originally thought to be of Miocene age. It was the basis for a personal communication by one of us (TR) cited in Crook (1975) that the tuatara was known from Miocene deposits.

Flora and Non-Tetrapod Fauna

In the following list of taxa represented from the Mimiomoko Pocket, D. Mildenhall identified the plants; R. H. Hoskins, the foraminifera; and I. W. Keyes, the chondricthyians.

Plants: Pyrrophyta (dinoflagellates), Eumycophyta (true fungi), Pteridophyta

Animals: Foraminiferida: Globigerina bulloides, G. ciperoensis, G. falconensis, G. foliata, G. quinqueloba, G. woodi decoraperta, Entosolenia sp., Trifarina bradyi, Globocassidulina subglobosa, Cibicides sp., C. molestus, Notorotalia sp.

Annelida: Terebellidae? (agglutinated tubes, not determinable to genus)

Cirripedia: Balanus sp.

Echinoidea: Arachnoides sp., Cidaroides? (small, stubby spines)

Mollusca: Turia waiauensis, Baryspira sp., Polinices sp.

Chordata: Chondrichthyes: Hypoprion sp., Deania sp., ?Dalatias sp., Pristiophorus lanceolatus, ?Trygon sp., Myliobatus sp.

Osteichthyes: Teleostei

Systematics of Tetrapods

Class Reptilia
Subclass Lepidosauria
Order Rhynchocephalia
Family Sphenodontidae

Sphenodon

Sphenodon punctatus

Material: Two right (Rep. 401–402) and two left (Rep. 403-404) humeri, nearly complete but lacking epiphyses. Proximal half of left humerus lacking epiphysis, Rep. 405. Right femur, nearly complete but lacking epiphysis, Rep. 406. Left tibia, nearly complete but lacking epiphysis, Rep. 407. Distal half of right fibula, lacking only epiphysis, Rep. 408. Partial left scapula and coracoid, Rep. 409. Partial right coracoid, Rep. 410. Partial left scapula, Rep. 411. Two teeth on bone fragment, Rep. 412. Centrum of anterior cervical vertebra, Rep. 413. Two centra of caudal vertebrae, Rep. 414, 415. Proximal one-third of left rib, Rep. 416. First phalanges, Rep. 417, 418. Second through fourth phalanges, Rep. 419-430. Terminal phalanx, Rep. 431.

Measurements (in millimetres):

	Length of Diaphysis or Midshaft	Minimum Diameter of Diaphysis or Midshaft
Humerii		
Rep. 401	21.5	2.0
Rep. 402	>21.9	2.8
Rep. 403	>20.7	2.3
Rep. 404	21.8	2.3
Rep. 405	-	1.7
Femur		
Rep. 406	35.9	2.7
Tibia		
Rep. 407	23.8	2.3
Fibula		
Rep. 408	-	1.3

	Centrum Length	Dorsoventral Diameter of Centrum, Anterior End
Anterior Cervical Vertebra		
Rep. 413	5.0	5.0
Caudal Vertebrae		
Rep. 414	-	1.5
Rep. 415	-	1.3

Second to Fourth Phalanges	Length	First Phalanges	Length
Rep. 419	4.2	Rep. 417	>7.3
Rep. 420	>2.9	Rep. 418	>5.7
Rep. 421	>3.1		
Rep. 422	5.0	Terminal phalanx	
Rep. 423	>4.5	Rep. 431	3.4
Rep. 424	4.1		
Rep. 425	5.1		
Rep. 426	4.6		
Rep. 427	>4.3		
Rep. 428	>3.8		
Rep. 429	>3.5		
Rep. 430	4.6		

Diagnosis (Same as for genus, only known species): Larger than any other member of the Sphenodontidae except the genera *Monjurosuchus* and *Anisodontosaurus* (Broili, 1925; Endo, 1940; Gilmore, 1909; Hoffstetter, 1955; Welles, 1947). Distinguished from *Monjurosuchus* by having amphicoelous rather than opisthocoelous dorsal vertebrae and broad based rather than acutely conical teeth (Endo, 1940). Distinguished from *Anisodontosaurus* by posterior mandibular teeth that are asymmetrical in lateral view, the anterior side of the teeth being nearly vertical and the posterior sloping at about 45° rather than both sides sloping at about the same degree and the inclination of the sides increasing progressively anteriorly from 45° to vertical by the middle of the tooth row (Welles, 1947). Distinguished from all Cenozoic reptiles except Gekkonidae and some dorsals of Chelonia by amphicoelous vertebra perforated for passage of notochord in adult (Romer, 1956). Distinguished from amphicoelous dorsals of Chelonia by width of centrum greater than, rather than less than one-half the length. Distinguished from amphicoelous Gekkonidae by absence of paired foramina subcentralia (Hoffstetter and Gasc, 1969). Distinguished from all other Cenozoic reptiles and rhynchosaurs by the presence of an entepicondylar foramen on the humerus (Romer, 1956).

Discussion: *Sphenodon punctatus* is the sole surviving rhynchocephalian and structurally, one of the most primitive living reptiles. Many of the osteological features of this species are duplicated nowhere else among living reptiles but are common to entire orders of extinct ones.

No single bone of *Sphenodon punctatus* can be distinguished from the same element in every other reptile. Only by examining different bones known to be parts of one individual can a specimen be unequivocally diagnosed as belonging to this species. However, it is possible to distinguish isolated vertebrae and humerii of *S. punctatus* from all other known Cenozoic reptiles. In light of the fact that Mimiomoko Pocket is probably late Quaternary in age and certainly no older than Miocene, a more limited diagnosis from Cenozoic reptiles is useful, for it is unlikely that this material represents a late Cenzoic example of some taxon previously thought restricted to the Mesozoic or Paleozoic.

Order Squamata
Suborder Lacertilia
Infraorder Indet.

Material: Proximal ends of two right humeri, Rep. 432, 433. Distal ends of two humeri, right and left respectively, Rep. 434, 435. Tooth in right maxilla fragment, Rep. 436. Teeth in bone fragments, Rep. 437, 442. Right mandible fragment with bases of the most anterior seven teeth preserved, Rep. 438.

Discussion: The presence of epiphyses on the humeri fragments supports the assignment of these specimens to the lizards. Other than lizards, only in *Sphenodon punctatus* and crocodiles, plus some other rare cases, are epiphyses developed among reptiles (Romer, 1956). These specimens are much too small to be referrable to either adult *S. punctatus* or crocodiles (see Fig. 2.3).

Infraorder Nyctisauria (Gekkota)
Family Gekkonidae
Genus *et* species indet.

Material: Two dorsal vertebrae, Rep. 439, 440. Caudal vertebrae, Rep. 441.

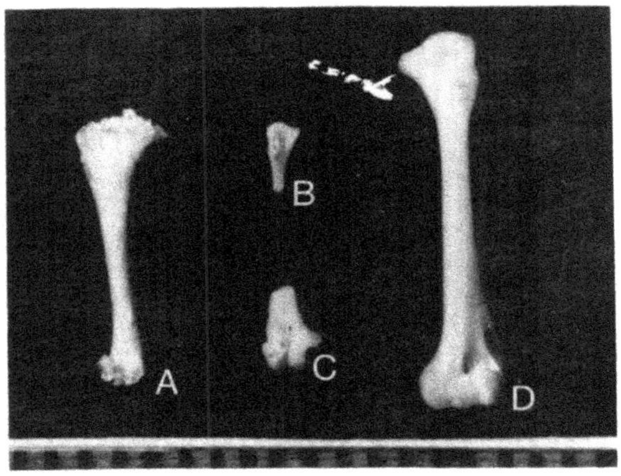

Fig. 2.3 A. Midshaft of right humerus of Hoplodactylus duvauceli, modern specimen.

B. Proximal end of right humerus of small lizard, Mimiomoko Pocket, Rep. 432.

C. Distal end of right humerus of small lizard, Mimiomoko Pocket, Rep. 434.

D. Left humerus of Leiolopisma infrapunctatum, modern specimen. Scale, each division equals 1 mm.

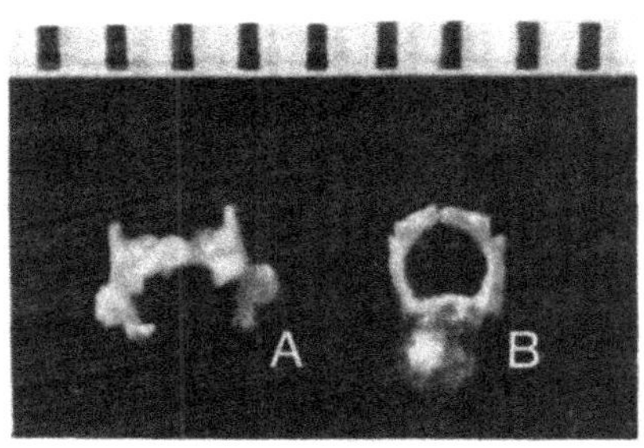

Fig. 2.4 A. Neural arch of modern specimen of Hoplodactylus duvauceli, posterior view. B. Posterior view of Gecko vertebra from Mimiomoko Pocket, Rep. 439. Neural arch of H. duvauceli figured here is from the same individual as the humerus midshaft illustrated in fig. 3A. Closeness of the neural canals in size indicates the similarity in size of the H. duvauceli specimen and the Mimiomoko Gecko figured, whereas the two limb fragments from Mimiomoko are from a much smaller adult animal than this specimen of H. duvauceli. Scale, each division equals 1 mm.

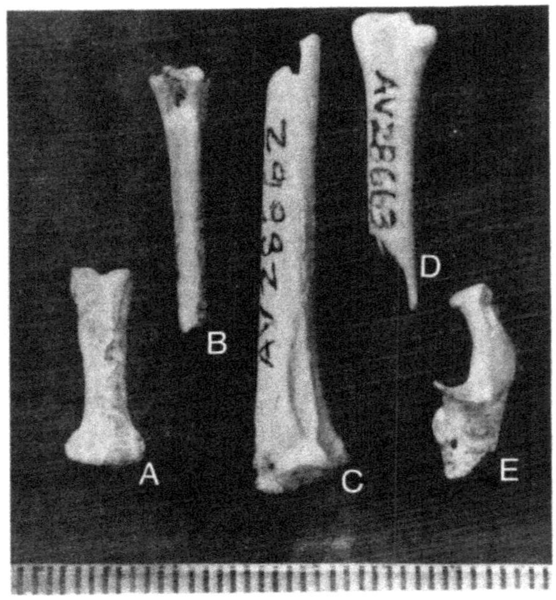

Fig. 2.5 A, cf. Apterygidae, phalanx, AV28667; Procellariidae: B, tarsometatarsus, AV28665; C, ulna (proximal) AV28662; D, ulna (distal), AV28663; and E, coracoid, AV28671. Scale, each division equals 1 mm.

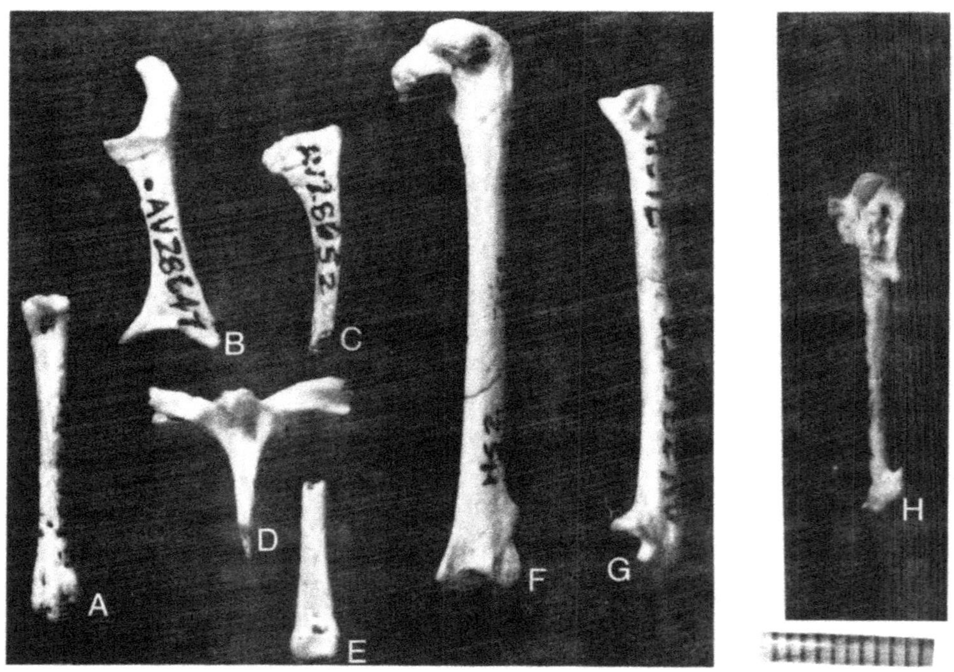

Fig. 2.6 <u>Pelecanoides</u> sp. from Mimiomoko Pocket: A, tarsometatarsus, AV28650; B, coracoid, AV28647; C, scapula, AV28652; D, sternum, AV28657; E, tibiotarsus (distal), AV28654; F, humerus AV28645; G, ulna, AV28653; and H, carpometacarpus, AV28648. Scale, each division equals 1 mm.

Measurements (in millimetres):

	Neural Canal Height	Neural Canal Width	Centrum Length	Centrum Width	Dorsoventral Diameter of Centrum Anterior End
Dorsal Vertebrae					
Rep. 439	1.0	1.3	1.5	0.6	0.7
Rep. 440	0.8	1.0	1.3	0.9	-
Caudal Vertebrae					
Rep. 441	-	-	1.4	1.0	0.8

Diagnosis: Many members of the family including this material distinguished from all other Cenozoic reptiles except some turtles and Sphenodon punctatus by amphicoelous vertebrae perforated for passage of notochord in adult (Romer, 1956). Distinguished from S. punctatus by presence of paired foramina subcentralia on amphicoelous forms (Hoffstetter and Gasc, 1969). Distinguished from amphicoelous dorsals of Chelonia by width of centrum greater than rather less than one-half the length.

Discussion: The lizard humeri mentioned previously are much too small to be conspecific with these gecko vertebrae. Owing to their fragmentary nature, it has not been possible to give meaningful measurements of these humeri. But comparison with a complete humerus of a modern skink (Leiolopisma infrapunctatus) together with a midshaft of a gecko (Hoplodactylus duvauceli) indicates how small the fossils are, probably no more than half the length of the modern specimens before being broken (see Fig. 2.3). The largest dorsal vertebrae associated with the modern gecko humerus figured here were only slightly larger than the Mimiomoko Pocket dorsal vertebrae (see Fig. 2.4). Because the epiphyses are fused to the diaphyses on the fossil humeri, they presumably belonged to adult individuals. Therefore, at Mimiomoko Pocket, there are gecko vertebrae the size of a modern specimen of H. duvauceli and humeri of a much smaller adult lizard.

Order ?cf. Apterygiformes
Family ?cf. Apterygidae

Material: ?Left phalanx 1, digit III (AV28667) (see fig. 2.5).

Measurements (in millimetres): Length (maximum), 14.6; maximum proximal width, > 6.6; maximum distal width, 4.6; maximum proximal depth, > 4.9; depth, internal condyle, 3.8; depth, external condyle, 3.9.

Diagnosis: Firm identification of this single phalanx is far from certain, but after careful comparisons across a broad range of avian families, AV28667 was most closely allied to the Apterygidae for the following reasons: in proximal view, proximal end of phalanx has convex sides where broadest part occurs somewhere between dorsal and ventral borders, not along either dorsal or ventral border; in ventral view, ventral shaft surface not grooved, in fact over most of extent convex ventrally; area for attachment of flexor tendon on ventral surface at proximal end bulbous, protruding ventrally with two short, distinct ridges either side of midline present; in ventral view, proximal border of phalanx V-shaped, with proximalmost extension just about at mediolateral midpoint; phalanx expanded at both proximal and distal ends, with narrower shaft in between; external condyle extends slightly further distal than internal condyle.

Discussion: Some differences of unknown significance can also be seen between AV28667 and phalanx 1, digit III of the living kiwi *Apteryx australis*: AV28667 has a prominence, presumably for tendonal or ligamental attachment, along the medial part of the shaft near the proximal end, which *A. australis* (NMV B9127) does not possess: AV28667 also flares more broadly at the proximal end and has a more nearly parallel-sided shaft between distal and proximal ends; excavations on the medial and lateral sides of the internal and external condyles respectively are decidedly shallower in AV28667; and AV 28667 is only slightly greater than half the size of *A. australis* (where phalanx length = 23.2 mm for NMV B9127).

Further comparisons to other kiwi species (*A. oweni*, *A. haasti*) need be made before any final comment or assignment of this specimen should be attempted. On the basis of a single phalanx, however, any final assignment would have to be tentative, even once the above comparisons had been made.

Order Procellariiformes
Family Procellariidae

Material: Proximal end of right ulna (AV28662); distal end, right ulna (AV28663); dorsal fragment of right coracoid (AV28671), including most of bone to just ventral of coracoidal fenestra; proximal end of right tarsometatarsus with an abraded hypotarsus (AV28665) (see Fig. 2.5).

Measurements (in millimetres): Ulnae. AV28662, maximum width across cotylae, 6.2; maximum depth across cotylae and olecranon, 6.3; maximum depth of internal cotyla, 4.5; maximum depth of external cotyla, 3.5. AV28663, maximum width of distal end measured parallel with appropriate shaft surface, 6.6; maximum width across condyles, 4.1; depth of internal condyle, ≯ 3.6; depth of external condyle, ≯ 4.6; width of shaft just proximal to expansion for distal articular surfaces, 4.0; depth just proximal to expansion for distal articular surfaces, 2.9. Coracoid. AV28671, length from proximal most extension to procoracoid, 8.1; maximum width across procoracoid, ≯ 5.2; anteriorposterior depth of coracoidal shaft at dorsal end, 4.4; anteroposterior depth of coracoidal shaft just dorsal to expansion for procoracoid, 3.7. Tarsometatarsus. AV28665, maximum width of proximal end, 4.5; maximum depth of internal calcaneal ridge from base of intercotylar process, ≯4.7; length of internal calcaneal ridge @ 4.4.

Diagnosis and Discussion: Because of the incompleteness of the two ulnae of procellariforms present in the Mimiomoko sample, which might even be opposite ends of the same bone, and the lack of comparative material of *Pagodroma*, *Priocella*, *Bulweria* among others, assignment of these specimens (AV28662 and AV28663) has been made only to family. Both ulnae represent birds the size of or slightly larger than *Puffinus gavia*, the Fluttering Shearwater, whereas the tarsometatarsus (AV28665) is from a much smaller procellarid. The coracoid fragment (AV28671) is slightly smaller than, although perhaps within the size range of, P. gavia. Thus at least two, possibly three, taxa in the Procellariidae are present in the Mimiomoko avifauna.

ULNAE. The two ulna fragments were placed in Procellariidae because they possessed the following combination of character states: ulna not mediolaterally compressed, unlike that in Pelecanoididae; shaft triangular in cross-section near proximal end; olecranon elongate but projecting anconally, not proximally, and triangular in shape when viewed proximally, located only slightly mediad of mediolateral midpoint of proximal end; palmar surface of shaft near proximal end with two moderately deep channels, separated by a distinct ridge; cotylar surfaces oriented nearly perpendicular to palmar surface of shaft, not at all approaching same plane; proximal end of shaft (viewed palmarly) not expanded much to accommodate articular surfaces, thus not broadly splayed; shaft nearly straight, not highly curved in any view; on distal end, prominent process present lateral to tendinal groove, presumably for further stabilization of the tendon passing across this area; internal and external condyles not broadly separated from one another; carpal tuberosity an elongate, prominent process; distal end of shaft somewhat palmoanconally compressed (when viewed distally); in lateral view, little expansion over shaft width occurs at distal end; external condyle quite "squared off", not rounded, with about an equal, slight amount of expansion in both palmar and anconal directions; tendinal pit deep and located just distad of process on lateral side of tendinal pit.

CORACOID. Although only a small fragment of the coracoid (AV28671) was preserved, it can be assigned to Procellaridae because of the following combination of character states: shaft of coracoid between procoracoid and dorsal end, when viewed medially, not deep anteroposteriorly and of equal depth from procoracoid to head, not increasing in dimensions in either direction; ventral border of coracohumeral surface, along which lies brachial tuberosity, forming straight line that is tilted slightly posteriorwards and not undercut; procoracoid well developed, pointing only slightly dorsally, but mainly medially, not primarily dorsally or anteriorly; procoracoid delicate and V-shaped, not broad or squared-off; coracoidal fenestra present and separated from, yet close to, medial margin of shaft; brachial tuberosity not elongate, pointing primarily medially; procoracoid and brachial tuberosity extending about an equal distance mediad; dorsal end of coracoid not noticeably pneumatized; area between brachial tuberosity and procoracoid broadly C-shaped with separation of two structures wide, neither closely apposed, and forming a nearly closed-O; glenoid facet decidedly more extensive in area than scapular facet, semicircular in shape, with straight side bordering medial margin of coracoid; glenoid facet not protruding abruptly beyond lateral margin of coracoidal shaft; in anterior view, dorsal end of coracoid not bulbous, but only slightly broader than shaft that occurs between procoracoid and dorsal end.

Of all the procellariid material available for comparison, the coracoid of Pterodroma seems most similar to AV28671. It should be noted again, however, that Priocella, Pagodroma, and Bulweria, all forms now known in the New Zealand region, were not available for comparison, so any of the following comments may need to be revised in light of further study involving these genera.

With coracoids of Pterodroma, AV28671 shares the following character states: distinct and prominent muscle scar present along medial border of coracoidal shaft just medial to coracoidal fenestra; in medial view, dorsalmost projection of coracohumeral surface at anterior end a distinct bulge with surface tilted gently ventrad towards posterior end; in same view, medial surface of shaft between coracohumeral surface and procoracoid not deeply excavated; viewed dorsally, coracohumeral surface compact, boxy, with medial and lateral margins parallel, posterior margin perpendicular to these, while anterior margin forms acute angles at intersection with medial and lateral margins; posterior end of coracohumeral surface protrudes as a distinct process; about size of Pterodroma mollis; slightly smaller, but perhaps even within the size range of Puffinus gavia.

TARSOMETATARSUS. The tarsometatarsus fragment (AV28665) was assigned to Procellariidae because it possessed the following combination of character states: three calcaneal ridges present, internal longer than external two; shaft deep, with marked channelling on anterior part; channelling present on posterior surface of shaft, but much shallower than that on anterior; internal calcaneal ridge set along internal border of shaft, not displaced towards mid-line of shaft; intercotylar prominence projecting far proximal of cotylae; proximal end not played broadly. Because the tarsometatarsus is only represented by the proximal moiety, and even that is rather worn, particularly in the area of the hypotarsus, it seems rather pointlett to attempt assignment of this form, which is about the size of Halobaena caerulea (but definitely not that taxon), to any category more specific than Procellariidae.

Family Pelecanoididae

Pelecanoides sp.

Material: Sterna (AV28657, AV28658), fragments of anterior part including coracoidal suici and most of the anterior margin of the keel with remainder of element completely eroded; scapulae, right (AV28651) and left (AV28652); coracoids, right (AV28646, AV28647), both lacking sternocoracoidal processes; humeri, right (AV28644, AV28645); ?radius, left (AV28661); ulnae, right (AV28653) and left (AV28659); carpometacarpii, right (AV28648) and left (AV28649); femur, right (AV28664), lacking proximal end but most of shaft and distal end present; ?tibiotarsus, right (AV28654); tarsometatarsus, right (AV28650) (see Fig. 2.6).

Measurements (in millimetres): See Table 2.1.

Diagnosis: the following diagnosis is appropriate at both familial and generic level, because Pelecanoides is the only genus in that family.

HUMERUS. Head palmoanconally compressed, not inflated or even smoothly rounded (P,H);[1] head deeply undercut in two areas that are separated by a pointed projection from the head instead of having a single, large undercut (P,H); bicipital crest forming an angle approaching 90° with internal margin of shaft, rather than a large, obtuse angle (P,H); deltoid crest not prominent, low (P,H); shaft markedly mediolaterally compressed, particularly near distal

1. P, Procellariidae; H Hydrobatidae. Initial given in parentheses after a character indicates those family(ies) which differ(s) in that character from the taxon being diagnosed.

Table 2.1 Measurements (in mm) of Pelecanoides from Mimiomoko Pocket, north Canterbury, N.Z. and modern species

Element/Measurement	Fossil Material		Recent Material			
			P. urinatrix		P. georgicus	
CORACOID	AV28646	AV28647	Range	n	Range	n
1. Length	23.7	21.4	22.9-25.8	14	21.2-22.8	3
2. Width across procoracoid	-	5.5	5.5-6.2	14	5.2-5.8	3
SCAPULA	AV28651	AV28652				
1. Width across head	6.3	6.2	6.4-7.0	14	5.9-6.7	2
HUMERUS	AV28644	AV28645				
1. Total length	44.0	41.8	41.7-45.6	16	38.6-41.7	3
2. Proximal width	8.8	8.7	9.2-11.8	16	8.6-9.3	3
3. Distal width	4.3	4.2	4.4-4.8	16	4.1-4.6	3
ULNA	AV28659	AV28653				
1. Total length	34.5	35.0	33.8-37.1	16	29.2-31.8	3
2. Proximal width	4.0	4.1	4.0-4.5	16	3.8-4.1	3
3. Distal width	4.8	5.1	4.7-5.7	16	4.3-4.6	3
CARPOMETACARPUS	AV28648	AV28649				
1. Length	24.7	24.1	23.7-26.7	15	21.4-22.2	3
2. Proximal width (across carpal trochleae to metacarpal I process)	5.5	5.8	5.7-6.3	15	5.5-5.6	3
TARSOMETATARSUS	AV28650					
1. Total length	24.3		24.9-28.8	13	21.9-23.4	3
2. Proximal width	3.8		4.2-5.1	13	3.9-4.2	3
3. Distal width	3.7		4.2-4.8	13	3.7-4.1	3
4. Minimum shaft width	1.8		1.7-2.3	13	1.4-1.8	3

end (P. H); viewed distally, distal end square, not rectangular in outline (P.H); brachial depression shallow, not deep (P); development of attachment for pronator brevis slight (H); olecranal fossa deeply excavated, forming deep pit in distal end (F);[2] entepicondyle projects far ventrally (F).

ULNA. Prominent process projecting dorsal to elecranon and lying between cotyla (P, H); entire bone markedly mediolaterally compressed, rather than being more gently rounded in cross-section (P, H); robust process projecting dorsodistally and lying lateral to tendinal groove (less well developed in most P, H).

CARPOMETACARPUS. Process of metacarpal I elongate proximodistally but abbreviate dorsoventrally, giving a "snub-nosed" appearance (P, H); deep, elongate excavation present on internal surface of head just dorsal to attachment for pisiform process (P, H); attachment of pisiform process not prominent, only slightly indicated (prominent in most Pu);[3] lacking or with only slight development of indentation in dorsal margin of internal carpal trochlea (F); entire proximal end and shaft markedly compressed dorsoventrally (P, H).

TARSOMETATARSUS. Intercotylar prominence not as well developed as that in most Procellariidae; posterior surface of shaft planar and oriented at nearly right angle to lateral and medial shaft surfaces, not tiled otherwise or rounded (P); distal end does not flare much beyond nearly parallel medial and lateral margins of shaft (most P, although Daption approaches this condition); posterior surface between trochleae highly excavated (most P).

STERNUM. Deep excavation present at base of manubrial spine (P, H); flange present along ventral margin of coracoidal sulcus at lateral edge (H); in ventral view, angle formed at manubrial spine by margins of two coracoidal sulci large, obtuse, approaching 180°, not 90° or less (meaning that ventral ends of coracoids articulating here are directed mediolaterally, rather than anteromedially-posterolaterally (P)).

CORACOID. Brachial tuberosity not protruding far medially, and thus head of coracoid more in line with main shaft axis (P, H); procoracoid not protruding far dorsally, and thus entire medial margin of shaft and head between brachial tuberosity and procoracoid much less highly curved (P, H); coracoid much more slender and elongate than in Hydrobatidae, not flaring into broad sternocoracoidal process as far proximally as in Hydrobatidae, thus flaring only very near distal end, and producing a straight lateral margin over much of coracoidal length (H); width of sternal facet about equal to width across proximal end of bone at level of procoracoid, not decidedly more (H).

SCAPULA. Proximal articular facet dumbell-shaped with slight expansions at both medial and lateral ends, rather than being broadest laterally (H) or having a marked medial constriction as well as a marked lateral expansion (Pu).

Discussion: Although several bones (femur, tibiotarsus, and radius) other than those diagnosed above have been tentatively assigned to Pelecanoides, they have not been diagnosed due to incompleteness and thus some uncertainty of their identity. The single femur lacks a proximal end and thus is difficult to assign with any high certainty. As with most of the hind limb elements, the

1. F, Fulmarinae: see footnote 1 on previous page.
3. Pu, Puffininae: see footnote 1 on previous page.

tibiotarsus is rather generalized and assigned to the Pelecanoididae primarily on the basis of size and proportions, a most unsatisfactory basis.

Assignment of the Pelecanoides from Mimiomoko Pocket to any one of the living species, based on the modern samples available in this study is problematical. Three of the living species listed in Peters (1931) were unavailable for comparison. No consistent qualitative differences can be found to allow distinction of the two modern New Zealand species from one another or the Mimiomoko form(s). Based on a limited modern sample, Pelecanoides urinatrix and P. georgicus can be distinguished in many cases on the basis of size, P. georgicus being consistently smaller. For some bony elements there is overlap in size, where only specimens on either extreme can be confidently assigned to one or the other species (see Table 2.1). As Table 2.1 illustrates, the Mimiomoko Pelecanoides do not consistently fit into either species. Certain measurements fall only within the range of P. urinatrix (coracoidal length, ulnar length, ulnar distal and proximal widths, carpometacarpal length). Other measurements fall only within the range of P. georgicus (scapular width, proximal and distal humeral widths, distal tarsometatarsal width). Yet some measurements fall within the range of overlap of P. georgicus and P. urinatrix (width across procoracoid, humeral length, proximal ulnar width, proximal carpometacarpal width, minimum tarsometatarsal shaft width). One measurement (tarsometatarsal length) lies intermediate between those of the two extant species, and another (proximal tarsometatarsal width) is smaller than measurements of both species.

Measurements that are only in the range of the larger Pelecanoides urinatrix tend to be the length measurements of bones, whereas width measurements tend to more closely approximate P. georgicus, thus, perhaps suggesting the presence of a third species with more slender elements, but tending to be about the same overall size as P. urinatrix. Perhaps, however, if modern samples were larger, the Mimiomoko form would simply reside in that intermediate "no-man's land" where P. urinatrix and P. georgicus cannot be distinguished. For the moment, then, the Mimiomoko material will be assigned only to Pelecanoides sp.

Family Rallidae
Gallirallus cf. minor (? = australis)

Material: Left carpometacarpus (AV28672); left radius (AV28656); left proximal and distal fragments of ulna (AV28660 and AV28655) (See Fig. 2.7).

Measurements (in millimetres): Carpometacarpus. See Table 2.2. Radius. Total length, 28.6; width of proximal end, 3.9; width of distal end, 3.4; maximum palmoanconal depth of proximal end, 2.3; minimum shaft width, 1.2. Ulnae Fragments. Proximal end, maximum width, 4.4; maximum depth 4.1. Distal end, maximum width, 4.4; maximum depth, 4.1.

Diagnosis: CARPOMETACARPUS. Placed in the Rallidae because of the following combination of character states: neither trochlea extending markedly further proximal than the others; trochleae not compressed proximodistally with respect to depth of proximal end (= distance across internal condyle and process of metacarpal I); deep pit present between and at distal

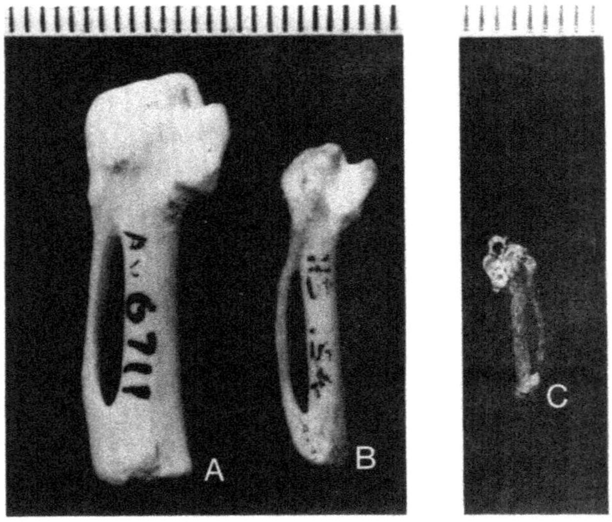

Fig. 2.7 Carpometacarpi of fossil Rallidae from the New Zealand region. A, *Diaphorapteryx hawkinsi*, AV6711; B, *Gallirallus* cf. minor, AV28672, from Mimiomoko Pocket; and C, *Capellirallus karam* AV20615. Scale, each division equals 1 mm.

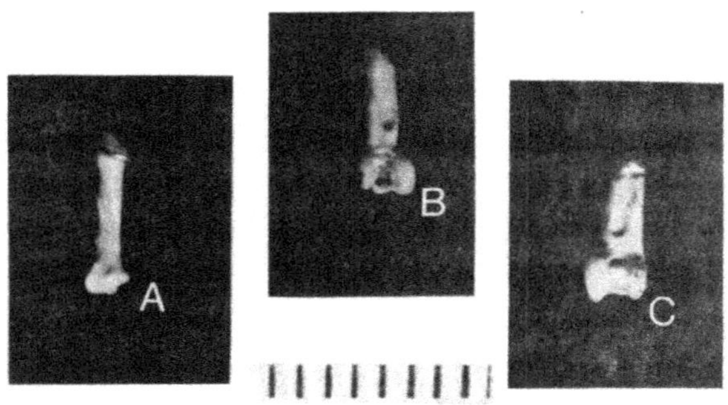

Fig. 2.8 Passeriform birds from Mimiomoko Pocket: A, Ulna (distal), AV28669. Distal tibiotarsi: B, AV28668; and C, AV28670. Scale, each division equals 1 mm.

bases of external and internal trochleae, just proximal to where metacarpal III joins remainder of bone; metacarpal III originating proximally primarily off metacarpal II rather than from base of carpal trochleae; process of metacarpal I broad (or elongate) proximodistally, not narrow and not tremendously elongate; ligmental attachment of pisiform process set back towards metacarpal III, not set towards base of metacarpal I process; intermetacarpal tuberosity that in some avian groups approaches or joins metacarpal III lacking; in cross-section, metacarpal II has rounded outlines, not angular; facet for digit II extending further distad than that for digit III; metacarpal III closely opposed to metacarpal II, not broadly bowed away from it, thus making the intermetacarpal space relatively narrow; metacarpals II and III fused distally over broad area (elongate proximodistally).

RADIUS. Considered rallid because of the following combination of character states: in proximal view, shape of humeral cotyla rectangular; cotyla compressed palmoanconally, not mediolaterally; small capital process present, triangular in shape. Bicipital tubercle located very near proximal end, merging directly into it, not displaced distally; distal end slightly, not broadly splayed; splaying, although slight, beginning near proximodistal midpoint of shaft so that shaft not parallel-sided over distal half; distal end not highly curved palmarly; in palmar view, distal end not deeply undercut, distal margin of radius rounded (curved), not straight; in distal view, scapholunar facet not extremely slender palmoanconally, particularly inflated on side of ligamental prominence, while either side tapers to a point; expansion palmarly and anconally about equal; shaft of radius rather rounded, not highly compressed; shaft short and straight, not significantly curved in any direction.

ULNAE. Assigned to the Rallidae because of the following combination of character states: Proximal end. Olecranon a broad-based, triangular process, not narrow; in proximal view, proximal end triangular in shape; external cotyla not projecting abruptly laterad, margins nearly, if not entirely, straight, not irregular in outline; external and internal cotyla extending about same distance palmarly, one not projecting decidedly further than the other; no proximally projecting processes present on proximal end, palmar to olecranon; in lateral view, external cotyla highly recurved proximally, giving hook-like appearance to cotylar margin; in palmar view, proximal articular surface somewhat undercut, shaft not gradually merging into proximal end; prominence for anterior articular ligament not well developed; proximal end of shaft significantly mediolaterally compressed. Distal end. Carpal tuberosity low and rounded, not slender and prominent, not forming narrow process that projects distally; distal part of shaft compressed somewhat mediolaterally; external condyle short proximodistally and lacking any processes on lateral side; in lateral view, external condyle smoothly rounded, not flattened distally; distinct tendinal pit present on palmar edge of external condyle; no marked excavations present on palmar surface of shaft at proximal base of internal condyle; in distal view, distal end triangular in shape, gradually diverging from a carpal tuberosity apex, lacking irregular outlines or abrupt changes in directions of the margins; in distal view, groove between internal and external condyles, particularly on anconal surface, not deep.

Discussion: When compared with bones of other rails, both fossil and recent, from New Zealand and nearby islands, AV28672 appears to most closely approach the morphology of Gallirallus. The Mimiomoko carpometacarpus is a short, robust element (see Table 2.2) with an enlarged proximal end relative to the remainder of the bone. This combination of characters rules out inclusion of this form in Porphyrio (P. porphyrio), Rallus (R. pectorallis, philippensis), Porzana (P. pusilla, P. tabuensis), Gallinula (G. tenebrosa, G. chloropus), Fulica (F. atra, F. chathamensis [=Nesophalaris prisca]), Tribonyx ventralis, and probably T. hodgeni (names of rails of New Zealand follow Olson, 1975). On the other hand, the Mimiomoko carpometacarpus is not nearly so robust as is that of Diaphorapteryx hawkinsi, which is only slightly more elongate. The Mimiomoko bone is twice the size of Capellirallus karamu and further differs in having a more angular outline of the internal carpal trochlea, not being smoothly curved over much of its circumference, having a much more prominent proximal metacarpal I process, and having metacarpal III originate off the shaft of metacarpal II rather than smoothly grading into the base of the carpal trochleae. It differs from Notornis mantelli in being less than half the size, in that the proximal end of the Mimiomoko bone is much less dorsoventrally compressed, and in that the pisiform process is much more prominent. Although approximately the same size, the Mimiomoko Gallirallus differs from the Chatham Island G. (= Nesolimnas) dieffanbachii in that metacarpal III originates proximally off metacarpal II and not off the base of the carpal trochleae and is thus relatively shorter; metacarpal II is more robust; the carpal trochleae appear to be more enlarged relative to the rest of the bone; and the facet for digit II is much more dorsoventrally compressed.

In all the above-mentioned characters, the carpometacarpii of Gallirallus australis and that from Mimiomoko Pocket are identical. The only differences between the two that can be noticed are: AV28672 is somewhat smaller than any of the G. australis in our measured sample, and the long axis of the metacarpal I process of AV28672 is not shifted as far proximally as in most G. australis.

Because of the modifications of the forelimb associated with loss of flight so common in rails, bones of the wing are of limited use in determining relationships within many of the Rallidae. Certainly the rallid carpometacarpus and radius from Mimiomoko Pocket indicates a bird of at least reduced flying ability. The carpometacarpus shows the greatest number of similarities to Gallirallus australis, but is smaller than any we have been able to measure (see Table 2.2). Perhaps if Gallirallus minor is indeed distinct from G. australis (see Olson, 1975, for a discussion of this problem), the Mimiomoko rallid belongs in the former species. This question, as Olson (1975) points out, is yet to be resolved, however. Thus, for the present, until more comprehensive work is carried out on New Zealand fossil rails, assignment of the Mimiomoko rail material will tentatively be made to Gallirallus cf. minor.

Order Passeriformes

Material: Distal end of left ulna (AV28666); two distal ends of right tibiotarsi (AV28669, AV28670); distal end of left tibiotarsus (AV28668) (see Fig. 2.8).

Table 2.2 Measurements (in mm) of Rallidae from Mimiomoko Pocket, north Canterbury, N.Z., and extant or fossil rails from the Australasian, mainly New Zealand, region

Taxa	Measurements of Carpometacarpi				
		Proximal width	Width across distal end (from tuberosity of metacarpal II to facet for digit III)	Length of metacarpal I	Minimum width of metacarpal II
AV28672	22.2	6.5	≯3.6	4.3	2.3
<u>Capellirallus karamu</u> (AV20615) (r)	@11.1	3.4	-	@2.8	@1.2
<u>Diaphorapteryx hawkinsi</u> (n=3)	25.1-28.1	9.5-10.6	6.2-6.8	6.2-6.9	3.3-3.6
<u>Fulica atra</u> (n=14 for length; 1 for others NMV W4573 (l)	39.1-45.0	8.1	4.4	5.2	2.8
<u>Fulica chathamensis</u> (CSIRO-BS1657) (l)	46.9	10.7	6.4	6.7	3.8
<u>Gallinula tenebrosa</u> (NMV W6821) (l)	38.4	7.8	4.1	4.8	2.4
<u>Gallirallus australis</u> (n=7)	25.8-27.9	6.9-7.6	4.3-4.8	5.4-6.7	2.6-2.8
<u>Gallirallus dieffenbachii</u> (CSIRO-BS1660) (r) [2]	21.8	≯6.1	@3.9	> 3.6	1.9
<u>Notornis mantelli</u> (DM17177) (cast, l)	45.4	10.9	5.9	7.4	3.9
<u>Porphyrio porphyrio</u> (NMV W6813) (l)	50.9	9.2	5.4	6.2	3.2

(contd.)

Table 2.2 (contd.)

	Length	Proximal width	Width across distal end (from tuberosity of metacarpal II to facet for digit III)	Length of metacarpal I	Minimum width of metacarpal II
Porzana pusilla (NMV W5552) (l)	13.7	2.9	1.8	2.0	1.1
Porzana tabuensis (NMV W6866) (l)	15.2	3.2	1.8	1.8	1.2
Rallus pectoralis (NMV B11094) (l)	19.1	4.3	2.4	2.6	1.4
Rallus philippensis (NMV B5234) (r)	27.3	5.7	3.3	3.6	2.1
Tribonyx ventralis (NMV W6814) (l)	40.2	8.0	4.6	4.9	2.9

1 Left Element
2 Right Element

Measurements: See Table 2.3.

Diagnosis: The following diagnoses are appropriate only to the ordinal level. More refined diagnoses were not attempted owing to lack of availability of a complete suite of skeletal material of New Zealand passeriform families. The combination of character states, and not necessarily any single character, is diagnostic.

ULNA. Small size; papillae of secondaries well developed as promiment knobs; carpal tuberosity slender, coming to a point internally, not squared off as in Alcedinidae or bulky as in Cuculidae, and not projecting primarily distally as in Turnicidae; distal radial impression deep; shaft near distal end straight, lacking any curvature between it and distal end; in distal view, distal articular surface neither extremely deep nor mediolaterally compressed; lateral to external condyle lies short flange producing deep channel between it and external condyle for passage of tendon.

TIBIOTARSI. Small size; condyles deep and of subequal depth and length; in lateral and medial views, margins of condyles describing a half circle; condyles subparallel to long axis of shaft, not angled, and not displaced to one side of shaft; in distal view, condyles nearly subparallel, only slightly diverging anteriorly; condyles projecting far anterior to shaft but only slightly posterior; supratendinal bridge "broad" proximodistally (elongate), lying nearly, if not, perpendicular to long axis of shaft; intercondylar fossa well excavated but raised area present connecting proximal ends of condyles on anterior surface; distal opening of tendinal canal located centrally, not noticeably laterally or medially displaced, opening just proximal to bar connecting condyles.

Discussion: As many as three different kinds of passeriforms may be represented by the 3 tibiotarsi, but much better comparative material is needed to ascertain that those few differences noted cannot be matched by variability within one species. AV28668 has a more slender shaft and is the smallest of the three tibiotarsi; the long axes of the distal condyles are twisted at a slight angle laterad of the long axis of the shaft, thus not lying parallel with it; it thus differs from AV28670 with which it shares a broad (proximodistally) supratendinal bridge. AV28670 and AV28669, although very similar in size, differ in that AV28670 has a very broad supratendinal bridge while AV28669 has a much narrower one. Additionally, the tendinal canal is centrally located in AV28669 but displaced laterally in AV28670.

In size AV28668 is comparable to that of Zosterops lateralis (the Silvereye), while AV28669 and AV28670 are somewhat larger. The ulna (AV28666), on the contrary, is markedly smaller than that of Z. lateralis. Thus, unless limb proportions of the fossil form are distinctly different than those exhibited by Z. lateralis (and they well could be), as many as 3 or 4 different species of passeriforms could be represented in the Mimiomoko sample.

Bias of Mimiomoko Sample

The sample of bones from Mimiomoko Pocket by no means represents all, or even nearly all, elements present in a vertebrate skeleton. To demonstrate this, Table 2.4 is a tabulation of the tetrapod sample by major categories of elements. Owing to the small sample size, only broad groupings of elements were tabulated. Many elements used in this tabulation were too fragmentary

Table 2.3 Measurements (in mm) of Passeriformes from Mimiomoko Pocket, north Canterbury, New Zealand

ULNA	AV28666		
1. Width across condyles	1.6		
2. Depth of external condyle	1.2		
3. Width of shaft at base of external condyle	0.8		
TIBIOTARSUS	AV28668	AV28669	AV28670
1. Width across distal end of condyles	-	2.4	$\ngtr$ 2.2
2. Depth of internal condyle	@ 1.9	@ 2.3	-
3. Depth of external condyle	-	$\ngtr$ 2.3	-
4. Length of internal condyle	-	1.7	-
5. Length of external condyle	-	2.0	1.9

Table 2.4 Tetrapod Elements Represented at Mimiomoko Pocket

	Rib	Skull and Mandible	Vertebrae	Limb Girdles	Limbs	Phalanges
Reptilia						
Number	3	4	7	4	21	25
% of reptiles	4.7	6.3	10.9	6.3	32.8	39.1
Bird						
Number	0	1	6	16	28	29
% of birds	0	1.3	7.5	20.0	35.0	36.2

to be identified more precisely other than as bird or reptile, hence the discrepancy between the number of specimens listed in the systematics section and enumerated here. Of the 199 bones or bone fragments in the tetrapod sample, 55 could not be identified to either Class or element.

Perusal of Table 2.4 reveals a marked bias towards rod-shaped over spherical or disc-shaped elements. This is strikingly evident in the case of the limb girdle elements. Among the reptiles, these elements are primarily disc-shaped and occur infrequently in the sample. By contrast, birds have many rod-shaped limb girdle elements such as the clavicle, scapula, and coracoid and these occur three times as frequently as in the reptilian case. Such post-mortem sorting on the basis of shape was most likely owing to mechanical action by either flowing water or wind.

Conclusions

Except for the possible presence of a kiwi, the tetrapod sample from Mimiomoko Pocket could be duplicated on Stephens Island in Cook Strait, New Zealand, today. Before the drastic alteration of the mainland New Zealand fauna by the exotic rodents and carnivores introduced by man, this same suite occurred there as well.

The sample appears to represent an accumulation in a shallow cave or burrow. The best represented taxa in the sample are the very ones to be expected in burrows where the tuatara and seabirds live alongside one another at the present time. Between the death of the tetrapods and their final burial, either wind or flowing water brought about mechanical sorting of the bones so that rod-shaped elements were preferentially preserved over spherical or disc-shaped ones.

Tuatara had disappeared from mainland New Zealand by 200 years ago. With this in mind together with the fact that none of the tetrapods introduced by man are represented in the collection from Mimiomoko Pocket, it is likely to be still older, possibly more than one thousand years. The high concentration of collagen in the bones argues against a significantly greater age than that.

Mimiomoko Pocket was originally believed to be comformable within the Miocene Double Corner Shell-beds. Primarily the presence of an unexpectedly large concentration of collagen in the tetrapod bones makes this interpretation now unacceptable. Without that critical piece of evidence, the site would still be regarded as Miocene. Unfortunately, the tetrapods themselves as yet do not provide a safeguard to prevent a similar error in the future. Without some well dated pre-Pleistocene fossil tetrapod assemblages as standards for comparison, it is unreasonable to rule out a pre-Pleistocene age for a modern-looking suite of terrestrial vertebrates. Small tetrapods in New Zealand may have been allied with living genera by the Miocene or even earlier in the Tertiary. Modern genera of birds and reptiles elsewhere in the world frequently have records extending well back into the Tertiary, quite in contrast to the mammals where only a few living genera are known before the Pliocene.

Afterward

The original impetus for two of us (TR & PR) to work on this project was the possibility that it represented the first pre-Pliocene terrestrial vertebrate fauna in New Zealand. New Zealand has a unique potential to test the ideas of MacArthur and Wilson (1967) on the relationship between island area and faunal diversity over an extended period of geologic time. The almost continuous record there of terrestrial sediments from the Cretaceous to the Holocene has the potential for yielding tetrapod remains (Rich, 1975).

The limited diversity of the New Zealand tetrapod fauna is owing in part, no doubt, to its long isolation from other land masses since the Cretaceous (Griffiths and Varne, 1972). However, a second factor could have had a long term effect on the diversity of the fauna. During the Landon (Oligocene), a widespread marine transgression occurred that probably reduced New Zealand to an archipelago of less than one-third its present area (Stevens, 1973). Indirect evidence of a decrease in diversity at this time is furnished by the marked turnover in the New Zealand flora during the Oligocene (Fleming, 1963). In light of this "Atlantis Factor" in New Zealand's history, beyond assuming that the highly endemic groups were present, only the discovery of pre-transgression fossil assemblages will provide an adequate basis for a characterisation of the tetrapod fauna there prior to the Oligocene.

Acknowledgements

Many persons and institutions assisted in various phases of this project. Sir Charles Fleming, New Zealand Geological Survey, Lower Hutt gave his unstinting support through all phases of this project, supplying much useful advice, technical assistance and comparative materials. Comparative materials was also supplied by Dr. Ian Crook, New Zealand Wildlife Service, Wellington; Mr. Alan McEvey, National Museum of Victoria, Melbourne; Dr. Jerry Van Tets, Wildlife Research, C.S.I.R.O. Canberra; and Dr. Anthony H. Whitaker, Ecology Division, D.S.I.R., Lower Hutt.

Technical assistance was provided by several persons identified in the appropriate places of the text. To them we wish to express our thanks.

Dr. Roger Duff, Director of the Canterbury Museum, Christchurch, made available the facilities of the institution, and many members of the staff provided many needed services, which greatly facilitated the work.

Mr. and Mrs. Allister Schroeder and family, Amberley, were most gracious hosts during the time when the collections were made, permitting the use of their motel for the somewhat unusual demands of processing of the matrix samples as well as assisting in the fieldwork.

The fieldwork related to this project was supported by a grant no. 1276 from the National Geographic Society. Subsequent study was carried out in part under the auspices of the Australian-American Educational Foundation and the M. A. Ingram Trust.

Mr. Robert Suddarth, Geosciences Department, Texas Tech University, drafted Fig. 2.2.

Mrs. Mary Lee Vickers and Miss Kayelene Voege skillfully typed the manuscript.

BIBLIOGRAPHY

Berggren, W. A. 1972. A Cenozoic time-scale—some implications for regional geology and paleobiogeography. Lethaia 5:195-215.

Broili, F. 1925. Beobachtungen an der Gattung Homoeosaurus H. v. Meyer. Sitz.-Ber. bayer. Akad. Wiss. 1925, Heft 2:81-121.

Crook, I. G. 1975. The Tuatara. In: G. Kuschel, (ed.) Biogeography and Ecology in New Zealand. W. Junk, The Hague. 331-352.

Endo, R. 1940. A new genus of Thecodontia from the Lycoptera Beds in Manchoukuo. Bull. Cent. Nat. Mus. Manchoukuo 2:1-14.

Fleming, C. A. 1963. Paleontology and southern biogeography. In: Tenth Pacific Science Congress. Bishop Museum Press, Honolulu. 369-385.

Gilmore, C. W. 1909. A new rhynchocephalian reptile from the Jurassic of Wyoming, with notes on the fauna of "Quarry 9." Proc. U.S. Nat. Mus. 37:36-52.

Gregg, D. R. 1959. Stratigraphy of the Lower Waipara Gorge, North Canterbury. New Zealand Jour. Geol. Geophys. 2:501-527.

Griffiths, J. R. and R. Varne. 1972. Evolution of the Tasman Sea, Macquarie Ridge and Alpine Fault. Nature Phys. Sci. 235:83-86.

Hawkins, D. N. 1957. Beyond the Waimakariri: A Regional History. Whitcombe and Tombs, Christchurch.

Hoffstetter, R. 1955. Rhynchocephalia. In: J. Piveteau, (ed.) Traite de paleontologie, Tome V. Masson, Paris. 556-576.

Hoffstetter, R. and J. P. Gasc. 1969. Vertebrae and ribs of modern reptiles. In: C. Gans, (ed.) Biology of the Reptilia. Vol. 1. Morphology A. Academic Press, New York and London. 201-310.

Kinsky, F. C. 1970. Annotated Checklist of the Birds of New Zealand. A. H. and A. W. Reed, Wellington.

MacArthur, R. H. and E. O. Wilson. 1967. The theory of island biogeography. Princeton Univ. Press, New Jersey.

McKenna, M. C. 1965. Collecting microvertebrate fossils by washing and screening. In: B. Kummel and D. Raup, (eds.) Handbook of Paleontological Techniques. W. H. Freeman and Co., San Francisco. 193-203.

Olson, S. L. 1975. A review of the extinct rails of the New Zealand region (Aves: Rallidae). Nat. Mus. New Zealand, Rec. 1:63-79.

Peters, J. L. 1931. Checklist of Birds of the World, 1. Harvard University Press, Cambridge.

Rich, T. H. V. 1975. Potential pre-Pleistocene fossil tetrapod sites in New Zealand. Mauri Ora 3:45-54.

Romer, A. S. 1956. Osteology of the Reptiles. Univ. Chicago Press, Chicago.

Stevens, G. R. 1973. The palaeogeographic history of New Zealand. New Zealand Entomologist 5:230-239.

Welles, S. P. 1947. Vertebrates from the upper Moenkopi Formation of northern Arizona. Univ. Calif. Publ., Bull. Dept. Geol. Sci. 27:241-289.

Weston, R. J., C. A. Repenning, and C. A. Fleming. 1973. Modern age of supposed Pliocene seal, *Arctocephalus caninus* Berry (= *Phocarctos hookeri* Gray), from New Zealand. New Zealand Jour. Sci. 16:591-598.

AVIFAUNAL COMPOSITION OF SKELETAL MATERIAL COLLECTED FROM SKUA MIDDENS AND BEACHES AT SOME AUSTRALASIAN SUB-ANTARCTIC ISLANDS

G. F. van Tets

Introduction

Skeletal material of birds that have died recently from natural causes can help in evaluating the status of the species frequenting an area. It can also indicate the food of predators, give a measure of relative abundance and habitat preference, and reveal the presence of species not otherwise recorded.

With the assistance of members of the 1972/73 Auckland Islands, 1973/74 Macquarie Island and 1975/76 Campbell Island Expeditions I collected skeletal material of birds at Enderby Island, Disappointment Island and Sugar Loaf Rocks in the Auckland Islands; Hasselborough Bay, Macquarie Island and at Campbell Island and nearby Dent Island. The material has been deposited in the collection of the National Museum of New Zealand, Wellington, and the Australian National Wildlife Collection, the Division of Wildlife Research, CSIRO, Canberra. Logistic support for reaching the islands was provided by the Wildlife Service and the Department of Lands and Survey of New Zealand, and by the Antarctic Division of Australia. The islands and their birds are described in Bailey and Sorenson (1962—Campbell Island); Law and Burstall (1956—Macquarie Island): and Yaldwyn (1975—Auckland Islands). Further papers on the birds of the islands are in preparation. The nomenclature is based on Condon (1975) and Kinsky (1970). Methods for processing the material were as described in Mestjens et al. (1975) and Scarlett (1972).

Results

Most of the material (see Table 3.1), was of birds that had been eaten by skuas although it was not possible to determine how many of these had actually been killed by skuas. Colour-banding of skuas has indicated that some of them have a daily foraging range of at least 8 km. Some of the material was exposed by erosion after burial in sand dunes and all the material had been exposed to sun and rain. If it had been left in situ it would have disintegrated within about a year and may therefore be mostly regarded as less than a year old.

Penguins

The distribution of penguin material reflects the nesting distribution of the species represented. No material was found of species which are presently rare in the islands, e.g. the Erect-crested Penguin (*Eudyptes sclateri*). Most of the King Penguin bones were found, sub-fossil, in an eroding sand dune at the Macquarie Island Base Camp. Material of at least seven Rockhopper Penguins was found at Enderby Island where the species did not breed and was not seen ashore at the time of the 1972/73 Expedition

Albatrosses

Considering the presence of huge nesting colonies of several species of albatrosses and mollymauks in the islands, very little material of them was found. The Royal Albatross material was found mainly beside their nest sites on Campbell Island and sub-fossil in the sand-dunes of Enderby Island. Beside skua nests on Campbell Island were found bones of very small albatross or mollymauk chicks which were indeterminable to genus and species.

Petrels

Except for species of which only traces were found the distribution of the material reflects the breeding distributions of species presently represented on each island. It was not possible to determine to species all the Giant Petrel material. Probably most of the material collected at Macquarie Island is of Macronectes giganteus, and that collected at the other islands of M. halli. Large numbers of White-headed Petrel, Sooty Shearwater and Common Diving Petrel were found to have been killed by skuas. No trace was found of the South Georgian Diving Petrel (Pelecanoides georgicus), which may have formerly bred on Enderby and Macquarie Islands, in sand-dunes.

The material of the Short-tailed Shearwater which was found at Macquarie Island may be of non-breeding birds foraging in the Southern Ocean during the austral summer (see Shuntov 1974). No trace of this species was found at Campbell and Auckland Islands, which are situated further north, but are presumably outside the foraging range of the birds breeding around Tasmania and south-eastern Australia.

Bones of prions were much more abundant in skua regurgitation pellets at Enderby than at the other islands. Broad-billed, and Medium-billed Prions, and the somewhat similar Blue Petrel, have not been previously reported in the Auckland Islands. The Mottled Petrel and the Little Shearwater have been reported breeding in the Auckland Islands but were not seen during the 1972/73 Expedition.

No material was found of the Cape Pigeon (Daption capense), which was found breeding in the Auckland Islands for the first time during the 1972/73 Expedition. It was suspected of breeding at Campbell Island during the 1975/76 Expedition. It was commonly seen following ships and fish schools around the islands.

Shags

The shag material mainly consisted of loose bones on beaches and in nesting colonies. The species were Leucocarbo colensoi at Enderby Island, L. campbelli at Campbell Island, and L. albiventer at Macquarie Island. Material was not found of the Black Cormorant, (Phalacrocorax carbo) which occurs in the islands as a vagrant, nor of the Little Pied Cormorant (P. melanoleucos), of which there was a small breeding population on Campbell Island.

Ducks

Considering the small populations surviving of the Auckland Island Teal (Anas a. aucklandica) and the Campbell Island Teal (Anas a. nesiotis), a rather large number of fresh remains were found to have been eaten by skuas. The material of the teal has been used in the preparation of a revision of the Australasian Teal that is being drafted by Russ, Robertson and van Tets. Previously known material of the Campbell Island Teal consisted of only two museum skins.

Grey Duck, Mallard and their hybrids occur in small numbers on most of the ponds and inlets of the islands, but only a couple of bones of them were found.

Rails

No material was found of the Auckland Island Rail (Rallus muelleri) which may still be extant on Adams Island and elsewhere in the Auckland Islands, but was not found during the 1972/73 Expedition.

Only four bones of at least two birds were found of the Weka, (Gallirallus australis), at Macquarie Island where it was frequently seen and heard during the 1973/74 Expedition.

Waders

The only wader material found was of the Island Snipe at Enderby Island. This species was seen again on Enderby Island during the 1972/73 Expedition after not having been sighted there since 1891 (Taylor in Yaldwyn 1975:239).

Waders are uncommon in the islands and only at Enderby Island were they regularly seen. Most numerous were Turnstones (Arenaria interpres) in small flocks of up to forty birds along the shores, and Banded Dotterels (Charadrius bicinctus), dispersed in pairs on the upland heaths.

Skuas, Gulls and Terns

Relatively little material was found of gulls and terns. No material was found of the two gull species at Campbell Island where they occur in the inlets and along the shores. It was not possible to identify the tern material to species. The Antarctic Tern (Sterna vittata) breeds in the islands and was the species most often seen. The Arctic Tern (S. paradisaea) is a rare visitor to the islands, and the White-fronted Tern (Sterna striata) breeds in the Auckland Islands and is a rare visitor further south. All three species are similar in size.

Land-birds

Of ten species of land-birds seen at Enderby Island during the 1972/73 Expedition, material of only three species was found in skua regurgitation pellets. In particular it was surprising that no material was found of the Red-crowned Parakeet (Cyanoramphus novaezelandiae) which was common on Enderby Island.

DISCUSSION

As a survey method the collection of skeletal material of birds has a different bias in the qualitative representation of the species composition of an area from that which is obtained by visual or aural censusses, or by capture or collection of living birds.

In this study the collection of skeletal material failed to record the presence of several species of waders and landbirds. In some localities it missed the presence of gulls and ducks. It did, however, record the presence of three species of petrels at the Auckland Islands which had not been recorded there before, and two species which were not otherwise found during the 1972/73 Expedition. The differences may be due mainly to the propensity of the species to be preyed on by skuas, to wash up dead on beaches and to be found dead in open, exposed situations.

The results of this study may provide some help in the evaluation of the representativeness of the avifaunal composition of material found in prehistoric deposits.

BIBLIOGRAPHY

Baily, A. M., and Sorensen, J. H., 1962. Subantarctic Campbell Island. Proc. No. 10 Denver Museum of Natural History.

Condon, H. T., 1975. Checklist of the Birds of Australia, I. Non-passerines. RAOU Melbourne.

Kinsky, F. C. (Convener), 1970. Annotated Checklist of the Birds of New Zealand. A. H. & A. W. Reed Wellington.

Law, P. G. and Burstall, T., 1956. Macquarie Island. A.N.A.R.E. Interim Reports 14.

Scarlett, R. J., 1972. Bones for the New Zealand Archaeologist. Canterbury Mus. Bull. 4.

Shuntov, V. P., 1974. Sea Birds and the Biological Structure of the Ocean. National Science Foundation Washington. (Translated from Russian).

Vestjens, W. J. M., D'Andria, A. H. and van Tets, G. F., 1975. Notes on the Preparation of Osteological Specimens. CSIRO Division of Wildlife Research, Tech. Memo No. 10.

Yaldwyn, J. C. (Ed.), 1975. Preliminary Report of the Auckland Islands Expedition 1972-1973. Dept. of Lands and Survey, Wellington.

Table 3.1 Skeletal material of birds found at Auckland, Campbell and Macquarie Islands. X = material of one or two birds, XX = material of three to nine birds, and XXX = material of ten or more birds.

Species	Auckland Islands			Campbell Island		Macquarie Island
	Enderby Island	Disappointment Island	Sugar Loaf Rocks	Campbell Island	Dent Island	Hasselborough Bay
Aptenodytes patagonicus King Penguin						XXX
Megadyptes antipodes Yellow-eyed Penguin	XXX			XX		
Pygoscelis papua Gentoo Penguin						XX
Eudyptes chrysocome Rockhopper Penguin	XX			XXX		XXX
E. chrysolophus schlegeli Royal Penguin						XXX
Diomedia exulans Wandering Albatross	X					
D. epomophora Royal Albatross	XX			XX		
D. cauta Shy Mollymauk	XX					

Table 3.1 (Cont.)

Species	Auckland Islands			Campbell Island		Macquarie Island
	Enderby Island	Disappointment Island	Sugar Loaf Rocks	Campbell Island	Dent Island	Hasselborough Bay
Phoebetria palpebrata Light-mantled Albatross				X		
Macronectes sp. Giant Petrel	XX	XX			X	XX
Fulmarus glacialoides Antarctic Fulmar						X
Pterodroma lessoni White-headed Petrel	XXX	XXX	XXX	X		X
P. inexpectata Mottled Petrel	XX					
Halobaena caerulea Blue Petrel	X					
Pachyptila vittata Broad-billed Prion	X					
P. salvini Medium-billed Prion	XX			X		
P. desolata Antarctic Prion	XXX	X	X	XX		XX

Table 3.1 (cont.)

Species	Auckland Islands			Campbell Island		Macquarie Island
	Enderby Island	Disappointment Island	Sugar Loaf Rocks	Campbell Island	Dent Island	Hasselborough Bay
Procellaria aequinoctialis White-chinned Petrel	XX	XX	X		X	
Puffinus griseus Sooty Shearwater	XXX	XXX	X	XX	XXX	X
P. tenuirostris Short-tailed shearwater						XX
P. assimilis Little Shearwater	XX					
Garrodia nereis Grey-backed Storm-Petrel	X					
Pelagodroma marina White-faced Storm-Petrel	X					
Fregetta tropica Black-bellied Storm-Petrel	XXX			X	X	
Pelecanoides urinatrix Common Diving-Petrel	XXX		X	XXX	XXX	
Leucocarbo sp. Island Shag	XXX			XXX		XXX
Anas platyrhynchos Mallard				X		

Table 3.1 (cont.)

Species	Auckland Islands			Campbell Island		Macquarie Island
	Enderby Island	Disappointment Island	Sugar Loaf Rocks	Campbell Island	Dent Island	Hasselborough Bay
A. superciliosa Grey (Black) Duck	X					
A. aucklandica ssp. Island Duck	XX	XX	X		XX	
Gallirallus australis Weka						X
Coenocorypha aucklandica Island Snipe	XX					
Stercorarius skua lonnbergi Southern Skua	XX	X		X	X	XX
Larus dominicanus Southern Black-backed Gull	XX					X
L. scopulinus Red-billed Gull	XX					
Sterna sp. Tern	X		X	X		
Turdus merula Blackbird	XX			X		
Anthornis melanura Bellbird	X					
Sturnus vulgaris Starling	XX					

THE BREEDING ECOLOGY OF MOAS

G. E. Hamel

From their skeletal anatomy it is apparent that moas occupied in New Zealand the ecological niches of large herbivores and of the larger cursorial birds, i.e. birds whose locomotion is mostly or wholly by walking. The concept of niche is useful because within it an organism's morphology, physiology and behaviour are seen as an intergrated system directed towards enabling the animal to make a living in nature in the face of competition from other species. Niche should be distinguished from habitat which is the place where an organism lives, whereas its niche is its occupation or way of life (Odum 1959).

According to niche theory a particular morphology such as that of a cursorial bird will tend to be linked to particular breeding adaptations such as the type of nesting dispersion and clutch size. For the prehistorian, breeding ecology is of interest not only because of its relationship to population dynamics, but also for the reason that the animals taken by hunters and gatherers are more vulnerable during the breeding season than at other times. Hence any deductions about moa breeding ecology from what we know of their bones, set within the context of niche theory, may be a profitable exercise for New Zealand prehistorians.

Lack (1968) has documented the ways in which breeding adaptations in birds are integrated in the major ecological groupings of littoral, marine, freshwater, wading and cursorial birds. He compiled data for most of the bird families of the world, looking for regularities between feeding habits and breeding patterns, as well as seeking functional explanations of these regularities. In this paper I shall bring together the available empirical data on moa nesting and, using these within the framework of Lack's deductions, suggest relationships between moa breeding patterns and Polynesian hunting techniques.

Breeding dispersion in cursorial birds

Among the cursorial birds, Lack includes 430 species belonging to 25 families. Two of the most important orders of cursorial birds are the Galliformes, which includes pheasants, turkeys and hens, and the Charadriformes, which includes the plovers. The seven families of ratites (Table 4.1) which are now considered to constitute a monophyletic group (Lack 1968:99), are all cursorial birds. Six of these families still have living representatives but the family to which the moas belonged is wholly extinct. Other New Zealand cursorial birds are the kiwi, weka, takahe, New Zealand dotterel, banded dotterel, wrybill, shore plover, the extinct New Zealand quail and several

Table 4.1 The genera of ratites

Cassowary	Casuarius
Emu	Dromiceius
Kiwi	Apteryx
Moas	Dinornis Pachyornis Euryapteryx Emeus Anomalopteryx Megalapteryx
Ostrich	Struthio
Rhea	Rhea
Tinamous	Crypturellus Nothoprocta Nothocercus Rhynchotus

Table 4.2 Scientific names of New Zealand birds mentioned in the text

Banded dotterel	Charadrius bicinctus
Bellbird	Anthornis melanura
Black-backed gull	Larus dominicanus
Extinct eagle	Harpagornis moorei
New Zealand dotterel	Charadrius obscurus
New Zealand falcon	Falco novaeseelandiae
New Zealand quail	Coturnix novaezealandiae
Shore plover	Thinornis novaeseelandiae
Takahe	Notornis mantelli
Tui	Prosthemadera novaeseelandiae
Weka	Gallirallus australis
Wrybill	Anarhynchus frontalis

non-breeding migrant dotterels (see Table 4.2).

Most cursorial species are solitary nesters, the main function of this dispersion being concealment of the nest and the sitting bird from predators. The few cursorial birds which are colonial (pratincoles, sand grouse and a seed-snipe) breed in loose colonies in which the nests are dispersed and the cryptic eggs hard to find (Lack 1968:100). The chicks are nidifugous, i.e. leaving the nest soon after hatching and often picking up food for themselves from their first day. Lack argues that these adaptations are the best strategy for raising the maximum number of young given a cursorial habit, ground nesting, a plant diet and the presence of egg and chick predators. Lack cites plant diet as a factor because herbivorous birds generally require larger volumes of food than insect eaters and so have a logistical problem in feeding their chicks. They solve this either by a change of diet (tuis and bellbirds feed insects to their nestlings), or by the nidifugous adaptation in which the newly hatched chick is able to move itself to the food supply instead of having the food brought to it. All the living ratites are solitary nesters and nidifugous, even the kiwi, which is not primarily a herbivore.

Moas as herbivores

Morphologically moas were well adapted to living on the ground. Besides being flightless and heavy boned, they had the metatarsi shortened and strengthened almost to the proportions typical of the Galliformes (Owen 1879:107), presumably for effectively scratching in the soil for food, rather than just for weight bearing. The herbivorous habit is evident in the heavy crushing bill, the long neck and legs to aid browsing, the overall pachydermal proportions and from the few gizzard contents competently analyzed. Identifiable gizzard contents have been found in only a few swamp deposits and most have contained fruits of forest and shrubland species (Oliver 1949, Falla 1941, Gregg 1972). Four of the gizzards from Pyramid Valley contained seeds of matai (Podocarpus spicatus), a tree which rarely fruits within five metres of the ground, but which does drop fruiting twigs very readily. After strong winds in summer and autumn, the ground under a matai tree is littered with twigs, and a moa scratching there would find a layer of half-rotting leaves and seeds, equivalent to a poor quality silage, as well as a dense mat of noduliferous roots of the matai and of any other nearby podocarps. Such an area represents a relatively high-calorie focus for a herbivore. It has also been plausibly argued that the unusually high proportion of divaricating shrubs in New Zealand is an evolutionary response to browsing by moas who are thought to have fed in a different way from mammals, probably pulling and breaking more than nipping and chewing (Greenwood and Atkinson 1977). As herbivores then, moas could have effectively browsed shrubs and the lower branches of trees, the taller species reaching up to four or five metres above the ground. They could also have been efficient scratchers of the litter layer on the forest floor.

Breeding dispersion in moas

The empirical evidence that moas were solitary nesters is slim. Ambrose (1970) described natural deposits of moa bone (mostly Megalapteryx), egg shell, droppings and nesting material in small rock shelters at Benmore, Waitaki Valley. Each rock shelter could have held at the most only a few nesting birds at one time. Hartree (in Falla 1962) was able to recognize the remains of solitary nests of Anomalopteryx on ledges and in sheltered cavities in limestone country that once carried fairly heavy forest in Hawkes Bay.

As herbivorous cursorial birds it is likely that the moa chicks left the nest soon after hatching. In a country lacking small mammals it does not seem likely that birds as large as moas would have solved the logistics of feeding chicks by supplying them with animal protein. In the emu, the chicks are led about by the male who establishes a small territory with plenty of open ground as dense undergrowth is difficult for the young emu chicks to scramble through (Eastman 1969). The bones of emu chicks are well ossified when they hatch and the chicks leave the nest as soon as the last of the clutch has hatched. Hector (1871) was able to compare the dried embryo of a moa chick in a whole egg found at Cromwell with an emu chick at the same stage of development and found that the patterns of ossification of the bones were very similar. If, as this suggests, the moa chick was nidifugous, moa adults may also have had to establish small chick feeding territories in a recognizable type of open forest, a behaviour pattern which could have been learnt and utilized by the human hunter.

Though mammalian predators were absent from New Zealand before man arrived, we cannot assume that there was no predation pressure on nesting moas. The extinct eagle, the New Zealand falcon, the black-backed gull and the weka (Table 4.2) were all possible chick predators, the last two being capable egg predators also. Only the eagle was large enough to have attacked the sitting adult. In the evolution of nest and chick dispersion then, it seems likely that moas were subjected to similar pressures and evolved similar adaptations to cope with those pressures as other cursorial birds.

Clutch size in cursorial birds

Lack has developed the argument that in nidicolous birds, where the young are fed in the nest, clutch size has evolved through natural selection to correspond with that brood-size from which on an average most surviving young are produced, the limit normally being set by the amount of food that the parents can collect for their nestlings (Lack 1968:165). Nidifugous birds are released in part from this limitation and raise some of the largest clutches known. A male rhea has been known to incubate 60 eggs, laid in his nest by several females, and hens of various species of the Phasianidae (pheasants and partridges) lay up to 17 eggs in a clutch. At the other extreme, there are several cursorial species which lay only one egg, the kiwi being a well-known example. Since fledging success is not affected by the food which the parents can bring to the brood, Lack suggests that the critical factor may be the food available to the laying bird, which she may use to produce either one large or several smaller eggs. In general, eggs of nidifugous birds are larger than those of nidicolous species, since the young must be larger at hatching so that they can fend for themselves (Lack 1968:199).

Ostriches, emus and rheas which live on the open plains lay average clutches of 8 to 30 eggs and tend to be polygamous, with several hens laying in anest guarded and incubated by one male. The forest ratites, however, (the cassowaries, kiwis and some tinamous) lay small clutches of one to eight eggs and tend to be monogamous. The functional value of this correlation between clutch size and habitat presumably has to do with the hunting techniques of chick predators.

The little empirical evidence available on clutch size in moas suggests that they follow the forest ratite pattern. In the nests of Anomalopteryx found by Hartree in Hawkes Bay, there tended to be the remains of only one egg or one chick (in Falla 1962). An entire skeleton of Dinornis robustus found in a cave near Tiger Hill, Alexandra was in a crouching position over the bones of four chicks (Ewen 1895). Also the eggs of moas seem to have been larger relative to body size than those of ostriches, suggesting that the moa female may have put the available energy into laying fewer, larger eggs than do ostriches. Lack estimated that large moa eggs would have weighed about 7.0 kg whereas ostrich eggs weigh only about 1.6 kg (Lack 1968:208). Certainly one would expect predation pressure in the New Zealand forests to be less than on the African savannah, and hence it would be better strategy for a moa to use the available energy in hatching fewer and larger chicks who would have a better chance of survival.

Discusion

Though the factors of solitary nesting, small clutches and nidifugous chicks suggest that moa adults, chicks and eggs would not be an easy or predictable harvest for Polynesian man to gather during the breeding season, there must have been sufficient regularities in moa breeding behaviour to enable man to gather their eggs. Moa egg shell is relatively common in moa hunter sites. The evidence from Hawkes Bay and Benmore shows that at least some moas favoured sheltered crevices and rock shelters as nesting sites. Such locations tend to be limited in number in most New Zealand lowlands where moa populations were largest, and those accessible to moas would also have been accessible to man. Knowing the favoured sites and the timing of nesting, it may have been a relatively successful hunting technique to visit cliffed and rocky areas at the right season. Faithfulness to nesting sites and to particular types of terrain as chick rearing territories could have made moas relatively vulnerable to human predation and could have been part of the pattern of events leading to their extinction.

REFERENCES

Ambrose, W., 1970. Archaeology and rock drawings from the Waitaki Gorge. Rec. Canterbury Mus. 8:383-437.

Eastman, M., 1969. The life of the Emu. Angus and Robertson, Sydney.

Ewen, C. A., 1895. On the discovery of moa remains on Riverton Beach. Trans. N. Z. Inst. 28:651-654.

Falla, R. A., 1941. The Avian Remains (Pyramid Valley). Rec. Canterbury Mus. 4:339-353.

Falla, R. A., 1962. The moa, zoological and archaeological. N. Z. Arch. Assn. Newsletter 5:189-191.

Greenwood, R. M. and Atkinson, I. A. E., 1977. Evolution of divaricating plants in New Zealand in relation to moa browsing. Proceedings of the New Zealand Ecological Society. 24:21-29.

Gregg, D. R., 1972. Holocene stratigraphy and moas at Pyramid Valley, North Canterbury, New Zealand. Rec. Canterbury Mus. 9:151-158.

Hector, J., 1871. On recent moa remains in New Zealand. Trans. N. Z. Inst. 4:110-120.

Lack, D., 1968. Ecological adaptations for breeding in birds. Metheun and Co., London.

Odum, E. P. , 1959. Fundamentals of Ecology. Saunders, Philadelphia.

Oliver, W. R. B., 1949. The moas of New Zealand and Australia. Dom. Mus. Bull. No.15.

Owen, R., 1879. Memoirs of the extinct wingless birds of New Zealand. John von Voorst, London.

TAMENESS AND MYSTICAL ASSOCIATIONS OF WILD BIRDS

Ralph Bulmer

One of Ron Scarlett's many engaging qualities is his boundless interest in all aspects of natural and human history. He is in no sense the narrow specialist in osteological remains; for him the living men and living birds are a constant subject of observation, interaction, and enthusiastic comment. The author received two letters from Ron describing his participation in David Cole's 1967 excavations in the Eastern Highlands of Papua New Guinea. These portray briefly but most graphically individual Highlands workmen on the site and inhabitants of the nearby village, among whom he clearly found kindred spirits. They also describe some of the familiar birds and mammals of the grasslands and gardens, as observed in life, or, in some cases, on their way to the local dinner menu. In presenting this essay to him, I hope it may refresh his memories of the experiences he had then and creatures he encountered.

Landsborough Thomson (1964:802-3) defines tameness as "toleration by birds of the close presence of human beings, in some instances amounting to a positive tendency to seek such presence".

There are no really tame wild birds in the Papua New Guinea Highlands, for obvious reasons. Adult humans (sometimes only women) are prepared to eat most, and in some societies all, bird species, and even where dietary restrictions and prohibitions of a mystical kind on interference give a measure of protection, any markedly tame individual bird would still be at high risk from children, dogs and, in recent decades, cats. Furthermore, Papua New Guinea Highlanders do not share the widespread custom of the English-speaking peoples of feeding wild birds—except occasionally as a device to facilitate their capture.

Nevertheless different Highlands species characteristically show very different degrees of toleration of man. The relatively tame Pied Chats (Saxicola caprata) in the sweet potato gardens, Willie Wagtails (Rhipidura leucophrys) by the streams and around the homesteads, and Friendly Fantails (Rhipidura albolimbata) and White-winged Thicket Flycatchers (Peneothello sigillatus) in the forest, are all prepared to come perilously close to humans, whereas the Tawny Grassbirds (Megalurus timoriensis) in the grass at garden edges, Papuan Lories (Charmosyna papou) and Slatey Thicket Flycatchers (Peneothello cyanus) in the forest and Mountain Ducks (Salvadorina waigiuensis) in the streams and rivers, are extremely shy.

The Kalam people of the Kaironk Valley disdain to eat adults of any of the small insectivores around the homesteads and in the gardens, on account of what they regard as their dirty feeding habits, but there is no simple correlation between this protection and the tameness of the species concerned, since these include not only the Pied Chat and the Willie Wagtail, which are moderately tame, but the Tawny Grassbird and Schach Shrike (Lanius schach), which are not. It is, however, the case that the eggs and nestlings of the Chat and Willie Wagtail are also given some protection, for mystical reasons; whereas children are not discouraged from taking and eating eggs and young of the Grassbird and the Shrike. Conversely, the White-winged Thicket Flycatcher and Friendly Fantail of the forest are freely killed and eaten, except that a serious hunter would not normally bother to shoot at them, in part, in the case of the Fantails, because of the risk that they will disturb more shy and valuable game. The behaviour of the Thicket Flycatcher, known in Kalam as wlmeñ saky, the "Foolish (or mad) Pied Chat" (it closely resembles the Pied Chat in plumage), which spontaneously approaches hunters, seems totally non-adaptive. Yet somehow the bird survives and remains common, perhaps in part because it is only casually hunted, and mainly by small boys.

In contrast, the natural shyness of some of the forest birds most prized for their plumage or their flesh, is clearly highly adaptive. These include the lories, especially the Papuan Lory, very highly valued for its plumes; certain of the birds of paradise; certain of the pigeons and doves, especially the fruit doves (Ptilinopus rivoli etc); and the hawks. But the chink in the hawks' armour is the boldness (which Landsborough Thomson (idem) appropriately distinguishes from tameness) with which at least some species (e.g. the goshawks Accipiter fasciatus and A. melanochlamys) return to the nest to defend their young against intruders. Of water birds, the shyness of that fascinating creature the Mountain Duck is also quite evidently its salvation; while that of the Torrent-Lark (Grallina bruijni), as compared with that of its cousin the Australian Magpie-Lark (G. cyanoleuca), may or may not be adaptive, but certainly has relevance to the special mystical status the Kalam accord it.

My thesis is that, for the Kalam, as I believe many other societies, both unusually tame species and unusually shy species, provided that they are readily identifiable, tend to be given special mystical values, which in turn affect man's interaction with them. The origins of such behaviour present interesting problems, though ones I can only speculate about in a quite amateur way. But also interesting are the human reactions and interpretations that such behaviour provokes; and the adaptiveness, for both birds and humans, of these particular avian-human interactions.

Of the various varieties of tameness, [1] the one I find most intriguing is that of birds which show, in Landsborough Thomson's phrase, "a positive tendency to seek (human) presence". My Kalam collaborator, Saem Majnep (Majnep & Bulmer, 1977), describes in some detail Kalam reaction to the five local species that do this: the Pied Chat, Willie Wagtail, Friendly Fantail and White-winged Thicket Flycatcher already mentioned, and the Mountain Mouse-Warbler (Crateroscelis robusta). Of all these the Pied Chat is much the most important, as I believe it is quite widely in highland New Guinea. This bird is present at all seasons in their gardens and, to a lesser extent, in the grasslands, provided that a few bushes remain there; and it is conspicuous by its visibility and by its song. It usually nests in the sides of banks and ditches at the edges of the gardens, but sometimes actually in among the sweet potato crop. Apart from the conspicuous singing of the male bird, generally from a free-standing bush or dead tree, but sometimes also delivered partly in flight, the birds also occasionally approach human beings with alarm notes or short challenging bursts of singing, when they intrude into their territory. When one does so, this is interpreted as the ghost of a dead father or brother bringing messages, either good news of impending visits of kin, or warnings of the approach of enemies. Kalam say, "It is our radio". I must stress that these approaches are not every-day occurences. If they were, I suspect that the messages would be devalued. Unfortunately my field notes do not indicate to what extent they are seasonal and correlated with the breeding cycle, but their frequency can be gauged from the fact that I have only once in two years' fieldwork received a message from a Chat. That was on a somewhat embarrassing occasion when I was squatting in a latrine hut and a hen bird appeared on the door-screen and briefly addressed me. Having no Kalam interpreter with me, I was unable to decode the message.

Like the English Robin, the Chat feeds on insects disturbed by the gardener, and it will approach pigs closely, as well as humans. Kalam interpret this as a ghost taking a helpful interest in the welfare of the pigs. Similarly, the protection afforded the Willie Wagtail, which also associates with pigs, is in part rationalised in terms of this being a good bird which helps look after these valuable beasts. But the Willie Wagtail's repetitive song does not, it seems, lend itself to interpretation as human utterance. Both these species are very successful invaders of the man-made environment of the Highlands valleys; the Chat, which has a wide distribution in Asia, being presumably a rather recent immigrant to New Guinea. While their presence must basically reflect both species' capacity to exploit the changed physical environment, the densities of their populations in certain areas, such as the Kaironk, must, I think, also reflect human patronage. Reciprocally, whether or not these two birds really "help the pigs", they do, presumably, along with the other partially protected small insectivores in the gardens, perform a significant role in keeping insect populations, including garden pests, in some kind of equilibrium.

In the forest it is the Mouse-Warbler that has something of the same significance as the Pied Chat, rather than the Thicket Flycatcher that the Kalam call the "Foolish Pied Chat". This nondescript little brownish bird skulks in the undergrowth and does not openly approach man. However it tends to remain close, if largely hidden, and to call in response to intruders. Its buzzing calls and brief but varied bursts of whistled song (Rand & Gilliard, 1967: 338; Diamond, 1972:233) can be interpreted by Kalam either as messages from ghosts, informing them that a game mammal is nearby or caught in a trap, or as an interfering warning to game animals that the hunter is approaching.

The fantails, especially the Friendly Fantail (so named, very appropriately, by Tom Gilliard (Rand & Gilliard, 1967:387)) are even more ambiguous; their over-familiarity and alleged capacity for disturbing bigger and more important birds with their alarm notes and antics, are not welcomed, and they may be identified with witches. But their dance-like posturings and movements, with spreading wings and tail, are admired, and their songs, in the Kalam version, are incorporated into male beauty ritual.

The poor White-winged Thicket Flycatcher has, in spite of its endearing attempts at familiarity, nothing going for it. Kalam regard its behaviour as rather pathetic, as evidence by the epithet "foolish" (or perhaps "mad" or "wild") in its name. I suspect that what the unfortunate bird lacks is a repertoire of calls that lend themselves to translation into Kalam.

In relation to humans, or at least Kalam humans, who are prepared to eat almost any form of vertebrate life from the forest, the behaviour of all these forest species seems non-adaptive. Yet the ones I have described are common. I hazard the guess that what they are doing is responding to man in the same way as they respond to other and more continuously resident large fauna (cassowaries, perhaps wallabies, and latterly, pigs) and may perhaps relate to the disturbance of insects by these creatures. I have been led to this speculation by watching New Zealand Fantails (Rhipidura fuliginosa) interact with cattle, by reading David Lack's (1953:195) interpretation of the genesis of European Robin tameness in the advantages forest-dwelling birds obtain in winter from disturbance of ground cover by man and perhaps other animals, and by my brief reading of the literature on the tameness of such New Zealand forest species as the New Zealand Robin (Petroica australis), (e.g. Buller, 1967:27-8).

The other variety of tameness that I will refer to is manifested in the classic case by the European Swallow (Hirundo rustica), cohabiting seasonally with man, and reciprocating his patronage by a remarkable degree of indifference to his presence, though not usually approaching him or soliciting his attention. To some degree the Greater Wood Swallow (Artamus maximus) enjoys this kind of relationship with the Kalam. This handsome bird nests, often in very visible locations, high in the dead casuarina trees in Kalam gardens, sometimes quite close to houses. As far as I know it is not specifically protected, or restricted as food; but many Kalam men and youths are superb tree-climbers, and if motivated to do so could undoubtedly rob almost every nest. Yet they do not do so. Nor do they chop down trees to get the fledglings, as I have seen them do in the forest to get mammals and even, on one occasion, fledgling Dusky Lories (Pseudeos fuscata). Their basic attitudes to Wood Swallows are very positive. According to Saem Majnep (Majnep and Bulmer, 1977:106-8) the very conspicuous family life of this species epitomises domestic virtue. Good housewives who abundantly and unselfishly provide for their families and kin are likened to this bird; and a ritual performed over a newly-married couple involves burning of Wood Swallow feathers in the fire by which they sit, and presentation to the bride of the beak and other plumage of the bird, for her to keep in her net-bag. The bird's alleged capacity to distinguish unfamiliar from familiar humans as they approach, and react with excitement to them, is interpreted as a warning of danger to people who live nearby. This also is in its favour.

Although this is not, I think, reported in the literature, my suspicion is that the Greater Wood Swallow also nests in dead trees within the forest. But whether or not this is so, its very successful colonisation of the garden lands of the Highlands may, I would argue, reflect human patronage as well as the convenience of the changes man has made to the landscape.

As in many other regions, in the Highlands many more birds are shy than tame. I can here refer only briefly to the special interpretations that Kalam put on the behaviour of a few of these species. The lories, several species of which are common in the forest but generally very shy, are in many cases just birds, but can also be the souls of men (i.e. adult males), either as they leave their bodies in dreams or, if I understand correctly, after death. The long-tailed birds of paradise, Stephanies (Astrapia stephaniae) and Sickle-bills (Epimachus fastosus and E. meyeri) are similarly, though apparently not so consistently, identified with the souls of women.

Lack (1956:110) points out that "... (European)... bird books written earlier than a century ago started with the eagle, for he was the King of Birds. But about the time when monarchs were tumbled from their thrones by black-coated republican intellectuals, the bird books followed suit, banishing the eagles to a back page as primitive and giving leadership to the crows". Saem Majnep, in his listing of Kaironk birds, gives first place to the lories, but allows the eagle and his relatives a position of almost equal prominence, at the rear of the procession. There is no doubt that the shyness and elusiveness of both groups greatly enhance Kalam interest in them.

In contrast, two species which combine shyness with intermittent but fleeting close approaches to humans are identified as witches. These are the Torrent-lark and the Slatey Thicket Flycatcher (which is remarkably different in its behaviour from its congenor, the White-winged Thicket Flycatcher). Both have loud and sometimes startling calls, so once again voice as well as physical movement is significant. People are genuinely scared by their encounters with these birds. The elusive Mountain Duck is also associated with witches and witchcraft.

Finally I must note that any bird, but expecially a rare and shy bird, seen or heard in an unusual place, tends to be given a mystical identification by Kalam. When I last revisited the Kaironk, in April-May 1976, my greatest friend, Wpc of Gobnem, the old Big Man and former war leader, had died four months previously. On my first day I sadly walked up the hill, through his gardens, to visit his grave. As we came close to his settlement—he had been buried very close to his house—a bird called. My companions and I argued as to what it was. I thought it was a Shrike, for there was one in the trees below the house, and Shrikes are great mimics, producing an astonishing range of calls. No, said the Councillor, it was a kabay kl (female or immature male Superb Bird of Paradise, Lophorina superba). This seemed to me highly doubtful, as this species is not normally found outside the secondary forest, the boundary of which was nearly half a mile away, and even there it is quite rare. But later the Councillor said, "Didn't you realise, I think that was the Old Man greeting you".

There are, of course, many other dimensions of birds' behaviour which, for any particular human society, in part detemine the attitudes towards them and mystical values placed upon them. In particular, as I have indicated in this paper, birds' calls and songs are very frequently given significance by man. But what I have attempted here is merely to suggest that, important as adaption to human induced change to the material environment so clearly is to an understanding of the status of bird populations today, the remarkable capacity of many species of birds to adapt to man himself may also be a feature of some significance. And on the human side, there are perhaps general tendencies, present in many, perhaps most human societies, to respond as patrons to those birds that succeed in initiating human-like interaction with man, and to develop special attitudes and values in respect of both conspicuously tame and conspicuously shy, but identifiable, bird species (cf. Armstrong 1958).

ACKNOWLEDGEMENTS

I am grateful to E. G. Turbott, D. S. Blake and J. M. Diamond for helpful suggestions on a number of points; to Atholl Anderson for his editorial patience, for this paper was planned and promised two years before it was written; and to Christine Helliwell for getting me to complete it in time for the 1979 ANZAAS Congress, where it was delivered to the 'Ethno-ornithology in Melanesia" session of Section 25 on 24 January, and thus just in time to be included in this volume.

NOTE

1. As applied to the behaviour of wild creatures, "tameness" is a curious, anthropocentric concept, and to date its main interest to ornithologists appears to lie in the interesting problems it poses regarding the interaction of genetic factors and learned traits in bird behaviour (e.g. Huxley 1947 and other authors cited by Thomson 1964). A systematic treatment of this topic is not here attempted, but if it were, the following factors might need consideration: quality, intensity and immediate contexts of interaction or close coexistence with man; distribution of such behaviours in the species and in local populations; apparent or putative motivations (e.g. access to food or nesting sites; curiosity; defense of territory, nests or young); adaptive functions (again, advantageous access to otherwise unexploitable food, shelter of breeding sites); etiology (e.g. pristine fearlessness of many bird populations that have not previously encountered man or any human-sized terrestrial predator, or, that in certain contexts of their life, have never been forced to recognise that he is a threat; transference to man of behaviour earlier developed in relation to other large animals); extent to which, deliberately or otherwise, this behaviour is stimulated by man; and the extent to which such human patronage is elicited by the bird's own behaviour and morphological character.

REFERENCES

Armstrong, E. A., 1958. The Folklore of Birds. Collins, London.

Buller, W. L. , 1967. Birds of New Zealand, (ed) E. G. Turbott. Whitcombe & Tombs, Christchurch.

Diamond, J. M. , 1972, Avifauna of the Eastern Highlands of New Guinea. Nuttall Ornithological Club, Cambridge (Mass.).

Huxley, J. S. , 1947, "Notes on the problem of geographical differences in the tameness of birds", Ibis 89:539-52.

Lack, D. , 1953, The Life of the Robin, Penguin. Harmondsworth. 3rd Edn.

Lack, D. , 1956, Swifts in a Tower. Methuen, London.

Majnep, I. S. & R. Bulmer, 1977, Birds of my Kalam Country. Auckland University Press & Oxford University Press, Auckland.

Rand, A. L. & E. T. Gilliard, 1967, Handbook of New Guinea Birds. Weidenfeld & Nicolson, London.

Thomson, A. L. , 1964, "Tameness", in A. L. Thomson (ed) A New Dictionary of Birds. Nelson, London.

AVIFAUNA AND MAN

R. J. Scarlett*

(The slightly revised text of a paper presented at the New Zealand Archaeological Association Conference, Dunedin, 17-23 May, 1969. Summary added by J. C. Yaldwyn)

At the 39th ANZAAS Congress in Melbourne, 1967, I presented a paper on "Moa and Man in New Zealand" (Scarlett 1974), and therefore I omit a detailed analysis of Moa remains at the sites discussed in the present paper, merely noting where they are present. The treatment given here is selective. From the many sites associated with Man from which I have now determined bird remains, I have selected some from each area, beginning in the far north and ending in Stewart Island.

At the outset, I will mention the difficulty, when dealing with "weathered out" middens in sand-dunes and beach areas, of being certain that every bone represented is contemporaneous with the midden, as birds still die on beaches, or are washed up as dead bodies. In the lists, I have included from such sites only those species of which there is reasonable certainty that their presence was due to the activity of Man. This can often be inferred from the state of the bone. Mammalian remains and fish are also listed. Systematic names are given only the first time a species is listed and follow those used by the Check-list Committee (1970). The reader is referred to "Bones for the New Zealand Archaeologist" (Scarlett 1972) for comments on the identification of archaeological bones in New Zealand.

I begin in Northland. From Spirits Bay in January 1961 Bob Jolly collected bones of:

Sooty Shearwater, *Puffinus griseus*
North Island Weka, *Gallirallus australis greyi*
Black-backed Gull, *Larus dominicanus*
North Island Kaka, *Nestor meridionalis septentrionalis*

as well as a few undetermined bones of other birds, Moa and seal.

Jack Grant-Mackie in January 1966 collected bones of:

Northern Blue Penguin, *Eudyptula minor* subsp.
Fluttering Shearwater, *Puffinus gavia*

To this list I added in February 1968:

Moa, *Euryapteryx exilis*
Seal, (probably Southern Fur Seal), ? *Arctocephalus forsteri*

---

*Published with Scarlett's general permission, but without him knowing where it would appear. Ed.

From Tom Bowling Bay Trevor Hosking in 1965, Jack Grant-Mackie and others in 1966 and 1967, and Jack and myself in February 1968, have gathered many bones. A composite list of species follows: those with an asterisk Jack and I also gathered at the adjacent Waikuku Beach.

*North Island Kiwi, Apteryx australis mantelli
Little Spotted Kiwi, Apteryx oweni (extinct in North Island)
*Northern Blue Penguin
Albatross, Diomedia sp. (uncertain if Royal or Wandering)
*Mollymawk, Diomedia cauta subsp.
Cook's Petrel, Pterodroma cooki
*Fairy Prion, Pachyptila turtur
Broad-billed Prion, Pachyptila vittata
Various other prions (Pacyptila spp.) not determinable to species
*Sooty Shearwater
Buller's Shearwater, Puffinus bulleri
*Fluttering Shearwater
*Short-tailed Shearwater, Puffinus tenuirostris
Little Shearwater (?North Island subsp.) Puffinus assimilis ? haurakiensis
Diving Petrel, Pelecanoides urinatrix
Gannet, Sula bassana serrator
Pied Shag, Phalacrocorax varius
Black Shag, Phalacrocorax carbo novaehollandiae
*White Heron, Egretta alba modesta
*Royal Spoonbill, Platalea leucorodia regia (association with midden uncertain)
Grey Duck, Anas superciliosa
Other ducks, not determined
*Large Harrier, the North Island form of Circus eylesi (extinct)
*N.Z. Quail, Coturnix novaezealandiae (recently extinct)
*North Island Weka
Little Weka, Gallirallus minor (extinct)
Falla's Rail, Capellirallus karamu (extinct)
North Island Takahe, Notornis mantelli mantelli (extinct)
*Variable Oystercatcher, Haematopus unicolor
Banded Dotterel, Charadrius bicinctus
*Black-backed Gull
*Red-billed Gull, Larus novaehollandiae scopulinus
Caspian Tern, Hydroprogne caspia
White-fronted Tern, Sterna striata
*N.Z. Pigeon, Hemiphaga novaeseelandiae
*Kakapo, Strigops habroptilus
North Island Kaka
Red-crowned Parakeet, Cyanoramphus novaezelandiae
Morepork, Ninox novaeseelandiae
Barn Owl, Tyto alba subsp. (several individuals, association with midden uncertain)
Bellbird, Anthornis melanura
Tui, Prosthemadera novaeseelandiae
*North Island Kokako, Callaeas cinerea wilsoni

*N.Z. Crow, Palaeocorax moriorum (extinct)
Moa, Dinornis struthoides
Moa, Euryapteryx exilis
Moa, Euryapteryx curtus
(other species of Moa have also been found at Tom Bowling Bay)
Polynesian Dog or Kuri, Canis familiaris
*Polynesian Rat or Kiore, Rattus exulans
*Tuatara, Sphenodon punctatus
Seal and fish not determined

This gives a total of 42 plus birds, other than Moa, associated with Man, for Tom Bowling Bay.

To those already listed for Waikuku Beach can be added:

Buller's Mollymawk, Diomedea bulleri
Harrier, Circus approximans gouldi
?Eastern Bar-tailed Godwit, ?Limosa lapponica baueri
?Skua, ?Stercorarius sp.
Black-fronted Tern, Chlidonias hybrida albostriatus

a total of 26 certain and two probable species.

From the important site of Houhora, N6/4, where for five enjoyable weeks I helped with the excavation, the number of species was less numerous. As I have not examined the bones, other than the Moa, in detail, I will mention only the New Zealand Swan Cygnus sumnerensis (the presence of this extinct bird at Houhora extended the known range greatly as the furthest north it had been found previously was in Hawkes Bay) and the extinct N.Z. Crow. The Moas at Houhora were Anomalopteryx didiformis (rare), Euryapteryx curtus and E. exilis (both abundant), E. geranoides (rare), Dinornis struthoides and D. giganteus (rare as bones, but many of the fish-hooks were made from Dinornis). C14 dates range from 556 ± 55 years B.P. (about 1394 A.D.) to 796 ± 56 years B.P. (about 1154 A.D.).

Moving down to the Coromandel Peninsula, the species represented at the various sites at Opito and Sarah's Gully are:

(O = Opito; S.G. = Sarah's Gully)

Little Spotted Kiwi (extinct in North Island)	O.	
Northern Blue Penguin	O.	S.G.
Albatross, Diomedea sp.	O.	S.G.
Mollymawk, Diomedea cauta subsp.	O.	S.G.
Mollymawk, Diomedea sp.	O.	S.G.
Grey-faced Petrel, Pterodroma macroptera gouldi		S.G.
Fairy Prion	O.	S.G.
Prion, Pachyptila sp.		S.G.
Flesh-footed Shearwater, Puffinus carneipes hullianus	O.	
Sooty Shearwater	O.	S.G.
Fluttering Shearwater	O.	

Other shearwaters, Puffinus spp.	O.	S.G.
White-faced Storm petrel, Pelagodroma marina maoriana	O.	
Northern Diving Petrel	O.	S.G.
Other petrels, not determined	O.	S.G.
Various shags, not specifically identified	O.	S.G.
Paradise Duck, Tadorna variegata		S.G.
Grey Duck	O.	S.G.
N.Z. Shoveler, Anas rhynchotis variegata	O.	
Duck, Anas sp.	O.	S.G.
North Island Weka	O.	S.G.
Stilt, Himantopus sp.		S.G.
Black-backed Gull	O.	S.G.
N.Z. Pigeon	O.	
North Island Kaka	O.	S.G.
Pipit, Anthus novaeseelandiae		S.G.
Tui		S.G.
North Island Kokako		S.G.
N.Z. Crow (extinct)	O.	S.G.
Moa, Dinornis giganteus	O.	S.G.
Moa, Dinornis struthoides	O.	S.G.
Moa, Euryapteryx geranoides	O.	
Moa, Euryapteryx exilis	O.	
Elephant Seal, Mirounga leoninus	O.	
Southern Fur Seal	O.	S.G.
Kuri	O.	
Kiore	O.	
Tuatara	O.	S.G.
Snapper, Chrysophrys auratus	O.	S.G.
Whale, not determined to genus	O.	

A total of 20 birds, other than Moa, for the Opito sites and 23 for Sarah's Gully. For Skipper's midden, Opito (N 40/3), we have a C14 date of 654 ± 50 years B.P. (about 1296 A.D.) and for the main site at Sarah's Gully, (N 40/9), about a mile away over a hill, dates of 621 ± 50 years B.P. (about 1329 A.D.) and 664 ± 50 years B.P. (about 1286 A.D.) showing that the two main sites were contemporaneous.

For Taranaki, I choose site N 128/3, Kaupokonui, excavated by Alistair Buist and associates. N 129/77, Ohawe, also excavated by Alistair and the adjacent Te Rangatapu sites, also contained much bird bone, including Moa, but time necessitates their omission.

North Island Kiwi
Little Spotted Kiwi (extinct in North Island)
Northern Blue Penguin
(?Fairy) Prion
Diving Petrel
(?Black) Shag
Shag, Phalacrocorax sp.
Spotted Shag, Stictocarbo punctatus punctatus
Grey Duck

Blue Duck, Hymenolaimus malacorhynchos
Four other ducks (one a teal or scaup) not identified to species
Large Harrier (extinct)
Giant Rail, Aptornis otidiformis (extinct)
Hodgen's Rail, Capellirallus hodgeni (extinct)
North Island Takahe (extinct)
Black-backed Gull
(?Red-billed) Gull (it is impossible to separate the Red and Black-billed Gulls on any bones other than the bill)
N.Z. Pigeon
Kakapo
North Island Kaka
Red-crowned Parakeet
Tui
Huia, Heteralocha acutirostris (recently extinct)
North Island Kokako
N.Z. Crow (extinct)
Moa, Dinornis giganteus
Moa, Dinornis struthoides
Moa, Dinornis gazella
Moa, Pachyornis mappini
Moa, Pachyornis septentrionalis
Moa, ?Euryapteryx geranoides
Kuri
Kiore
Southern Fur Seal
Fish (not identified)

A total of 28 birds, other than Moa. The Ohawe site also contained Takahe, Aptornis, and Huia. Note that fewer species of Petrels were found than in more northern sites. There is a C14 date for Kaupokonui of 570 years B.P. (about 1380 A.D.).

The final North Island site is N 160/50, Paremata, on the Wellington coast, excavated by Janet Davidson and E. D. Sinclair and a small amount by the late Walter Carkeek. The bones are in the National Museum, Wellington.

North Island Kiwi
Little Spotted Kiwi (extinct in North Island)
Northern Blue Penguin
Fiordland Crested Penguin, Eudyptes pachyrhynchus pachyrhynchus
(?Erect-crested) Penguin, Edyptes pachyrhynchus ?sclateri
Southern Crested Grebe, Podiceps cristatus australis (this is the third occurrence known to me of this in association with Man. Trevor Hosking had it in his Whakamoenga Cave site, Taupo (see Leahy 1976), and it was found at Marfell Beach, Lake Grassmere)
Mollymawk, Diomedia cauta subsp.
(? Buller's) Mollymawk
(?Broad-billed) Prion
Fluttering Shearwater (both P. gavia and the closely allied Hutton's Shearwater P. huttoni are found on the Wellington coast)

Shearwater, Puffinus sp.
Black Shag
(? Little) Shag, Phalacrocorax ? melanoleucos
Shag, Phalacrocorax sp.
Spotted Shag
N.Z. Swan, Cygnus sumnerensis (extinct)
N.Z. Goose, Cnemiornis calcitrans (extinct)
Grey Duck
Blue Duck
Duch (teal or scaup), Anas sp.
Duck, Anas sp.
Harrier
Large Harrier (extinct)
Falcon, Falco novaeseelandiae
Banded Rail, Rallus philippensis
North Island Weka
Little Weka (extinct)
(? Falla's) Rail (extinct)
North Island Takahe (extinct)
Giant Rail (extinct)
Oystercatcher, Haematopus sp.
(? Antarctic) Skua, Stercorarius skua ? maccormicki
Black-backed Gull
N.Z. Pigeon
Kakapo
North Island Kaka
Red-crowned Parakeet
Parakeet, Cyanoramphus sp.
Laughing Owl, Sceloglaux albifacies (probably extinct)
Tui
North Island Kokako
N.Z. Crow (extinct)
Moa, Eurapteryx gravis (represented by a " reel" possibly imported from Wairua Bar)
Moa, Euryapteryx geranoides
Moa, Dinornis giganteus
Moa, Dinornis struthoides
Moa, Dinornis novaezealandiae
Moa, Pachyornis mappini
Moa, Anomalopteryx didiformis
Kuri
Kiore
Southern Fur Seal
(probably other seals are represeted also)
Cetacean, not determined
Tuatara

Total birds, other than Moa, 43. There is a C14 date of 547 ± 48 years B.P. (about 1413 A.D.) for Paremata. Judging from the various levels reported by Janet Davidson, the site was probably occupied over a considerable period, and this single date is not to be taken as indicating either the earliest or latest occupation of the site.

Turning now to the South Island, Marfell Beach, S 29/8, is an area of several acres of sand dune at the north-east corner of Lake Grassmere, Marlborough. Artifacts of apparently early Maori manufacture were collected by C. H. Robson, as well as Moa bones, in the vicinity in the 1870s. Much of the coast-line west from Cape Campbell was occupied by both early (Moa-hunter) and later Maori settlements. In the 1940s Jim Eyles and Roger Duff, and later Elliot Dawson, Lew Gurr, and Jack and Ruby Britton collected bones from the dunes at Marfell Beach. In July 1952, I was, so far as I know, the first to uncover part of the actual ash midden layer. Briefly the sequence is as follows: a damp fine sea-gravel layer, on which when exposed by wind erosion, rotten skeletons of Moa are occasionally found. These crumble at the touch. Over this the dunes are formed. Between the gravel and the earliest occupation layer, occasional bones occur including those of *Dinornis robustus*, which apparently died out before man came to the district, as so far remains of this species have not been found in the human occupational levels. The midden varies, from a shallow 6 or 8 inches in the part where I first excavated (here the bones were in "pockets", with large areas of sterile ash between them), to a much more heavily concentrated deposit, two to three feet thick, with an uneven base, but a level top. Above this are dunes of varying heights now largely overgrown by Marram grass. These contain bone, *haangi* and artifacts. Some at least of the ovens are of Moa-hunter origin, as indicated by the Moa bones found around them.

In the following list I have included bones from the sand above the midden, which I consider to have been derived from the occupation layer by wind plantation.

South Island Kiwi, *Apteryx australis australis*
Northern Blue Penguin
Fiordland Crested Penguin
Southern Crested Grebe
N.Z. Dabchick, *Podiceps rufopectus*
Mollymawk, *Diomedia* sp.
White-headed Petrel, *Pterodroma lessoni*
Petrel, *Pterodroma* sp.
Fairy Prion
Prion, *Pachyptila* sp.
Sooty Shearwater
Fluttering Shearwater
Short-tailed Shearwater
Shearwater, *Puffinus* sp.
White-faced Storm petrel
Diving Petrel
N.Z. Pelican, *Pelicanus conspicillatus novaezealandiae* (extinct)
Black Shag
Little Shag, *Phalacrocorax melanoleucos brevirostris*
Shag, *Phalocrocorax* sp.
King Shag, *Leucocarbo carunculatus carunculatus*
Spotted Shag
N.Z. Swan (extinct)
N.Z. Goose (extinct)

Paradise Duck
Grey Duck
Grey Teal, *Anas gibberifrons gracilis*
Brown Teal, *Anas aucklandica chlorotis*
N.Z. Scaup, *Aythya novaeseelandiae*
Finsch's Duck, *Euryanas finschi* (extinct)
Australian Musk Duck, *Biziura lobata*
Merganser, *Mergus* sp. (extinct)
several other ducks, *Anas* spp.
N.Z. Eagle, *Harpagornis moorei* (extinct)
Harrier
Large Harrier, *Circus eylesi* (extinct)
Falcon
N.Z. Quail (recently extinct)
Banded Rail
Weka, *Gallirallus australis* subsp.
Little Weka (extinct)
Hodgen's Rail (extinct)
South Island Takahe, *Notornis mantelli hochstetteri*
Giant Rail (extinct)
several small rails, *Rallus* sp. (extinct)
N.Z. Coot, *Nesophalaris chathamensis* (extinct)
(?Asiatic) Whimbrel, *Numenius phaeopus* ? *variegatus*
Eastern Bar-tailed Godwit
Snipe, *Coenocorypha* sp.
Pied Stilt, *Himantopus himantopus leucocephalus*
Skua, *Stercorarius skua* subsp.
Black-backed Gull
Red-billed Gull
Caspian Tern
White-fronted Tern
Tern, *Sterna* sp.
N.Z. Pigeon
Kakapo
South Island Kaka, *Nestor meridionalis meridionalis*
Small Kaka, *Nestor* sp. (extinct)
Red-crowned Parakeet
Parakeet, *Cyanoramphus* sp.
Morepork
Laughing Owl (probably extinct)
South Island Robin, *Petroica australis australis*
Tui
Saddleback, *Philesturnus carunculatus*
South Island Kokako, *Callaeas cinerea cinerea*
South Island Thrush, *Turnagra capensis capensis*
N.Z. Crow (extinct)
Moa, *Emeus crassus*
Moa, *Euryapteryx gravis*
Moa, *Anomalopteryx didiformis*
Kuri

Kiore
Southern Fur Seal
Elephant Seal
Tuatara
Fish, unidentified

A total of 71 plus birds, other than Moa. The great number of water fowl is easily explained by the proximity of the Lake, but the numerous bush birds indicate extensive forest cover much closer to the site than is the case today.

By contrast with this extensive list, that from the only South Island West Coast site so far excavated, the Heaphy River encampment (S 7/1) is very short, which is rather surprising when one considers the large numbers of subfossil species in the adjacent limestone caves.

Kiwi, *Apteryx* sp.
Southern Blue Penguin, *Eudyptula minor minor*
Fiordland Crested Penguin
Mollymawk, *Diomedia cauta* subsp.
Allied Shearwater, *Puffinus assimilis* subsp. (one bone)
Pied Shag
(small) Shag, *Phalacrocorax* sp.
N.Z. Pigeon
South Island Kaka
Small Kaka (extinct)
Tui
Moa, *Anomalopteryx didiformis*
Moa, ?*Dinornis robustus*
Kuri
Southern Fur Seal

The absence of Rat in the midden was striking. There was plenty of it in the adjacent caves. Eleven birds present, apart from Moa. Our one C14 date from the site, 573 ± 70 years B.P. (about 1377 A.D.) which came from a shell sample, represents neither the earliest nor the latest phase of occupation, but, I think, somewhere about the middle of it.

Site S 49/20, Whaler's Bay Cave, Kaikoura, excavated by the Canterbury Museum Archaeological Society mainly in 1968, is of fairly late occupation. The faunal content was identified as follows:

South Island Kiwi
Little Spotted Kiwi
Blue Penguin, *Eupyptula minor* subsp.
Fiordland Crested Penguin
Albatross, *Diomedea* sp.
Buller's Mollymawk
Mollymawk, *Diomedea cauta* subsp.
Petrel, *Pterodroma* ? *hypoleuca* subsp.
White-chinned Petrel, *Procellaria aequinoctialis steadi*
Sooty Shearwater
(? Hutton's) Shearwater, *Puffinus* ? *huttoni*

Diving Petrel
(?Black) Shag
Little Shag
Spotted Shag
Swan, Cygnus sp. (this is a puzzle, it is neither the extinct C. sumnerensis nor the Black Swan C.atratus)
Paradise Duck
Grey Duck
Grey Teal
Harrier
N.Z. Quail (recently extinct)
Weka, Gallirallus australis subsp.
Pukeko, Porphyrio porphyrio melanotus
Black-backed Gull
(? Red-billed) Gull
Black-fronted Tern
White-fronted Tern
South Island Kaka
Laughing Owl (probably extinct)
South Island Robin
Tui
Moa, fragments of subfossil material, not identified to genus
Kuri
Kiore
Southern Fur Seal
Fish, various species

A total of 30 birds. Less certainly associated with the midden were bones of the Reef Heron, Egretta sacra collected by Owen Wilkes in 1963.

Site S 84/46, Bromley, Christchurch, a small Moa-hunter occupation area (I think fairly late) excavated between 1963 and 1965 by the Canterbury Museum Archaeological Society, had bones of the following species represented. The site lies on the north side of the Heathcote estuary, in dunes. Rather strangely, no ducks or petrels were present.

(?Blue or White-flippered) Penguin, Eudyptula sp.
(?Erect-crested) Penguin
Black Shag
Spotted Shag
N.Z. Quail (recently extinct)
Weka, Gallirallus australis subsp.
Australian Coot, Fulica atra australis
N.Z. Pigeon
?Fernbird, ?Bowdleria punctata
Moa, Euryapteryx gravis
Kuri
Southern Fur Seal
Cetacean, bones not determined as to genus

A total of nine birds other than Moa.

Site S 111/1, Dashing Rocks, Timaru, contained the following species:

Albatross, Diomedia sp.
(?Fluttering) Shearwater
Black Shag
Grey Duck
small Duck (a teal or scaup), Anas sp.
N.Z. Quail (recently extinct)
Small Kaka (extinct)
?Tui
Moa, too broken for identification
Kuri
Seal, not identified

A total of eight birds, other than Moa.

An inland rock shelter, now under the water of Benmore Dam, S 109/9, Shepherd's Creek II, Waitaki Gorge, was excavated by Graeme Mason and myself in late 1964. Jan Hjarnø assisted for two days.

Mottled Petrel, Pterodroma inexpectata
Paradise Duck
Grey Duck
Blue Duck
Duck, Anas sp.
Falcon
N.Z. Quail (recently extinct)
Weka, Gallirallus australis subsp.
Little Weka (extinct)
Black-backed Gull
Black-fronted Tern
White-fronted Tern
South Island Kaka
Small Kaka (extinct)
Red-crowned Parakeet
Parakeet, Cyanoramphus sp.
Laughing Owl (probably extinct)
Pipit
South Island Robin
Saddleback
South Island Kokako
South Island Thrush
Moa, Euryapteryx gravis, eggshell only, mainly in the moa nests we found. Possibly ante-dates Man, although it appeared to be contemporaneous with the lower occupation.
Moa, Megalapteryx didinus, bone and eggshell
Kuri
Kiore
Bat, species not yet determined
Fish, not determined

The paradise Duck, White-fronted Tern, and Black-backed Gull were still present during our visit. Even if some of the bush birds found were brought from elsewhere as is possible, the climate must have differed vastly from the arid, dessicated region it is today. Sparrow, Goldfinch, Song Thrush, and Blackbird had died in the rock shelter also, and their bones were present on the surface and just into layer 1. Bones of Skink and Gecko were also present, but they were plentiful as living animals in the rocks near the site, and their presence may have nothing to do with Man. It is possible that Man's occupation was seasonal and that the considerable depth of Euryapteryx nest material, which contained much dung and broken eggshell, was deposited during a period of the year when Man was absent.

Site S 136/2, Ototara, a few miles from the sea near Teschemaker's, North Otago, excavated by Michael Trotter, had bones of the following species:

Yellow-eyed Penguin, Megadyptes antipodes
Prion, Pachyptila sp.
Diving Petrel
(?Black) Shag
Shag, Phalacrocorax sp.
N.Z. Swan (extinct)
N.Z. Goose (extinct)
Paradise Duck
Grey Duck
Duck (teal), Anas sp.
Duck, Anas sp.
N.Z. Shoveler
Harrier
Falcon
N.Z. Quail (recently extinct)
Pukeko
South Island Takahe
Black-billed Gull
White-fronted Tern
N.Z. Pigeon
Red-crowned Parakeet
Morepork
Laughing Owl (probably extinct)
Moa, Euryapteryx gravis
Kuri
Kiore

A total of 23 birds, other than Moa. The C14 date for the site is 561 ± 32 years B.P. (about 1389 A.D.).

Duckworth Midden, Long Island, Dusky Sound, in what appears to have been an occasional site usedby coasting Maori, was situated on a rock platform. The midden was practically all along the lip, among trees which had since grown there, about 50 feet up from the sea, with forest below and sheltered by a high cliff at the back. It was found by John Hall-Jones on Nigel Duckworth's March 1965 trip, when a few bones were removed, and in March 1966, on Duckworth's second expedition, was excavated under my supervision.

The bones represented were:

Southern Blue Penguin
Fiordland Crested Penguin
Fairy Prion
Prion, Pachyptila sp.
Fluttering Shearwater
Diving Petrel
Pied Shag
Little Shag
N.Z. Quail (recently extinct)
Western Weka, Gallirallus australis australis
N.Z. Pigeon
South Island Kaka
South Island Kokako
Moa, Euryapteryx gravis, one piece of femur and two pieces of tibio-tarsus, probably a haunch taken round from Southland.
Kuri
Kiore
Southern Fur Seal
Fish, not identified

A total of 13 birds, excluding the Moa. Most of them, except the Quail, which became extinct last century, can be found in Dusky Sound today. From the appearance of the bones, they might have been killed a few days before excavation.

The Native Island midden, S 189/1, Stewart Island, has not yet been excavated. The following bones were surface collected by myself in 1954 and by Baden Norris in 1961 and 1963. They indicate the rich potential of the site.

Southern Blue Penguin
Yellow-eyed Penguin
Mollymawk, Diomedia cauta subsp.
(?Buller's) Mollymawk
Mollymawk, Diomedia sp.
Broad-billed Prion
Sooty Shearwater
Fluttering Shearwater
Southern Diving Petrel, Pelecanoides urinatrix chathamensis
Little Shag
Stewart Island Shag, Leucocarbo carunculatus chalconotus
Blue Shag, Stictocarbo punctatus steadi
N.Z. Swan (extinct)
(?Brown) Teal
Stewart Island Weka, Gallirallus australis scotti
Little Weka (extinct)
Black-backed Gull
Red-billed Gull
Small Kaka (extinct)
Laughing Owl (probably extinct)
N.Z. Crow (extinct)
Moa, Euryapteryx gravis
Kuri
Southern Fur Seal

Kuri
Southern Fur Seal

The birds total 21 species, plus three more not as yet identified.

The final site is Old Neck, S 189/4, Stewart Island. The following bones were surface collected from midden material weathered out from the sand dune by myself in 1954, Baden Norris in 1962, Olga Sanson in 1964 and by Les Lockerbie in 1959. I have not seen Les's excavated bones as yet, except the Moa, but expect that they will add still more species to the list.

Stewart Island Kiwi, *Apteryx australis lawryi*
Southern Blue Penguin
Yellow-eyed Penguin
Fiordland Crested Penguin
Mollymawk, *Diomedia* sp.
Fairy Prion
Broad-billed Prion
Sooty Shearwater
Fluttering Shearwater
Shearwater, *Puffinus* sp.
Southern Diving Petrel
Little Shag
Stewart Island Shag
Blue Shag
(?Brown) Teal
Duck, *Anas* sp.
Banded Rail
Stewart Island Weka
Little Weka (extinct)
Variable Oystercatcher
Black-backed Gull
N.Z. Pigeon
South Island Kaka
Small Kaka (extinct)
South Island Kokako
Moa, *Euryapteryx* n.sp., abundant, found also in the Murihiku area of the South Island
Kuri
Kiore
Southern Fur Seal
Sea Lion, *Otaria hookeri*

Lockerbie's excavated Moa contained some *Eurapteryx gravis*, abundant material of the new small *Euryapteryx* and one femur of *Anomalopteryx didiformis*.

The total of bird species, other than Moa, is 25, plus six others not as yet identified.

A bird rarely found in association with Man, the Giant Rail, *Aptornis otidiformis*, was recorded by von Haast at Moa Point Cave, Redcliffs. A few bones occurred in middens adjacent to the Cave, and Michael Trotter found some in his Katiki excavations. Thus we have a picture of man exploiting all the animal resources of his environment, including shellfish, not mentioned in this paper, but abundant in many of the sites. The complementary resources, such as fruits and berries, as well as vegetables, rarely leave traces, except in the form of pits, which in many cases were presumably for storage. At Sarah's Gully, for instance, we found comparatively shallow rectangular pits, with drains, and the bottoms covered with fine white shell sand from the beach, which probably served such a purpose. The bell shaped Rua are more obviously storage pits.

What is certain, from the faunal remains, is that the Polynesian population of these islands found almost every bird and animal edible. I have not attempted in this paper to deal with the proportions of the different birds in the respective sites, even where I have such data, which is not always the case. In Whaler's Bay Cave, for instance, there are far more Albatross and Mollymawk represented than any other bird. At Houhora, more Snapper appear to have been eaten than any other form of animal food. Various shags are plentiful in some sites. Cook's sailors found shags palatable food. In most middens, some species are represented by one or a few bones. The chances of survival of the bones of the smaller bush birds and tiny petrels, such as Pelecanoides, are less than those of the bigger birds. The presence of a Tui humerus indicated the use of at least one bird, and many more may have been eaten, without any trace surviving. At some sites, excavated at several periods, later 'digs' have produced a number of bones of species not represented at all in the first excavations. For such reasons, I consider in the present stage of our knowledge that statistics are rather unreliable for many sites.

SUMMARY

Bird, mammal and Tuatara bone identifications from 19 New Zealand archaeological sites (9 in the North Island and 10 in the South Island) are listed site by site. Fifteen moa species, 111 other bird species and subspecies, and five mammals are named in the list.

Extinct non-ratite birds are recorded as follows: N.Z. Pelican from a site near the South Island coastal Lake Grassmere; N.Z. Swan from North Auckland, Paremata near Wellington, the Lake Grassmere site, North Otago, and Stewart Island; N.Z. Goose (Cnemiornis) from Paremata, Lake Grassmere, and North Otago; Finsch's Duck (Euryanas) and Merganser from Lake Grassmere; Large Harrier (North Island form) from four sites between North Auckland and Paremata; Large Harrier (Circus eylesi) and N.Z. Eagle (Harpagornis) from Lake Grassmere; N.Z. Quail from North Auckland to Dusky Sound (nine sites); North Island Takahe from four sites between North Auckland and Paremata; Little Weka (Gallirallus minor) from North Auckland to Stewart Island (six sites); Giant Rail (Aptornis) from Taranaki, Paremata, and Lake Grassmere; Hodgen's Rail (Capellirallus) from Taranaki and Lake Grassmere; Falla's Rail (C. karamu) from North Auckland and possibly from Paremata; a small extinct rail and N.Z. Coot (Nesophalaris) from Lake Grassmere; small extinct Kaka from six South Island sites between Lake Grassmere and Stewart Island; Huia from Taranaki; and N.Z. Crow (Palaeocorax) from eight sites between North Auckland and Stewart Island.

Moa records are as follows: Anomalopteryx didiformis from North Auckland, Paremata, Lake Grassmere, West Coast South Island, and Stewart Island; Megalapteryx didinus from Waitaki Gorge; Pachyornis mappini from Taranaki and Paremata; P. septentrionalis from Taranaki; Emeus crassus from Lake Grassmere; the small Euryapteryx curtus from North Auckland; E. exilis from North Auckland and Coromandel; the larger E. geranoides from four North Island sites between North Auckland and Paremata; the large, characteristic "moa-hunter" species E. gravis from Paremata in the North Island (worked into a reel) and from seven South Island sites between Lake Grassmere and Stewart Island (represented by eggshell at Waitaki Gorge and by a single leg in Dusky Sound); an undescribed small species of Euryapteryx from Stewart Island; Dinornis gazella from Taranaki; D. struthoides from six North Island sites between North Auckland and Paremata; D. novaezealandiae from Paremata; D. giganteus from five North Island sites between North Auckland and Paremata; and D. robustus possibly from the West Coast of the South Island.

Other interesting records include: Little Spotted Kiwi (now extinct in the North Island) from four North Island sites between North Auckland and Paremata; crested penguins (now not breeding north of south Westland) from Paremata and Lake Grassmere as well as sites further south; Southern Crested Grebe (now breeding only in the South Island) from Paremata and Lake Grassmere; Mottled Petrel from the inland Waitaki Gorge site (did it breed nearby?); a swan different from the extinct N.Z. Swan and the Australian Black Swan from a cave site at Kaikoura; Australian Musk Duck (not known in N.Z. in post-European contact times) and South Island Takahe from Lake Grassmere; Australian Coot (only known as a straggler in N.Z. in post-European contact times) from a site near Christchurch; Kakapo from four North Island sites between North Auckland and Paremata, and from Lake Grassmere and North Otago in the South Island; Laughing Owl (probably extinct) from Paremata, and from five South Island sites between Lake Grassmere and Stewart Island; Barn Owl (the Australian subspecies is a rare straggler to N.Z.)from North Auckland; South Island Thrush (possibly extinct) from Lake Grassmere and Waitaki Gorge; and Tuatara remains from North Auckland, Coromandel, Paremata, and Lake Grassmere.

Bibliography

Checklist Committee (F. C. Kinsky convener), 1970. Annotated checklist of the birds of New Zealand including the birds of the Ross Dependency. A. H. & A. W. Reed (for Ornithological Society of N.Z.), Wellington.

Leahy, A., 1976. Whakamoenga Cave, Taupo, N 94/7, a report on the ecology, economy and stratigraphy. Rec. Auckland Mus. 13:29-75.

Scarlett, R. J., 1972. Bones for the New Zealand Archaeologist. Canterbury Mus. Bull. 4:1-69, 329 figs.

Scarlett, R. J., 1974. Moa and Man in New Zealand. Notornis 21:1-12.

SOME OSTEOGRAPHIC CONTEMPLATIONS ON MAORI AND KAKAPO IN EARLY WELLINGTON

Elliot W. Dawson

> "The finding of bird bones in middens does not necessarily mean that the birds were caught even within some miles of the site and any attempt at studying the ecology or distribution of various species of birds by examining the contents of middens is therefore unreliable. It sould be remembered that the Maori often carried large supplies of potted or preserved birds over great distances ..."
> W. Carkeek, 1966. "The Kapiti Coast", p.107.

Introduction

The review given by Williams in 1956 of past and present distribution of the Kakapo (*Strigops habroptilus* Gray), defined for the first time the limits of occurrence of this large ground-parrot as a living bird or as a fossil. The present contribution adds another locality to the distribution map given in that account, suggests that this indicates the survival of the Kakapo to a late date in the Wellington district in which it has been recorded only as subfossil bones in various cave deposits, suggests also first-hand knowledge of the Kakapo by certain Maoris up to 1846, affirms the importance of the correct interpretation of archaeological sites in which midden debris may overlie earlier natural deposits and notes, in addition, the biogeographical implications of former distributions of species now restricted to specialized habitats.

Historical

Some years ago, I identified in the Department of Palaeontology, British Museum (Natural History), some 10,000 bones of subfossil and extinct birds from New Zealand and the Chatham Islands, formerly part of the late Lord Rothschild's collection at Tring (Dawson, 1958). Most of these bones were acquired by Dr. Henry O. Forbes when he returned to England in 1893 after relinquishing his post as curator of the Canterbury Museum (Rothschild, 1907). No examination of this collection had been made since the Museum obtained it, much of the material being just as Forbes had left it, a great jumble of bones, many of which were labelled in Forbes' hand in only brief and tantalising detail interpretable only by those familiar with Forbes' biography. In passing, it may be mentioned that Forbes' prior career as a renowned traveller, ethnologist and naturalist in the Indonesian and New Guinea regions, and, particularly, the detail of the circumstances of his departure from the Canterbury Museum are the subject of current research by myself and by my associate Dr. R. F. Ellen (cf. Ellen 1978).

In a preliminary account of his collections, Forbes (1892) listed the sources of his material, amongst which were bones of birds collected "in the notorious Rauparaha's Kitchen-middens, by Mr. A. Hamilton". No specific indication was given of the kinds of birds occurring at this site nor of the geographic locality. Augustus Hamilton (1854-1913), Director of the Dominion Museum for 10 years (1903 to 1913), is famed for his Maori Art amongst many other achievements.

An obituary in the Wellington newspaper The Evening Post of 13 October 1913, page 8, said of him:

> "He was an insatiate student, a persistent collector and classifier. Up and down the two islands he has marched in his quest of facts about this country's ancient and modern fauna and flora, and in this great hunting ground no game was too big or too little for his capacious bag. The range of his search was from moa bones to moths, from ferns to coins, from postage stamps to Maori war canoes. Many of the trophies of this fervent chase will be in the National Museum by and by—material with a message about Nature's wondrous ways, and matter in which the workings of the Maori mind are embodied."

Augustus Hamilton was Registrar of Otago University for 13 years and had previously been a teacher in schools in Wellington and Hawkes Bay. During this period in which these bones were collected, namely 1888, he was teaching at Petane School, Hawkes Bay, and had come into contact with Forbes when the moa bone deposits in Te Aute swamp were discovered, also in 1888, and had sent collections to him at the Canterbury Museum which were eventually taken by Forbes to England finding their way eventually to the British Museum (Natural History) in which I was able to re-discover them.

Discussing Hamilton's prowess as a collector of Maori material, Skinner (1914:vi) said of him: "The foundations of his Maori collection were laid in the early years of his residence in New Zealand ... To this period belong his first field expeditions among the ancient village sites of the East Coast and about Porirua."

The Rothschild collection in London contains several hundred bones labelled (in cumulative detail) as being from Te Rauparaha's pa at Paremata, near Wellington, collected on 18 may 1888 by A. Hamilton. A number of species of birds are represented, from the so-called extinct swan (Cygnus sumnerensis (Forbes)) the extinct coot (Nesophalaris chathamensis (Forbes)—syn. Palaeolimnas, see Brodkorb and Dawson, 1962, and Olson, 1975), recorded by Dawson (1962) from this collection as the first North Island record, and the extinct raven (Palaeocorax moriorum Forbes) to the domestic fowl (Gallus sp.) and the goose (Anser sp.) as well as ducks, hawks, pigeons, wekas, kakas, kiwis, owls, Kokako, Huia, shags, gulls, petrels and penguins.

Discussion of Probable Sites

The site of "Rauparaha's Kitchen-middens" explored by Hamilton seems most likely to have been either near the "Porirua" pa (Angas, 1874b:247) close to the site of Thom's whaling station and to the remains of the Paremata fort on what is now called the Ngati Toa Domain, or possibly the "Taupo" pa on the present site of Plimmerton (see map in Cowan, 1922:110, and by Best et al., 1916). Illustrations of these pa have been given by Angas (1847a,b) and by Brees (1848). The Taupo pa was built by the Ngati Toa under Te Rangihaeata, nephew of Te Rauparaha, in 1843 and 1844 following the Wairau affray in April 1843 (Wakefield, 1845) and was occupied at least until the arrest of Te Rauparaha on 23 July 1846 (cf. Ward, 1912). It is likely, however, that both this area of Plimmerton beach and the Paremata flat (i.e. the Ngati Toa Domain) have been Maori occupation sites for a great many years, certainly long before the establishment of the whaling station (cf. Brees, 1848: figs 14, 29, 52) and that the midden debris may represent an accumulation of far greater antiquity than 1843. Angas (1847b, I: 247) visited this place and the impression is given that the Porirua pa then occupied by Te Rauparaha had been existing for some time, whereas the Taupo pa in which Te Rangihaeata lived was "a new and very substantial stockade." (Angas, 1847a: plate 48).

Best (1918:215) stated: "This seems to have been one of the oldest of native-village sites in the Porirua district, to judge from the fact that bones of the moa have been there found, also from the amount of village debris formerly visible. This settlement was situated on the eastern side of the railway-line, a little way north of the bridge." Other details of Maori occupation and archaeological notes were also given by Best (1914) in his series of articles entitled "Porirua: and they who settled it."

Christie, in an unpublished address delivered to the Historical Society of Victoria College [now Victoria University of Wellington] about 1914, remarked: "... the Paremata district has experienced many changes of occupant and the Paremata people have changed many times also." The flat west of what is now called Mana (formerly "Dolly Varden") has been shown clearly to be an early occupation site, and Adkin (1950:11-12) has discussed its age and probable history. Duff (1956:276) considered "Paremata, in Porirua Harbour [as among] the few North Island sites that can be satisfactorily demonstrated as of Moa-hunter occupation." This conclusion was based on Christie's unpublished manuscript quoted above (the date of which Duff erroneously gave as "1943", the time of his own personal communication with the author; internal evidence indicates a date rather later then 1914). However, Duff remarked that Christie's description of the mixing of material from various strata after wind action "clearly subjects the primary association of the Moa and human remains to the doubts raised by Archey." Adkin (1950:12, footnote) has extended Hamilton's (1896:181) reference to "an old pa at Paremata" to one "situated on the eastern side of Paremata Flat," but there is no evidence of the exact location from Hamilton's account although this, in fact, may be the site of which he collected the bones under discussion. Yaldwyn (1959) listed identifications of moa remains from this area. Because of the linking of Te Rauparaha, chief of the Ngati Toa, with the site visited by Hamilton, it might appear more probable that his "Paremata" pa was at Taupo Point, Plimmerton, with middens extending over the present Plimmerton beach, the area marked "C" in Adkin's (1921:152) figure 3, (cf. also his plate XXXV which shows the northern aspect of the pa site on Taupo Point). Beckett (1955) examined the midden of the Taupo pa during 1910-1914 and found moa egg shell, Tuatara jaw bones, and "very old moa bone fragments" as well as bird bones.

The Kakapo Bones

Among the bones from Hamilton's Paremata site are nine tibiotarsi and eight tarsometatarsi of the Kakapo (Strigops habroptilus). These were identified from amongst a mixed collection of limb bones of various birds, as mentioned above, and they were labelled in ink or pencil on the bones themselves: "Parematta, New Zealand," "Parematta Pa," or, in Forbes' writing, "Hamilton's pa." The measurements of the bones are shown in Table 7.1.

Tibiotarsus: Strigops habroptilus, Paremata

L	P	D	M
116.5	20.7	14.6	6.4
123.0	20.5	15.0	6.2
113.0	18.3	13.7	6.0
115.4	20.0	14.7	6.6
127.0	19.2	16.0	6.0
114.0	20.0	14.5	6.5
-	-	13.8	6.5
118.0	20.4	15.6	6.5
129.0	23.6	16.0	7.5

Tarsometatarsus: Strigops habroptilus, Paremata

L	P	D	M
52.8	16.2	19.2	6.8
51.3	18.2	19.6	6.5
45.7	14.5	15.6	6.0
-	-	19.0	7.0
47.5	14.7	16.4	5.8
50.5	15.1	17.3	6.6
48.3	14.7	16.4	5.9
-	-	18.3	6.9

Key: L = total length; P = breadth of posterior end; D = breadth of distal end; M = breadth of middle of shaft (as in Larson, 1930)

The bones seem to show a variation in size amongst themselves comparable with that found in collections of subfossil and archaeological Kakapo bones now available in New Zealand museums (particularly many added since Williams' (1960:226) record of the " rather fragmentary subfossil material...lodged in New Zealand museums....."). Skeletal remains of recent living birds are still not common in museum collections but, despite attempts at establishing subspecies based on skins (see Williams, 1960:219), no discrete populations or subspecific grouping can be seen readily amongst the bones examined. It might be thought unwise (although tempting in view of the large quantity of bones, especially crania and humeri, now available), to make detailed statistical analyses of bones from individual sites and expect to be able to equate them with those from other sites which may differ chronologically to a degree of significance that cannot be demonstrated except by unavailable precision dating techniques. However, a statistical examination was made of the following series of tibiotarsi from various localities as listed with tabulated measurements below:

1. the Paremata bones (British Museum (Nat.Hist.)); 2. part of a large series of bones from Ruakokopatuna Cave 1, near Martinborough (in which initial investigations yielded at least 74 crania, see Yaldwyn, (1956:3), collected by E. W. Dawson and J. C. Yaldwyn in 1952 (National Museum of N.Z.); 3. representatives of at least 26 Kakapos from the spelæological site Harwood Hole, Canaan, near Takaka, collected by O. R. Wilkes, 1960 (Canterbury Museum): 4. a series of isolated bones from North and South Island localities, including bones of a single recently dead bird collected by T. Hitchings in Fiordland, January 1955 (Canterbury Museum). The results are shown in Table 7.2.

Table 7.2 Tibiotarsus: Strigops habroptilus

		L	P	D	M
Paremata	($\bar{X}$)	119.4	20.3	14.8	6.4
	N =	(8)	(8)	(9)	(9)
Martinborough	($\bar{X}$)	125.1	19.6	16.3	6.8
	N =	(15)	(15)	(15)	(15)
Harwood Hole	($\bar{X}$)	115.0	17.4	14.1	5.8
	N =	(16)	(14)	(14)	(25)
South I. locs:					
AV 5957		133.0	20.8	15.5	6.7
AV 11144		120.8	19.9	15.4	6.9
AV 11145		108.3	16.8	13.2	5.9
AV 11392		111.7	16.3	12.9	5.7
AV 12647		121.0	-	13.2	6.9
AV 15042		118.6	20.2	14.6	6.4
AV 15706		138.5	18.9	14.6	6.1
North I. locs:					
AV 17353		106.3	17.3	12.8	6.4
AV 17354		106.9	18.8	13.7	6.7
Recent:					
AV 13782		127.0	20.1	15.4	6.3

Key: AV 5957, Pyramid Valley, 1939/40; AV 11144-5, Marfell Beach, Marlborough, 1952; AV 11392, Karamea cave, 1952; AV 12647, Mt. Somers, 1923; AV 15042, Pyramid Valley, 1957, AV 15706, Marfell Beach, 1958; AV 17353-4, Briar's cave, Te Kuiti, 1961; AV 13782, Fiordland, 1955.

The Paremata bones appear to fall within the size ranges revealed by this comparative material. Morphologically, they are indistinguishable from any skeletal material available for examination. However, statistically, some differences are evident between the populations represented by the samples measured as outlined in Table 7.3.

Table 7.3 Tibiotarsus: <u>Strigops habroptilus</u>

		N	OR	$\overline{X}$	S	V	CD
Martinborough	L	15	112.0-134.3	125.14 ± 1.75	6.79 ± 1.24	5.42 ± 0.99	
	P	15	16.2- 22.1	19.68 ± 0.48	1.88 ± 0.34	9.58 ± 1.74	
	D	15	14.9 - 17.8	16.37 ± 0.26	1.01 ± 0.18	6.19 ± 1.13	
	M	15	5.9. 7.9	6.83 ± 0.03	0.58 ± 0.03	0.38 ± 1.52	9.25
Paremata	L	8	113.0-129.0	119.48 ± 2.44	6.90 ± 1.72	5.77 ± 1.44	16.87
	P	8	18.3- 23.6	20.33 ± 0.54	1.53 ± 0.38	7.53 ± 1.88	
	D	9	13.7- 16.0	14.87 ± 0.26	0.79 ± 0.18	5.31 ± 1.25	
	M	9	6.0- 7.5	6.47 ± 0.11	0.35 ± 0.08	5.40 ± 1.27	7.62
Harwood	L	16	106.2-129.3	115.01 ± 1.82	7.28 ± 1.28	6.32 ± 1.12	
	P	14	16.2- 21.4	17.47 ± 0.36	1.38 ± 0.26	7.89 ± 1.49	
	D	14	12.9- 16.7	14.16 ± 0.31	1.19 ± 0.22	8.40 ± 1.58	
	M	25	5.5- 6.7	5.81 ± 0.09	0.47 ± 0.06	8.08 ± 1.14	

Key: N = number of specimens; OR = observed range; $\overline{X}$ = mean; S = standard deviation; V = coefficient of variation; CD = coefficient of divergence.

The coefficients of variability for the three samples have these ranges: L, 5.4—6.3; P, 7.5—9.5; D, 5.3—8.4; M, 5.4—8.0. The Paremata sample is 91.2% as variable as the one from Harwood Hole and 6.45% more variable than that from Martinborough which is 93.9% and 85.75% as variable as those from Paremata and Harwood Hole respectively. Overall, therefore, these figures indicate relatively uniform samples with good average values of variability. The relative difficulty of consistently measuring the dimensions P, D and M may be reflected in the rather higher variabilities shown. The coefficient of variability, incidentally, is a useful guide for the selection of characters that are relatively little variable within a taxonomic group and hence are of good taxonomic value. In this instance, the dimension 'L', on its own, may be as reliable a distinguishing character as any limb bone dimension.

The coefficients of divergence show that the Paremata bones differ less (c. 8%) in the character of total length from both the Martinborough and Harwood samples than the latter do from each other (c. 17%). Statistical t tests for the significance of the difference of means applied to each pair of samples show that between the Paremata and both Martinborough and Harwood means $P > 0.05$; hence no real difference can be established. However, between the Martinborough and Harwood means, $P < 0.01$, a difference considered significant at this level. Accordingly, the two subfossil cave populations (North Island and South Island) differ from each other significantly whereas the Paremata sample, perhaps from an unknown provenance, has statistical features in common with both.

The bones from Paremata give no clue by their gross appearance and colour as to whether they are subfossil or archaeological in origin since they have been cleaned and varnished. Although they are generally well-preserved and of hard texture, they may have been, prior to treatment, no different in appearance from the subfossil bones of small birds from the pumice sands (Taupo outwash, see Adkin, 1950:24) at Walkers Hill gully situated to the south of the site of the Taupo pa (Dawson, 1952:260) and now largely built upon.

The bones form part of the collection of the Department of Palaeontology in the British Museum (Natural History), London. They bear the registration numbers A 5574—5579, A 11829—11831 (tibiotarsi) and A 11832—11839 (tarsometatarsi).

Significance of the Bones

From Williams' maps and comments, and from other museum specimens seen, it appears that the Kakapo was known as a living bird in the North Island, pre-1901, only from the central area (Williams, 1956, 1960). There are no published records, apart from the subfossil bones of the Martinborough caves, of its occurrence in the Wellington district.

For 17 bones to be collected on one visit to such a site, the birds must have moderately abundant as experience has shown in other places (Dawson, 1949), and this find may represent an extension of the range of the Kakapo to the south of the central area within recent times (i.e. up to about 1843). The evidence collected by Buller (1888:181-3) is that, by this time, the Kakapo was restricted in the North Island to the region near Lake Taupo, the Kaimanawa ranges and the upper reaches of the Wanganui River. Angas (1847b, I:244), despite the inaccuracy of his views on the rarity of preserved skins of the bird, said of it: "Another bird, indigenous to New Zealand, and now nearly, if not totally extinct, is the Kakapo." It seems unlikely, therefore, that there were any Kakapos at Paremata, even as traded preserved birds, at the time of Angas' visit.

If these bones are undoubted midden debris, this site may represent an accumulation of human-formed strata covering a wide range of time. The presence of fowl bones in Hamilton's collection certainly suggests midden debris of a late date. *Gallus* has been recorded in similar superficial deposits elsewhere in New Zealand, one of the most historic being at Waingongoro (Lydekker, 1891) where Moa-hunter and later Maori found a congenial habitation and accumulated a wide chronological range of debris. So far as my own examination extends, these are not the bones of the Jungle Fowl which might have reached New Zealand during early Polynesian voyaging. The bones of *Nesophalaris*, *Palaeocorax* and *Cygnus*, on the other hand, suggest an intermingling of material from two or more distinct strata as commonly seen in other wind-swept coastal sites.

However, it may be that these particular birds, whose remains were found at Paremata, were caught in the central area of the North Island and were traded southwards, since it is known that the old-time Maori valued the Kakapo for its meat and its feathers if the opportunity of taking it arose. Best (1942: 211, 216, 312) has outlined the methods of hunting, skinning and preservation of the Kakapo, noting that expeditions were made to the South Island to procure skins. Potted or preserved birds (*huahua manu*) less their bones were used as barter. He remarked (*ibid*:319): "The bones of the birds, together with the trail, were cooked and enjoyed by the fowlers and their assistants." If this was always so, however, the occurrence of the bones at Paremata, if midden debris, might suggest that certain Maoris in this area had first-hand dealings with the Kakapo as hunters of the birds rather than as receivers of traded *huahua manu*.

Conclusion

The importance of the correct recognition of stratification in such deposits must be stressed. Stratification in these places is undeniable and should always be appreciated. A distinction between primary and secondary associations formed from these deposits, especially with reference to the intermingling of moa and midden debris, was pointed out by Archey (1941), and, indeed, had already been recognised in the Chatham Islands, much earlier, by Forbes (1893) himself and, later, by Hill (1914) in Hawkes Bay. The possibility of intermingling at Hamilton's site cannot be dismissed. Field experience teaches one, nevertheless, that it is not always easy to know what material may have been derived from a particular layer especially if the original stratification is disturbed or destroyed and a careful *in situ* study is essential. In this respect, the model and suggestions for coastal dune site interpretation outlined by Coutts (1972) are to be commended for further application. Best's (1918) remarks, already quoted, on the antiquity of Paremata were based largely on a lack of awareness that the moa bones found there were not necessarily in association with midden debris, and the some conclusions have been reached quite often in other times and in other places. Every practising archaeologist in New Zealand knows about primary and secondary associations, of course, but, like many another well-known pitfall of observation and interpretation, some of us still flounder into it.

Closer examination of the bird remains from midden sites in New Zealand has led to the opinion that some of the birds thought to belong to a pre-human era did, in fact, live until Moa-hunter times (Duff, 1950, 1956) or that some of these sites, where, for example, bones of the extinct eagle (Harpagornis) occur, are indicative of an even earlier date in the Polynesian occupation of New Zealand (Falla, 1942). The Paremata site explored by Hamilton suggests both end points of such a time scale if all the bones are truly midden debris. In addition our current knowledge of the present-day specialized habitats of such birds as the Takahe and the Kakapo may imply differences in tolerance towards their environment in earlier times in contrast to a former widespread distribution of their present-day habitat and climatic association (cf. Dawson 1965). For example, Best (1942:215-6), discussing the habits and distribution of the Kakapo, stated: "In the South Island these birds seem to have frequented the plains in olden times, but in the North Island I have heard of them only as denizens of the forest..."

The Kakapo has occurred at Paremata, therefore, either in association with human activity or with earlier natural deposits which, in themselves, invite questions of faunal history and palaeoecology. In whatever way the bird's occurrence might be explained, Hamilton's find is still a matter of considerable interest. Careful exploration of such Maori occupation sites, indicative of both long history and recent historical events as at Paremata, will reveal many clues which, if properly recorded and carefully interpreted, will be of very great use to the archaeologist, the geologist and the ornithologist alike in providing a novel view of the early history of New Zealand.

ACKNOWLEDGEMENTS

My thanks are due to the Trustees of the British Museum (Natural History), London, and to Dr. W. E. Swinton, formerly of that institution, for allowing me to work on Forbes material; to the Directors of the National Museum of New Zealand and the Canterbury Museum for the use of comparative material; and, particularly, to Dr. R. K. Dell, National Museum, for his valued criticism of an earlier draft of this paper, and for always so willingly providing working space in his institution.

Note addedin proof: Davidson (1978) has recorded the Kakapo in bones collected during excavations at Paremata by E. D. Sinclair early in 1963. She regards the " Paremata Pa" as being the " Porirua Pa" herein described as lying close to Thom's whaling station.

BIBLIOGRAPHY

Adkin, G. L., 1921. Porirua Harbour: a study of the shore-line and other physiographic features. Trans. Proc. N.Z. Inst. 53:144-156.

Adkin, G. L., 1950. Supplementary data relating to the ancient Waitaha in the Horowhenua-Te Whanga-nui-a-Tara area, North Island, New Zealand. J. Polynes. Soc. 59:1-34.

Angas, G. F., 1847a. The New Zealanders illustrated. 60 pls & text. Thomas M'Lean. London.

Angas, G. F., 1847b. Savage life and scenes in Australia and New Zealand. Vol. I 339 pp. Smith and Elder. London.

Archey, G., 1941. The Moa. Bull. Auck. Inst. Mus.

Beckett, P., 1955. Taupo midden, Plimmerton, 1910-1914. J. Polynes. Soc. 64:161-162.

Best, E., 1901. Te Whanga-nui-a-Tara: Wellington in pre-Pakeha days. J. Polynes. Soc. 10:107-165.

Best, E., 1918. Shell-middens of the Porirua district. N.Z. Jl Sci. Technol. 1:212-217.

Best, E., 1942. Forest lore of the Maori... Mem. Polynes. Soc. 19 (Dom. Mus. Bull. 14).

Best, E., McLeod, H. N., & Mason, H. J. W., 1916. [Map of] Wellington country district showing native names. Scale, 1 inch to a mile. Wellington: N.Z. Dept. of Lands & Survey.

Brees, S. C., 1847. Pictorial illustrations of New Zealand. Pp. 1-36, 67 figs. Williams, London.

Brodkorb, P., & Dawson, E. W., 1962. Nomenclature of Quaternary coots from oceanic islands. Auk. 79:267-269.

Buller, W., 1888. A history of the birds of New Zealand. Vol.I: London.

Carkeek, W. C., 1966. The Kapiti coast. Maori history and place names. A. H. & A. W. Reed. Wellington.

Chapman, F., 1885. Notes on Moa remains in the Mackenzie Country and other localities. Trans. Proc. N.Z. Inst. 17:172-178.

Christie, H. M. n.d. The Paremata middens. A paper delivered to the Historical Society of Victoria College after 1914 [from internal evidence]. Unpublished typescript. [Copies in Canterbury Museum, National Museum, and the author's library].

Coutts, P., 1972. Interpretation of archaeological material eroded from coastal dune sites. J. Roy. Soc. N.Z. 2:407-412.

Davidson, J., 1978. Archaeological salvage excavations at Paremata, Wellington, New Zealand. National Museum of New Zealand Records 1:203-236.

Dawson, E. W., 1949. Sub-fossil bird remains from Lake Grassmere N.Z. Bird Notes. (Notornis: 3:132-133).

Dawson, E. W., 1952. Subfossil Nestor (Psittacidae) from New Zealand. Emu 52:259-272.

Dawson, E. W., 1958. Re-discoveries of the New Zealand sub-fossil birds named by H. O. Forbes. Ibis 100:232-237.

Dawson, E. W., 1962. A first record of the extinct New Zealand coot from the North Island. Notornis 10 (2): 84.

Dawson, E. W., 1965. Bird remains in archaeology. (Chapter 29) in: Brothwell, D. & Higgs, E., (Eds) Science in Archaeology: A comprehensive survey of progress and research. 2nd ed. Thames & Hudson. London: 279-293.

Duff, R. S., 1950. Moas and Man. (Part 2). Antiquity 24:72-83.

Duff, R., 1956. The Moa-hunter period of Maori culture. 2nd ed. Government Printer. Wellington.

Ellen, R. F., 1978. The contribution of H. O. Forbes to Indonesian ethnography: a biographical and bibliographical note. L'Archipel, 15.

Falla, R. A., 1942. Bird remains from Moa-hunter camps. Rec. Canterbury Mus. 5:43-49.

Forbes, H. O., 1892. Preliminary notice of additions to the extinct avifauna of New Zealand. Trans. Proc. N.Z. Inst. 24:185-189.

Forbes, H. O., 1893. The Chatham Islands and their story. Fortnightly Rev. n.s. 53:669-690.

Hamilton, A., 1896. The art workmanship of the Maori race in New Zealand. New Zealand Institute. Wellington.

Hill, H., 1914. The Moa—legendary, historical, and geological: why and when the Moa disappeared. Trans. N.Z. Inst. 46.

Larson, L. M., 1930. Osteology of the California road-runner Recent and Pleistocene. Univ. Calif. Publs. Zool. 32:409-428.

Lydekker, R., 1891. Catalogue of the fossil birds in the British Museum (Natural History). Trustees of the British Museum (Natural History). London.

Olson, S. L., 1975. A review of the extinct rails of the New Zealand region (Aves: Rallidae). Natnl. Mus. Rec. 1:63-79.

Rothschild, W., 1907. On extinct and vanishing birds. Proc. Fourth Int. Orn. Congr. 191-217.

Skinner, H. D., 1914. Obituary. Augustus Hamilton. Trans. Proc. N.Z. Inst. 46:v-vii.

Wakefield, E. J., 1845. Adventure in New Zealand, from 1839 to 1844. Vol. II. John Murray, London.

Ward, E., 1928. Early Wellington. Whitcombe & Tombs. Auckland.

Williams, G. R., 1956. The Kakapo (Strigops habroptilus, Gray). A review and re-appraisal of a near-extinct species. Notornis 7:29-56.

Williams, G. R., 1960. Distribution of specimens of the Kakapo, Strigops habroptilus Gray, in some museums throughout the world. Rec. Dominion Mus. 3:219-227.

Yaldwyn, J.C., 1956. A preliminary account of the sub-fossil avifauna of the Martinborough caves. Rec. Dominion Mus. 3:1-7.

Yaldwyn, J. C., 1959. Moa remains from the Wellington district. N.Z. Archaeol. Ass. Newsl. 2:20-25.

MAXIMIZING MINIMUM NUMBERS: AVIAN REMAINS FROM THE WASHPOOL MIDDEN SITE

B. Foss Leach

Introduction

During a programme of archaeological work in Palliser Bay from 1969 through 1971, excavations were conducted at about 25 prehistoric sites. A number of these contained midden refuse, and bird remains were fairly common. The largest quantity of avian fragments was recovered from the 65 m^2 excavated at a food processing area at the mouth of the Makotukutuku River, an area locally known as the Washpool (N168-9/22). The stratigraphy of the site is grouped into three cultural levels. Occupation at Level I was centred on about A.D. 1180, Level II at about A.D. 1345 and Level III at about A.D. 1650. The bird remains from this site are used here to document the changing relationships of prehistoric people to their natural environment in the area, and also the manner in which different parts of bird anatomy were treated by this community in selecting some parts as food and others for ornaments and so on. This study owes much to the work of Ron Scarlett who identified the several thousand bird bones recovered. A particular method was used for recording these identifications, and this should be described in full.

Recording Avian Bone Fragments

There are various ways of assessing the number of individuals from bone fragments (for example, Chaplin, 1971:70ff); all of these are fairly time consuming in the case of bird remains, owing to the large number of fragments which can be identified from any one individual bird. The approach suggested here has a number of advantages. On the one hand the method can be consistently applied between different species and this leads to a high degree of comparability. The second feature is that it allows a straightforward analysis of the parts of the body which were retained by prehistoric groups for preservation or immediate consumption. The figures for the different parts of the anatomy also provide an insight into butchering practices.

Figure 8.1 shows the common parts of each limb bone which may be identified. For each fragment an assessment is made whether the shaft portion covers more or less than the mid-shaft position. There are therefore seven different sorts of fragment for each bone.

1. C The complete bone.
2. P The proximal part with only a small part of the shaft.
3. PS The proximal part and more than half of the shaft.

4. S The shaft portion, including the mid-shaft area.

5. DS The distal portion and more than half of the shaft.

6. D The distal portion with only a small part of the shaft.

7. F Small fragments other than those above.

As the bones were identified they were entered on a cyclostyled form shown in Table 8.1. Space is allowed for nine major limb bones (including both sides), and a further eight bones which commonly occur in archaeological sites. The latter eight cannot be coded in any simple way, and common sense must dictate how these fragments are assessed. Of some importance is the sternum, and this has been divided into four fragment types. The rostrum is the small prominence at the anterior end of the sternum. This is often broken off in specimens and can be separately identified.

When a sample of bones is thus processed, Table 8.2 is filled in with the results. An example is given to explain the procedure. One 'minimum number' for the left femur is the addition of the figures in Table 8.1 under C plus P plus PS. Another is C plus D plus DS, and yet another is C plus PS plus DS plus S. Whichever is the greater of these sums is the 'minimum number' for the bone of that side. If only fragments (F) occur, then the minimum number at this point is one. These sums are calculated for the other side, and the final minimum number for the bone is simply the larger of the two numbers. This procedure is followed for each bone including the final eight (no simple rules can be formulated for these fragments). Finally, whichever is the largest number in the right hand column is the maximum minimum number.

Avian Remains at the Washpool

The bird bones from the Washpool Midden Site were processed according to the method outlined above. A minimum number of 273 birds was thus identified, belonging to 45 different species; the basic results are set out in Table 8.3. In Tables 8.4 and 8.5 the same data is organized into taxonomic orders, and in Table 8.6 into habitat types. Size differentiation details are presented in Table 8.7.

These figures show a clear emphasis on the medium and smaller forest birds, and in particular the very mobile flocking birds which characterize mixed forest conditions. In this medium size range are 192 individuals including *Callaeas cinerea wilsoni* (kokako), *Ninox novaeseelandiae* (morepork), *Prosthemadera novaeseelandiae* (tui), *Cyanoramphus novaezelandiae* and *C. auriceps* (parakeets), *Turnagra capensis tanagra* (thrush), *Philesturnus carunculatus rufusater* (saddleback), *Halcyon sancta vagans* (kingfisher), and *Anthornis m. melanura* (bellbird). The kokako, thrush and saddleback are now locally extinct. There are only 20 forest birds in the larger size range and most of them are *Hemiphaga novaeseelandiae* (wood pigeon), although there are also a few *Gallirallus australis greyi* (weka), *Nestor meridionalis septentrionalis* (kaka), and *Heteralocha acutirostris* (extinct huia). Small forest birds include *Petroica australis longipes* (robin), *Rhipidura fuliginosa placabilis* (fantail), and account for only 5 individuals.

Table 8.1 Recording form for avian identifications

Site: N168-9/22 Layer: Lens IIB Species: Cyanoramphus novaezelandiae

Bone	Fragments						
	C	P	PS	S	DS	D	F
L Carpometacarpus	1	1					
R Carpometacarpus	1	2					
L Ulna					1	1	
R Ulna	1	2					
L Radius							
R Radius							
L Humerus	2	6	3		2	4	
R Humerus	2	6	4			8	
L Scapula	2						
R Scapula		1					
L Coracoid	8		1		1	5	
R Coracoid	4	2	4			3	
L Femur	1	2	2				
R Femur		6	1			1	
L Tibiotarsus		3		1	2	8	
R Tibiotarsus		1	1	3	2	1	
L Tarsometatarsus	3	5			4	3	
R Tarsometatarsus	8	2	2	1	3		

	C	C minus rostrum	rostrum	F
Sternum	32	3	8	7

Mandibles Complete 3 Left side 4 Right side

Furcula Complete Left side Right side

Pelvis and Sacrum Complete Fragments 5 (Minimum No = 1)

Skull Fragments (describe) 16 Maxillae, 3 fragments

L Quadrate

R Quadrate

Phalanges

Vertebrae

Total number bones identified = 225

Table 8.2 Computation schema for minimum numbers

Bone	C+P+PS	C+D+DS	C+PS+DS+S	Minimum Per Side	Minimum Per bone	Maximum Min. No.
L Carpometacarpus	2	1	1	2	3	
R Carpometacarpus	3			3		
L Ulna		2	1	2	3	
R Ulna	3	1	1	3		
L Radius					0	
R Radius						
L Humerus	11	8	7	11	12	
R Humerus	12	10	6	12		
L Scapula	2	2	2	2	2	
R Scapula	1			1		
L Coracoid	9	14	10	14	14	
R Coracoid	10	7	8	10		
L Femur	5	1	3	5	7	
R Femur	7	1	1	7		
L Tibiotarsus	3	10	3	10	10	
R Tibiotarsus	2	3	6	6		
L Tarsometatarsus	8	10	7	10	14	
R Tarsometatarsus	12	11	14	14		
Sternum					40	40
Mandible					7	
Furcula					0	
Pelvis & Sacrum					1	
Quadrate					0	
Phalanges					0	
Vertebrae					0	
Skull					16	

There are a number of surprises in the species composition of this assemblage. One unexpected result is that the huia is only poorly represented. Huia were quite numerous, in the nineteenth century, in the Aorangi Mountains which form the hinterland of Palliser Bay, especially in the kanuka areas on spurs and ridges (Stidolph, 1971:17), and their tail feathers were highly prized as a trade item. The small numbers of these birds in the Palliser Bay middens may parallel the nineteenth century situation when they were apparently not eaten by the Maori (Best, 1942:221). Similarly, nineteenth century records indicate that kokako and the saddleback were not prized as food (Best, 1942:378, 380) and these two species are also poorly represented in the site. On the other hand, the robin was deliberately trapped by the nineteenth century Maori, and this bird is a more significant component of the midden. There are possible indications in these parallels of considerable conservatism in bird preferences over a period of seven centuries. At the same time, the weka is comparatively rare in the assemblage and this is unexpected since it was common on the coastal strip in the nineteenth century and also a favoured item of food (Stidolph, 1971). Similarly, swamp birds are uncommon in the assemblages; there is only one harrier hawk (Circus approximans gouldi), one swan (Cygnus sumnerensis), one banded rail (Rallus philippensis assimilis), and only a few ducks (Anas s. superciliosa, and Anas rhynchotis variegata). Several other species such as the white heron are occasional visitors to the Washpool but are not represented in the midden. Again, from the present open ponding condition of the river-mouth, bitterns, dabchicks, and crakes might be expected in the midden. Also birds common on shingle river beds such as the pied stilt, banded dotterel, paradise duck, and pipit are conspicuously absent in the site. These absences may indicate some environmental differences from today's conditions. In fact they may support the conclusion derived from the analysis of shellfish and land-snails that local hydrological conditions deteriorated markedly after a few centuries of human settlement (see B. F. Leach, 1976: Chapter 5). Another surprising feature is the relatively small number of sea birds such as black-backed gulls and shearwaters, the young of which can be easily taken from breeding colonies. It seems likely that modern colonies such as that at North Kawakawa, in Palliser Bay have not been long established. Judging by modern instances, a number of the sea birds which are present in the midden were probably taken as sea-wrecks after the passage of cyclonic storms. In this category are the albatross, shy mollymawk and petrels.

Moa are represented by ten individuals belonging to three species: Pachyornis mappini, Euryapteryx geranoides and E. gravis. It is also possible that some of the fragments could belong to Dinornis sp. The identifications were made by Ron Scarlett, and his overall opinion was that the bone was in fresh green condition when broken rather than from sub-fossil deposits. The remains are very fragmentary, and practically all are from the tibio-tarsus bone. Many of the bone artefacts from the site are made from moa bone, and all indications are that moa presence in the site was not the result of local exploitation of a living population. The most likely interpretation is that the bone was obtained from a population some distance from the Washpool, and was procured by trade or from hunting expeditions during visits elsewhere. A South Island source is a strong possibility since E. gravis was almost certainly not contemporary with man in the North Island. On the other hand it is the most abundant moa in South Island archaeological sites. This could not have been the only source of moa-bone, however, since E. geranoides , a strictly North Island moa, was also present. Clearly these people must have had access either directly or indirectly to a living population in the North Island as well.

The abundance of flocking, honey and berry-eating birds in this site indicates that the occupants of the Washpool had some particularly suitable catching methods. Ethnographic literature of the nineteenth century Maori shows that spears and nooses were used for taking the larger birds such as kaka, tui, and pigeon, and bird spears were a prominent artefact in the site. There are a number of smaller birds, however, which could not have been caught with spears, such as the thrush, robin, parakeet, bellbird, saddleback and fantail. Some quite different method therefore must be suggested, such as snares, set traps, and nooses, and these were also recorded during the nineteenth century. In particular, an effective method of catching parakeets was obviously well developed. Members of the order Psittaciformes are present throughout the high islands of Polynesia, and being social birds are attracted to a decoy. A knowledge of this behaviour, when combined with snares, can be used to advantage in catching the birds. Such a technique could well have been imported to New Zealand. Apart from tui, parakeet, and pigeon, most other birds were recovered in similarly low numbers, and this probably reflects chance catching rather than specially adapted methods.

By far the greatest number of birds was found in Level I, and the sharp decline in Level II may indicate either a decline in fowling or perhaps modification of the environment and reduction in the bird population. The site at Level II still possesses much midden material, so the decline in numbers may be significant. The remains of only two birds were present in Level III. The proportions of different species were compared from Level I to Level II to see if any significant differences could be detected which might shed light on this issue. When the raw data for the two levels is compared using the Chi-square test for two independent samples (qv. Reyment, 1971:53), the resulting difference is shown to be highly significant (see footnote to Table 8.3). If the data is reorganized into taxonomic orders (as in Tables 8.4 and 8.5) there are only three proportions which appear to change significantly between the two levels. The Passeriformes (such as tui, bellbird etc.) have dropped from 44.5% to 25.0%, and this is significant (p= .05, as determined by Rosenbaum's statistic qv, B. F. Leach, 1976: Appendix 35). The main increases detected are the Columbiformes (such as pigeons) from 3.2% to 13.9% (significant p less than .01), and the Procellariiformes (most of the sea birds) which have risen from 2.7% to 11.1% (significant p = .05). The overall difference however, is rather subtle and not easily described in environmental terms. When the frequencies are related to bird habitat categories (Table 8.6), there appears to be a rise in coastal birds at the expense of both forest and forest fringe dwellers, but the significance of the change cannot be proven with the smaller sample size of Level II. Finally, the change may be examined in terms of bird size (Table 8.7). There is a clear change in emphasis in Level II towards the larger birds (significant p less than .05), and this largely reflects the dwindling numbers of Passeriformes. These detected changes are consistent with the gradual loss of forest conditions close at hand.

Table 8.3 Bird remains at the Washpool Midden Site: Minimum numbers. Species arranged in decreasing abundance.

N.B.: Figures in brackets are percentages.

Species	Level I	Level II	Level III	Total
Cyanoramphus ? sp.	13 (5.7)	6(13.6)		
Cyanoramphus novaezelandiae	48(21.1)	5(11.4)		
Cyanoramphus auriceps (35.6)	20(8.8)	3(6.8)	(31.8)	95
Prosthemadera novaeseelandiae	77(33.9)	9(20.5)	1(50.0)	87
Hemiphaga novaeseelandiae	7(3.1)	5(11.4)		12
Passeriformes ? sp.	7(3.1)			7
Bird ? sp.	1(0.4)	4(9.1)	1(50.0)	6
Petroica australis longipes	4(1.8)			4
Nestor meridionalis septentrionalis	3(1.3)	1(2.3)		4
Turnagra capensis tanagra	3(1.3)			3
Gallirallus australis greyi	2(0.9)	1(2.3)		3
Capellirallus/Rallus ? sp.	3(1.3)			3
Coturnix novaezealandiae novaezealandiae	3(1.3)			3
Eudyptes pachyrhynchus pachyrhynchus	3(1.3)			3
Moa ? sp.	1(0.4)	2(4.5)		3
Philesturnus carunculatus rufusater	2(0.9)			2
Callaeas cinerea wilsoni	2(0.9)			2
Euryapteryx geranoides	2(0.9)			2
Euryapteryx gravis	2(0.9)			2
? Pachyornis mappini/Euryapteryx geranoides		2(4.5)		2
Anas ? sp.	2(0.9)			2
Anas superciliosa superciliosa	2(0.9)			2
Small petrel ? sp.	1(0.4)	1(2.3)		2
Eudyptula minor	2(0.9)			2
Puffinus ? gavia/huttoni	1(0.4)	1(2.3)		2
Rhipidura fuliginosa placabilis	1(0.4)			1
Anthornis melanura melanura	1(0.4)			1
Ninox novaeseelandiae novaeseelandiae	1(0.4)			1
Heteralocha acutirostris	1(0.4)			1
Pachyornis mappini	1(0.4)			1
Rallus philippensis assimilis	1(0.4)			1
Circus approximans gouldi	1(0.4)			1
Cygnus sumnerensis		1(2.3)		1
Halcyon sancta vagans	1(0.4)			1
Anas rhynchotis variegata	1(0.4)			1
Diomedea exulans/epomophora		1(2.3)		1
Diomedea cauta	1(0.4)			1

Table 8.3 Cont'd

Species	Level I	Level II	Level III	Total
Larùs dominicanus	1(0.4)			1
Phalacrocorax ? sp.	1(0.4)			1
Halobaena caerulea	1(0.4)			1
Petrel ? sp.		1(2.3)		1
Puffinus gavia	1(0.4)			1
Puffinus ? sp.	1(0.4)			1
Eudyptes pachyrhychus sclateri		1(2.3)		1
Eudyptes ? sp.	1(0.4)			1
Totals:	227(100.0)	44	2	273

Chi-square Level I to Level II = 79.1 with 44 degrees of freedom, which is a highly significant difference (p less than .005).

Table 8.4 Bird remains at the Washpool Midden Site—Species arranged into Taxonomic orders.

N. B.: Derived from the figures in Table 8.3.

Order	Level I	Level II	Level III	Total
Dinornithiformes	6	4	-	10
<u>Pachyornis</u> <u>mappini</u> <u>Euryapteryx</u> <u>geranoides</u> <u>Euryapteryx</u> <u>gravis</u> <u>?</u> <u>Pachyornis</u> <u>mappini</u> <u>Euryapteryx</u> <u>geranoides</u> Moa ? sp.				
Procellariiformes	6	4	-	10
<u>Diomedea</u> <u>?</u> <u>exulans/Epomophora</u> <u>Diomedea</u> <u>cauta</u> <u>Halobaena</u> <u>caerulea</u> Small petrel ? sp. Petrel ? sp. <u>Puffinus</u> <u>gavia</u> <u>Puffinus</u> <u>?</u> <u>gavia/huttoni</u> <u>Puffinus</u> <u>?</u> <u>sp</u>.				
Sphenisciformes	6	1	-	7
<u>Eudyptes</u> <u>pachyrhynchus</u> <u>pachyrhynchus</u> <u>Eudyptes</u> <u>pachyrhynchus</u> <u>sclateri</u> <u>Eudyptula</u> <u>minor</u> <u>Eudyptes</u> <u>?</u> <u>sp</u>.				
Pelecaniformes	1	-	-	1
<u>Phalacrocorax</u> <u>?</u> <u>sp</u>.				
Anseriformes	5	1	-	6
<u>Cygnus</u> <u>sumnerensis</u> <u>Anas</u> <u>rhynchotis</u> <u>variegata</u> <u>Anas</u> <u>?</u> <u>sp</u>. <u>Anas</u> <u>superciliosa</u> <u>superciliosa</u>				
Falconiformes	1	-	-	1
<u>Circus</u> <u>approximans</u> <u>gouldi</u>				
Galliformes	3	-	-	3
<u>Coturnix</u> <u>novaezealandia</u>				

Table 8.4 Cont'd

Order	Level I	Level II	Level III	Total
Gruiformes	6	1	-	7
Gallirallus australis greyi				
Capellirallus/Rallus ? sp.				
Rallus philippensis assimilis				
Charadriiformes	1	-	-	1
Larus dominicanus				
Columbiformes	7	5	-	12
Hemiphaga novaeseelandiae				
Psittaciformes	84	15	-	99
Nestor meridionalis septentrionalis				
Cyanoramphus novaezelandiae				
Cyanoramphus auriceps				
Cyanoramphus ? sp.				
Strigiformes	1	-	-	1
Ninox novaeseelandiae noveseelandiae				
Coraciiformes	1	-	-	1
Halcyon sancta vagans				
Passeriformes	98	9	1	108
Prosthemadera novaeseelandiae				
Passeriformes ? sp.				
Petroica australis longipes				
Turnagra capensis tanagra				
Philesturnus carunculatus rufusater				
Callaeas cinerea wilsoni				
Rhipidura fuliginosa placabilis				
Anthornis melanura melanura				
Heteralocha acutirostris				
Bird ? genus	1	4	1	6

Table 8.5 Bird remains from the Washpool Midden Site—Proportion in Different Orders (excluding Moa and Bird ? genus—percentages in brackets)

	Level I	Level II	Total
Procellariiformes (petrels etc.)	6(2.7)	4(11.1)	10
Sphenisciformes (penguins)	6(2.7)	1(2.8)	7
Pelicaniformes (shags)	1(0.5)	–	1
Anseriformes (ducks etc,)	5(2.3)	1(2.8)	6
Falconiformes (hawk etc,)	1(0.5)	–	1
Galliformes (quail etc,)	3(1.4)	–	3
Gruiformes (weka etc,)	6(2.7)	1(2.8)	7
Charadriiformes (gulls etc,)	1(0.5)	–	1
Columbiformes (pigeon etc,)	7(3.2)	5(13.9)	12
Psittaciformes (parakeets etc,)	84(38.2)	15(41.7)	99
Strigiformes (morepork etc,)	1(0.5)	–	1
Coraciiformes (kingfisher etc,)	1(0.5)	–	1
Passeriformes (tui etc,)	98(44.5)	9(25.0)	107
Totals:	220(100.0)	36(100.0)	256

Table 8.6 Bird remains from the Washpool Midden Site—Arranged into Habitat Types.

Habitat	Level I	Level II
Forest Dwellers	190(86.4%)	29(80.6%)
Columbiformes		
Psittaciformes		
Strigiformes		
Passeriformes		
Fringe Dwellers	16(7.3%)	2(5.6%)
Anseriformes		
Falconiformes		
Galliformes		
Gruiformes		
Coraciiformes		
Coastal Dwellers	14(6.4%)	5(13.9%)
Procellariiformes		
Sphenisciformes		
Pelecaniformes		
Charadriiformes		
Totals	220(100)	36(100)

Habitat	Level I	Level II
Forest and Fringe Dwellers	206(93.6%)	31(86.1%)
Coastal Dwellers	14(6.4%)	5(13.9%)
Totals	220(100)	36(100)

N.B.: None of the observed differences in proportions above are significant at the 5% level (tested by Rosenbaum's method q.v. B. F. Leach, 1976: App.35).

Table 8.7 Bird remains at the Washpool Midden Site—Arranged into two size groups.

Size Group	Level I	Level II
Larger Birds	41(18.6%)	13(36.1%)
Procellariiformes		
Sphenisciformes		
Pelecaniformes		
Anseriformes		
Falconiformes		
Galliformes		
Gruiformes (minus *R. philippensis assimilis*)		
Charadriiformes		
Columbiformes		
Nestor meridionalis		
Callaeas cinerea wilsoni		
Heteralocha acutirostris		
Smaller Birds	179(81.4%)	23(63.9%)
Rallus philippensis assimilis		
Cyanoramphus spp.		
Strigiformes		
Coraciiformes		
Passeriformes (minus *C. cinerea* and *H. acutirostris*		
Totals:	220	36

Preparation of the Bird Carcass

As shown above, the identification of bird remains followed a procedure which reveals details of the fate of the bird once it was caught. The recorded information is very extensive and only two examples are illustrated here. These are particularly interesting because they may indicate preservation practices, and also the relative importance of meat and feathers to the Washpool people. The examples chosen are the tui (Prosthemadera novaeseelandiae) and the two parakeet species (Cyanoramphus novaezelandiae and C. auriceps). In Tables 8.8 and 8.9 the minimum numbers are given for the major anatomical parts of each bird. As can be seen, the minimum numbers vary considerably for different parts of the body, but very little from the left to right side. The observed bilateral differences proved to be insignificant in the case of both tui and parakeet (see footnotes to Tables 8.8 and 8.9). For easy comparison of the results, the various minimum numbers were standardized as a proportion of the maximum minimum number, and these are plotted out schematically in Figures 8.2 and 8.3.

Evidently there are major differences in the various parts of the bird represented. One obvious possibility is that this reflects either differential survival or uneven ability to identify the discrete parts of the body. However, close inspection of the figures shows that neither of these claims c n be maintained. For example, tui are well represented by such fragile remains as the mandible, which is relatively difficult to speciate; and yet the same bird is very poorly represented by the relatively durable and diagnostic femur. Again, the similar minimum numbers for each side argue that the observed pattern has some additional significance. Some comments must be made, however, on particular bones for which the minimum numbers may be somewhat misleading.

The most difficult bones to assign to species are the phalanges, ribs, vertebrae, quadrate, scapula, furcula, carpometacarpus, radius, ulna and mandible. The carpometacarpus is not particularly diagnostic in shape, while the furcula, which is quite diagnostic, is very small and seldom recovered whole for these small species. Species identification is relatively easy for the tarsometatarsus, tibiotarsus, femur, humerus, pelvis, sternum, coracoid, and cranium, although complications can arise with the sternum, pelvis and cranium due to fragmentation. These factors relating to species identification, however, cannot account for the discrepancy between the pelvis and sternum, which are represented by 6% and 100% respectively in the case of parakeets, and 7% and 100% in the case of tui. Taking into account the few possibly unreliable figures, the main body area of each bird (Figs. 8.2 and 8.3) is conspicuously absent. The line of demarcation between what is well represented, and what is under-represented is very similar for the two species, but there are two notable differences, one in the head region, and the other in the lower leg area. In the case of the tui, the mandible remained at the site, but the cranium was removed with the body; in the case of the parakeet the exact reverse was found. It is precisely in these areas that the most valuable feathers on each bird are found. The male tui has a tuft of long curled white feathers under the mandible and these would be very easily removed by cutting the mandible and upper throat skin away in one slice, perhaps with a piece of obsidian. The skin could then be easily removed from the mandible. The beak area simply marks a convenient point for

Table 8.8 *Cyanoramphus* spp. from the Washpool Midden Site Level I (minimum numbers for different parts of the anatomy)

Anatomy		Minimum Number	% of Maximum Minimum Number
Carpometacarpus	L	3	4.4
	R	7	10.3
Ulna	L	4	5.9
	R	6	8.8
Radius	L	1	1.5
	R	0	0
Humerus	L	18	26.5
	R	23	33.8
Scapula	L	7	10.3
	R	5	7.4
Coracoid	L	25	36.8
	R	17	25.0
Femur	L	9	13.2
	R	10	14.7
Tibiotarsus	L	17	25.0
	R	12	17.6
Tarsometatarsus	L	25	36.8
	R	30	44.1
Quadrate	L	0	0
	R	0	0
Mandible	L	7	10.3
	R	7	10.3
Cranium and Maxilla		21	30.9
Furcula		1	1.5
Sternum		68	100.0
Vertebrae		7	10.3
Pelvis and sacrum		4	5.9
Phalanges		3	4.4
Ribs		1	1.5

Chi-square from Left to Right = 6.83, with 9 degrees of freedom. Therefore no significant bilateral asymmetry in identifications.

Table 8.9 Prosthemadera novaeseelandiae at the Washpool Midden Site Level I (minimum numbers for different parts of the anatomy)

Anatomy		Minimum Number	% of Maximum Minimum Number
Carpometacarpus	L	5	8.8
	R	6	10.5
Ulna	L	7	12.3
	R	8	14.0
Radius	L	8	14.0
	R	8	14.0
Humerus	L	27	47.4
	R	36	63.2
Scapula	L	15	26.3
	R	15	26.3
Coracoid	L	27	47.4
	R	30	52.6
Femur	L	3	5.3
	R	2	3.5
Tibiotarsus	L	49	86.0
	R	44	77.2
Tarsometatarsus	L	10	17.5
	R	18	31.6
Quadrate	L	2	3.5
	R	3	5.3
Mandible	L	20	35.1
	R	23	40.4
Cranium and Maxilla		3	5.3
Furcula		6	10.5
Sternum		57	100.0
Vertebrae		2	3.5
Pelvis and sacrum		4	7.0
Phalanges		3	5.3
Ribs		1	1.8

Chi-square from Left to Right = 3.68, with 10 degrees of freedom. Therefore no significant bilateral asymmetry in identification.

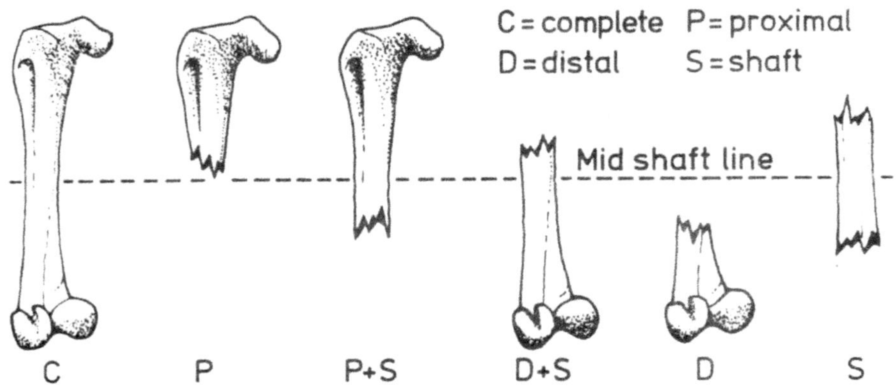

Fig. 8.1 Avian limb bone fragments recorded.

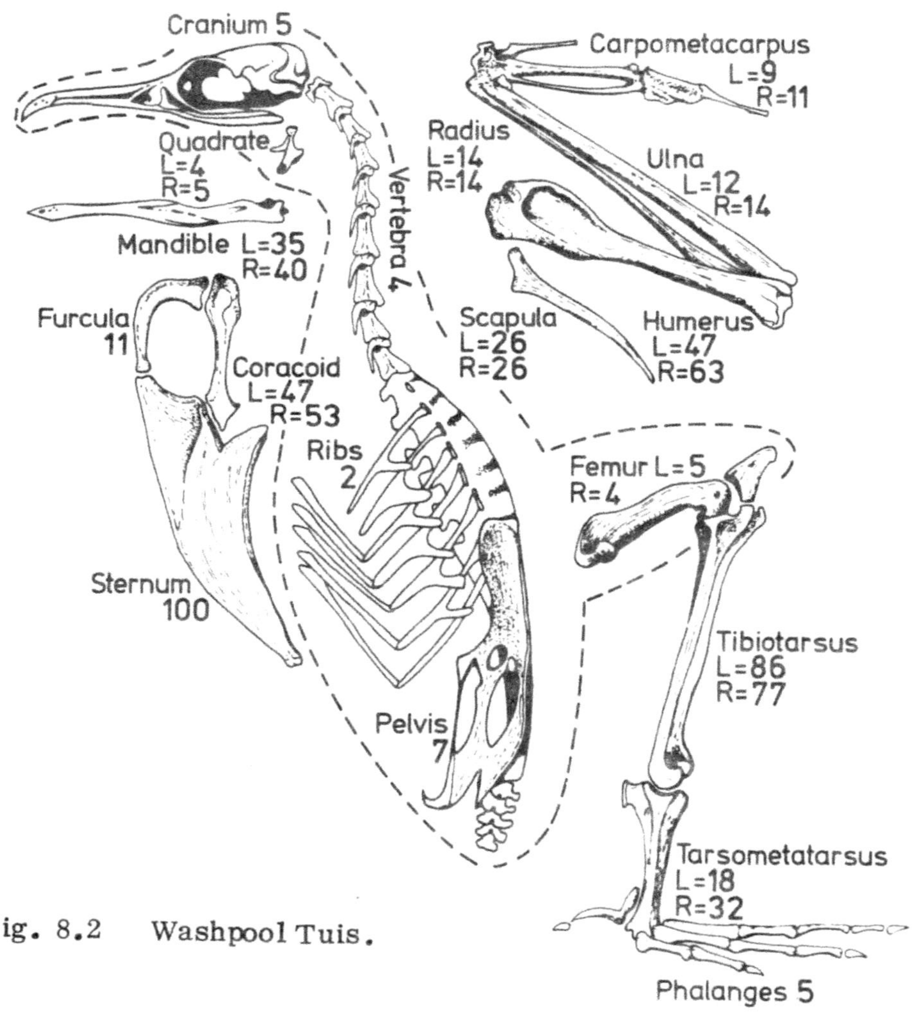

Fig. 8.2 Washpool Tuis.

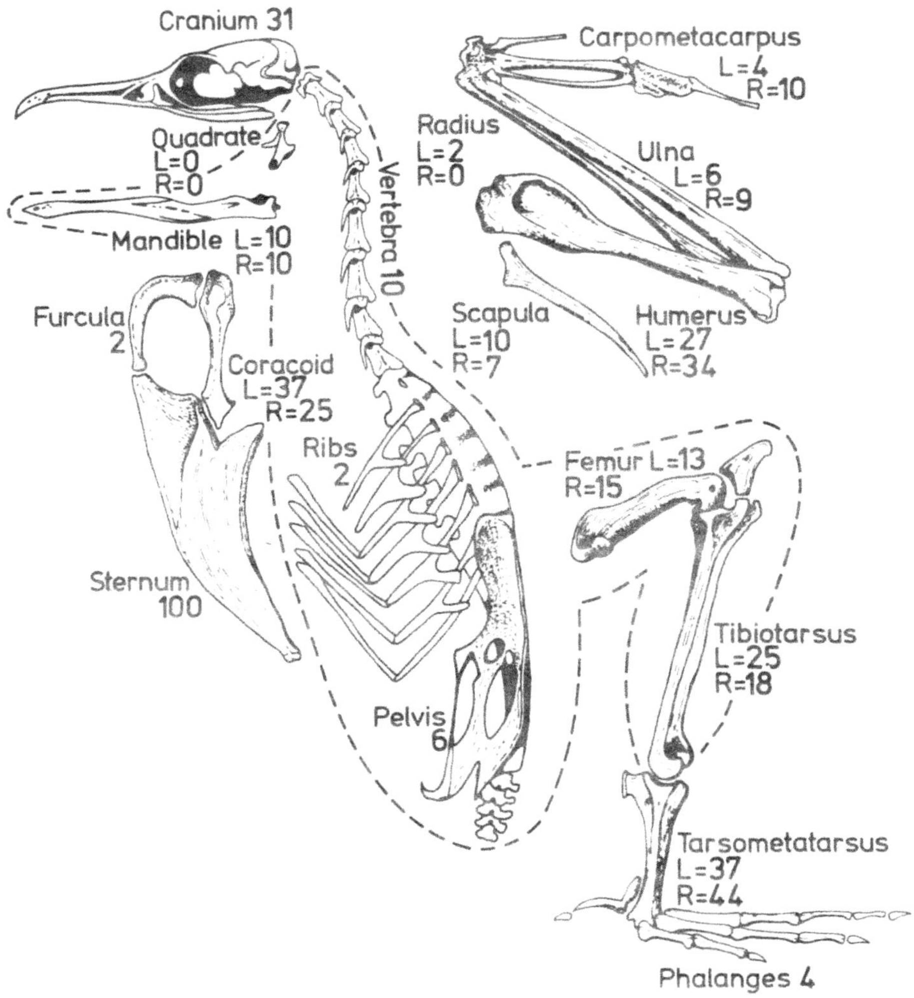

Fig. 8.3 Washpool Parakeets.

beginning the cut. The red and yellow-crowned parakeets, on the other hand, have what their common names imply, an area of red and yellow feathers on the top of their heads. Removal of these may have involved a similar process as in the tui (cutting at the back of the beak area) but this time removing the cranium from the rest of the body.

The second notable difference is in the lower leg region. The tui tibiotarsus (and the rest of the lower leg) stayed at the site, and the femora were removed with the body, whereas the demarcation line in the parakeet is a joint lower. Only the leg from the tarsometatarsus down remained at the site. This difference is explicable by reference to the outward appearance of each bird. The tui, a hopping bird, has practically no meat or feathers from the tibiotarsus down, and the obvious place to cut and discard the leg is at this point. The parakeet, however, is a walking bird, and has more muscle and feathers on the tibiotarsus; thus the obvious dividing line is one joint below that of the tui. It is suggested then that one of the operations involved in preparing these bird carcasses was to 'top and tail' them, as is done today, and the appropriate parts saved and discarded were determined by the presence of valuable feathers in the head area, and the presence of meat in the leg region.

The following interpretation is offered as a likely explanation of the observed patterns. When the tui were processed at the site the useless lower leg portion was cut off and discarded on to the midden. Then the lower beak (of the males) was sliced off with the piece of throat skin to which the white tuft of feathers was attached. Taking the mandible with the skin was for convenience of cutting only, for it was then cut from the throat skin and also thrown on to the midden. The wings were removed next, along with the sternum and the rest of the shoulder girdle. This contains a large portion of the meat of the bird and could only be removed with some tearing action as well as cutting. Presumably this part was eaten on the spot before discarding, for it also found its way on to the midden. The rest of the bird, which includes the drumsticks, and the remainder of the body cavity along with intestines, was removed from the site, perhaps preserved in fat for trading or eating elsewhere. The procedure applied to the parakeet must have been very similar except that the head was removed for its feathers, the skin presumably taken off separately, and the cranial bones discarded on the midden. The tibiotarsus was left on the body as having more useful meat than the tui.

Conclusion

The bird bones deposited at the Washpool midden site provide an insight into the nature of the surrounding environment and man's influence upon it over a period of about five hundred years, as well as the manner in which birds were treated as items of food and as a source of decorative feathers. Thus, these remains, in common with other types of economic debris, can form a prime source of information on natural history as well as culture history. Anyone who has worked with bird bones will realise what a tantalizing medium they present. Identifications are fairly difficult, and many must be made to achieve a 'minimum number' of even one. Even though the rewards of these labours always appear worthwhile in retrospect, it is important for archaeologists to be continually striving to maximize the return of information from this work. Ron Scarlett has contributed a great deal to this quest over the years by maintaining a high quality and an incomparable quantity of identifications for archaeologists. Above all, he has steadfastly demanded that even tiny fragments be included in any material sent to him. The documention of these fragments from the Washpool has helped uncover aspects of behaviour which would not have been possible by a more conventional analysis.

BIBLIOGRAPHY

Best, E., 1942. Forest Lore of the Maori. Dom. Mus. Bull. 14.

Chaplin, R. E., 1971. The Study of Animal Bones from Archaeological Sites. Seminar Press.

Leach, B. F., 1976. Prehistoric Communities in Palliser Bay, New Zealand. Unpublished Ph. D. Thesis, Anthropology Department, University of Otago.

Reyment, R. A., 1971. Introduction to Quantitative Paleoecology. Elsevier, Amsterdam.

Stidolph, R. H. D., 1971. The Birds Around Us. Hedleys Bookshop, Masterton.

ISLAND AND COASTAL FOWLING STRATEGIES OF THE PREHISTORIC MORIORI

D. G. Sutton

Introduction

A series of workers have collected bird bone from the Chatham Islands (Figure 9.1) since at least the 1860s. However, with the exception of unpublished material from the 1962-1963 expedition (Simmons, 1964), no previous collection has had the advantages of stratigraphical control and radiocarbon dating. Ten prehistoric sites have been excavated in the Chatham Islands since 1973. The analysis of midden material from all of these is now complete. Radiocarbon dates are available from five sites. They range in age from 1450-1620 A.D. at one standard deviation from the mean (Sutton, 1976; 1977). Unfortunately dates are not yet available from the two inland sites and the CHC seal bone midden (N.Z.A.A. Site No. C240/689) which are discussed in detail below. However, nitrogen determination (Smith, 1977) for CHC and artefacts present in all three sites indicate that they will fall within the 15th-16th century range.

The analysis of bird bone recovered from a broadly contemporary pattern of sites on Chatham Island is useful in at least two areas. First, the results contribute to the reconstruction of Morior subsistence economics (Sutton, n.d.) within the Durham study area (Figure 9.1). Secondly, it will assist attempts to establish the causes of bird extenctions and range reductions which have led down, evidently very recently, to the current impoverished native avifauna of the islands (vide Marshal, Scarlett and Sutton, n.d.).

Materials

The sites discussed below are shown in Figure 9.1. Figure 9.2 shows a schematic section through the coastal terrace (Hamel, 1977) near Waihora.

The CHA site (C240/681) is located on the southwest aspect of Ridge III Figure (9.1). It is small, discontinuous and shallow site which extends over 30 metres down a low spur. Test-pitting showed that it consisted on a number of small patches of midden, each near one or more scoop hearths. Two adjacent 5 by 5 m excavation areas were laid out over the largest midden. The southernmost revealed a single dense layer of bird bone midden. The other contained 4 small scoops.

The CHB site (C240/680) was found beside a small swamply basin on the top of Ridge III, 200 metres inland from CHA and approximately 900 m etres from the sea. Again two adjacent 5 by 5 m areas were laid out. They contained a single continuous deposit of midden up to 30 cm deep. It contained very few portable artefacts and no evidence of structures. A large area around the site was test-pitted in a search for house structures but none were found.

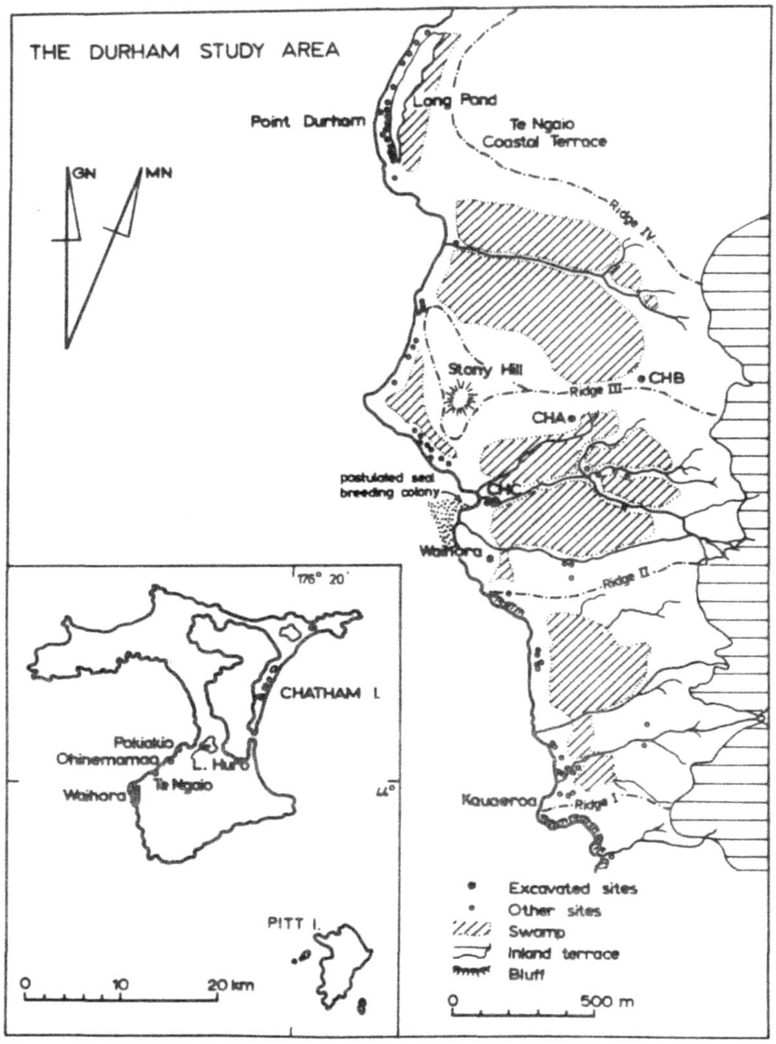

Fig. 9.1 The Durham Study Area.

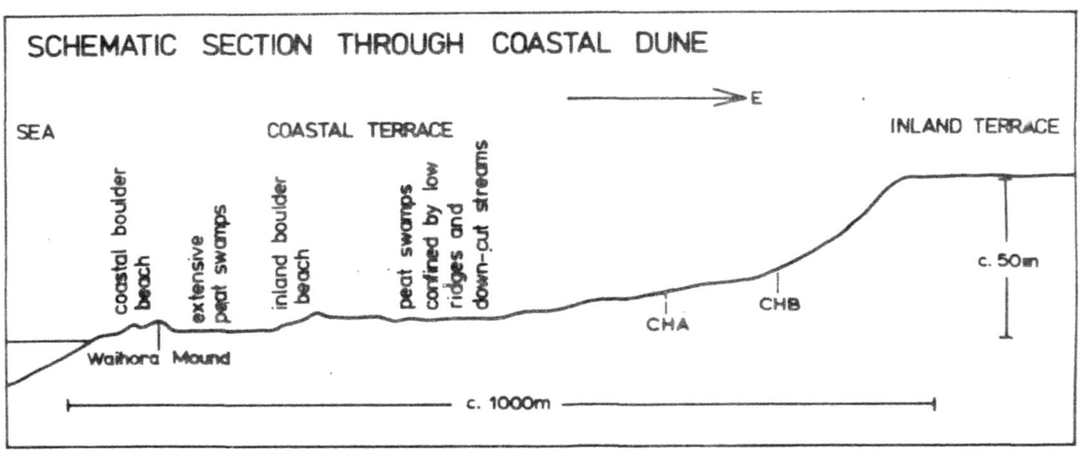

Fig. 9.2 Schematic Section through Coastal Terrace.

The coastal sites discussed below are on the CHC seal bone midden (Figure 9.1), excavated and reported by Smith (1977), and three small shellfish middens; Te Ngaio (C240/277), Ohinemamao (240/273) and Pokiokia (C240/266). They are located 4, 6 and 8 km respectively northeast of Waihora on the Southwest coast of Chatham Island (Figure 9.1). They were excavated and reported by McIlwraith (1976) and are radiocarbon dated to 1560, 1530 and 1530 respectively with a standard deviation of 40 years (Sutton, 1977).

Laboratory Analysis

The material recovered from these excavations was separated into primary components: shell, bone, charcoal and artefacts. A series of minor components was also separated out; crustacea, otoliths *et cetera*. The bone material was sorted into rat, bird, fish, marine mammal and human. The bird bone was steam cleaned and identified to anatomy.

The identification of this material to species level was undertaken by R. J. Scarlett and Y. M. Marshall. Minimum numbers *per* species were calculated using a method developed by B. F. Leach (this volume). The study of body parts representation *per* major species, discussed by Sutton and Marshall (1977) and mentioned below, also followed Leach's procedure.

The Assemblages

The assemblages from coastal and inland sites are summarised in Tables 9.1 and 9.2 respectively. The species identified are listed with the minimum number of birds represented *per* species.

Assemblages from the inland sites are very similar, although sample sizes are widely separated (n = 635 at CHB and n = 148 at CHA). The five most frequent represented species at both sites are:

Chatham Island Taiko
Chatham Island Pigeon
Chatham Island Petrel
Dieffenbach's Rail
Broad-billed Prion

They make up over 90% of both assemblages. The order in which they occur is almost identical. Moreover, the proportion of the total minimum number *per* assemblages which each species contributes is almost identical (Table 9.1). The only reversal is in the order of Pigeon and Diving Petrel. This may be explained by the location of CHB in a zone of broadleaf forest, as indicated by charcoal and landsnail data (Molloy, pers. comm., 1977; Wallace, 1977; 1979), contemporary vegetation ecology and historical records (Hamel, 1977; Begg, 1977). On the other hand, the CHA site was in a disturbed transitional vegetation containing fewer broadleaf tree species (Dodson and Kirk, n.d.) but nearer to the Stony Hill headland (Figure 9.2) where Diving Petrels would have bred.

The sets of minor species present in the two inland sites are also similar. All the nine major species represented in CHA are amongst the 26 species at CHB. The additional minor species at CHB are numerically incidental to the assemblage.

The coastal assemblages are quite distinctive (Table 9.2). Again a strong pattern is evident amongst the sites. They are all small and irregular in their species composition, and they are all unspecialised in that no one or a few species make up an overwhelming proportion of the assemblages. The species which dominate the inland middens are not strongly represented.

Habitat Types

The assemblages can be interpreted in terms of fowling strategies when the species lists are organised into habitat types. These categories include birds which are found most commonly in a specific habitat. Five habitat groups are discernible. They are:

- i Coastal Birds: shags, gulls, oystercatchers and terns (Orders Pelecaniformes and Charadriiformes) and some non-breeding visitors.
- ii Wetland Birds: all the duck species and the Extinct Swan (Anseriformes).
- iii Forest-edge and Bogshrubland Birds: rails and snipes (Gruiformes and Charadriiformes).
- iv Broadleaf Forest Birds: piegon (Columbiformes), Kaka and Parakeets (Psittaciformes) and Tui (Passeriformes).
- v Marine Birds: there are species which come ashore for short intervals at various seasons of the year to breed and/or moult. They are the penguins and petrels (Sphenisciformes and Procellariiformes).

The percentage of the 2 inland and 4 coastal assemblages which are drawn from each of these habitat types are shown in Figure 9.3.

The operation of three separate fowling strategies is evident from the inland sites. The clearly dominant one is "muttonbirding" in which fledglings of small to medium sized petrels are taken from burrows or crevices during the latter stages of their pre-flight growth period. A commerical version of this ancient (Jones, 1971; Bowdler, 1974; Sutton and Marshall, n.d.) and very efficient strategy is used on Flesh-footed, Sooty and other shearwaters in New Zealand (Richdale, 1948; Oliver, 1955) and on the Short-tailed Shearwater in the Bass Strait Islands (Simpson, 1972). It is interesting to note here that Jones' (1971:550) procedure of counting minimum numbers of birds on the basis of "...pre-maxillae, dentaries, and posterior parts of the mandibles" may have resulted in an under estimation of the importance of birds throught the Rocky Cape sequence, possibly by a factor of at least 4-5 times (vide Sutton and Marshall, 1977: Figure 6 and 7; Sutton and Marshall, n.d.: Figure 6).

An examination of the degree of osteological maturity evident in the most frequently ocuurring bones of the Taiko, Diving Petrels and Broad-billed Prions at CHB proves that a large proportion of them were taken as fledglings. At least 65.5% of the Taiko scapulae, 35.6% of scapulae and coracoids of Diving Petrel and 16.6% of the tiny coracoid of the Prion survived as immature

TABLE 9.1:
COMPOSITION OF INLAND BIRD BONE ASSEMBLAGES
(CHA, L.1. and CHB, L.1.)

Species are arranged in numerical order as they occur in CHB and then taxonomic order. Nomenclature after Kinsky (et al., 1970), except where more recent revisions apply.

Species (Total 31 spp.)	Numbers per Species		% Assemblage		Total per Species
	CHB	CHA	CHB	CHA	
Chatham Island Taiko (Pterodroma magentae)	339	75	53.39	50.68	414
C.I. Pigeon (Hemiphaga novaeseelandiae chathamensis)	95	14	14.96	9.46	109
Southern Diving Petrel (Pelecanoides urinatrix chathamensis)	69	32	10.87	21.62	101
Dieffenbach's Rail (Rallus philippensis dieffenbachi)	40	9	6.30	6.08	49
Broad-billed Prion (Pachyptila v. vittata)	30	4	4.72	2.70	34
Extinct Giant Rail (Diaphorapteryx hawkinsi)	11	3	1.73	2.03	14
White-faced Storm Petrel (Pelagodroma marina maoriana)	6	3	0.94	2.03	9
Fluttering Shearwater (Puffinus g. gavia)	4	2	0.63	1.35	6
Pitt Island Shag (Stictocarbo punctatus featherstoni)	4	1	0.63	0.68	5
Fairy Prion (Pachyptila turtur)	3	-	0.47	-	3
Grey Teal (Anas gibberfrons gracilis)	3	-	0.47	-	3
New Zealand Shoveler (Anas rhynchotis variegata)	3	1	0.47	0.68	4
Weka (Gallirallus autralis)	3	-	0.47	-	3

Species (Total 31 spp.)	Numbers per Species CHB	CHA	% Assemblage CHB	CHA	Total per Species
C.I. Red-crowned Parakeet (Cyanoramphus novaezelandiae chathamensis)	3	1	0.47	0.68	4
Chatham Island Petrel (Pterodroma hypoleuca axillaris)	2	-	0.31	-	2
Grey Duck (Anas s. superciliosa)	2	1	0.31	0.68	3
Extinct Weka (Gallirallus minor)	2	-	0.31	-	2
White-fronted Tern (Sterna striata)	2	-	0.31	-	2
C.I. Tui (Prosthemadera novaeseelandiae chathamensis)	2		0.31	-	2
C.I. Blue Penguin (Eudyptula minor chathamensis)	1	-	0.16	-	1
Erect-crested Penguin (Eudyptes pachyrhynchus sclateri)	1	-	0.16	-	1
Northern Royal Albatross (Diomedea epomophora sanfordi)	1	-	0.16	-	1
C.I. Mollymawk (Diomedea cauta eremita)	1	1	0.16	0.68	2
Sooty Shearwater (Puffinus griseus)	1		0.16	-	1
Short-tailed Shearwater (Puffinus tenuirostris)	1	-	0.16	-	1
Paradise Duck (Tadorna variegata)	1	-	0.16	-	1
New Zealand Scaup (Aythya novaeseelandiae)	1	-	0.16	-	1
Extinct Coot (Fulica c. chathamensis)	1	-	0.16	--	1
C.I. Snipe (Coenocorphya aucklandica pusilla)	1	-	0.16	-	1
Red-Billed Gull (Larus novaehollandiae scopulinus)	1	-	0.16	-	1
Kaka (Nestor meridionalis)	1	1	0.16	0.68	2
	635	148	100.15	100.13	783

Table 9.2 Coastal Bird Bone Assemblage (CHC, Te Ngaio, Ohinemamao and Pokiakio)

Species are arranged in taxonomic order. Nomenclature after Kinsky (et al., 1970) except where more recent revisions apply.

Species (Total 40 spp.)	CHC No.	CHC %	Te Ngaio No.	Te Ngaio %	Ohinemamao No.	Ohinemamao %	Pokiakio No.	Pokiakio %	Total per species
King Penguin (Aptenodytes patagonicus)	1	5.0	-	-		-		-	1
C.I. Blue Penguin (Eudyptula minor chathamensis)	1	5.0	1	6.25	3	12.50	1	2.44	6
Erect-crested Penguin (Eudyptes pachyrhynchus sclateri)	1	5.0	-	-	1	4.17	-	-	2
Northern Royal Albatross (Diomedea epomophora sanfordi)	-	-	-	-	-	-	1	2.44	1
C.I. Mollymawk (Diomedea cauta eremita)	1	5.0	-	-	-	-	1	2.44	2
C.I. Taiko (Pterodroma magentae)	-	-	1	6.25	1	4.17	1	2.44	3
C.I. Petrel (Pterodroma hypoleuca axillaris)	-	-	-	-	-	-	1	2.44	1
Broad-billed Prion (Pachyptila v. vittata)	1	5.0	1	6.25	-	-		-	2
Fairy Prion (Pachyptila turtur)	-	-	1	6.25	1	4.17	1	2.44	3
Sooty Shearwater (Puffinus griseus)	-	-	1	6.25	1	4.17	2	4.88	4
Fluttering Shearwater (Puffinus g. gavia)	1	5.0		-	-	-	-	-	1
Short-tailed Shearwater (Puffinus tenuirostris)	1	5.0	-	-	-	-	-	-	1
White-faced Storm Petrel (Pelagodrmoa marina maoriana)	-	-	-	-	-	-	1	2.44	1

Table 9.2 (contd.)

Species	CHC No.	CHC %	Te Ngaio No.	Te Ngaio %	Ohinemamao No.	Ohinemamao %	Pokiakio No.	Pokiakio %	Total per species
Southern Diving Petrel (Pelecanoides urinatrix chathamensis)	2	10.0	2	12.50	2	8.33	8	19.51	14
C.I. Shag (Leucocarbo carunculatus onslowi)	-	-	-	-	-	-	1	2.44	1
Pitt Island Shag (Stictocarbo punctatus featherstoni)	7	35.0	-	-	1	4.17	4	9.76	12
Extinct Swan (Cygnus sumnerensis)	-	-	-	-	-	-	1	2.44	1
Paradise Duck (Tadorna variegata)	-	-	-	-	-	-	1	2.44	1
Grey Duck (Anas s. superciliosa)	-	-	-	-	1	4.17	1	2.44	2
Grey Teal (Anas gibberfrons gracilis)	-	-	1	6.25	1	4.17	-	-	2
New Zealand Shoveler (Anas rhynchotis variegata)	-	-	1	6.25	-	-	-	-	1
N.Z. Scaup (Aythya novaeseelandiae)	-	-	-	-	-	-	1	2.44	1
Harrier (Circus approximans subspp.)	-	-	-	-	-	-	1	2.44	1
Extinct Giant Rail (Diaphorapteryx hawkinsi)	-	-	-	-	1	4.17	-	-	1
Dieffenbach's Rail (Rallus philippensis dieffenbachi)	2	10.0	1	6.25	2	8.33	1	2.44	6
Extinct Weka (Gallirallus minor)	-	-	1	6.25	-	-	1	2.44	2
Weka (Gallirallus australis)	1	5.0	-	-	-	-	-	-	1

Table 9.2 (contd.)

Species	CHC		Te Ngaio		Ohinemamao		Pokiakio		Total per species
	No.	%	No.	%	No.	%	No.	%	
Extinct Coot (Fulica c. chathamensis)	-	-	1	6.25	-	-	-	-	1
C.I. Oystercatcher (Haematopus chathamensis)	-	-	1	6.25	-	-	1	2.44	2
Extinct Snipe (Coenocorphya chathamica)	-	-	-	-	1	4.17	-	-	1
C.I. Snipe (Coenocorphya aucklandica pusilla)	1	5.0	-	-	1	4.17	-	-	2
Southern Skua (stercorarius skua lonnbergi)	-	-	-	-	-	-	1	2.44	1
Southern Black backed Gull (Larus dominicanus)	-	-	-	-	-	-	3	7.32	3
Red-billed Gull (Larus novaehollandiae scopulinus)	-	-	1	6.25	3	12.50	3	7.32	7
White-fronted Tern (Sterna striata)	-	-	-	-	-	-	1	2.44	1
C.I. Pigeon (emiphaga novaeseelandiae chathamensis)	-	-	-	-	1	4.17	1	2.44	2
Kaka (Nestor meridionalis)	-	-	-	-	-	-	1	2.44	1
C.I. Red-crowned Parakeet (Cyanoramphus novaezelandiae chathamensis)	-	-	1	6.25	1	4.17	-	-	2
C.I. Yellow-crowned Parakeet (Cyanoramphys auriceps forbesi)	-	-	1	6.25	1	4.17	-	-	2

Table 9.2 (contd.)

Species	CHC		Te Ngaio		Ohinemamao		Pokiakio		Total per species
	No.	%	No.	%	No.	%	No.	%	
C. I. Tui (Prosthemadera novaeseelandiae chathamensis)	-	-	-	-	1	4.17	1	2.44	2
Totals	20	100.0	16	100.00	24	100.04	41	100.03	101

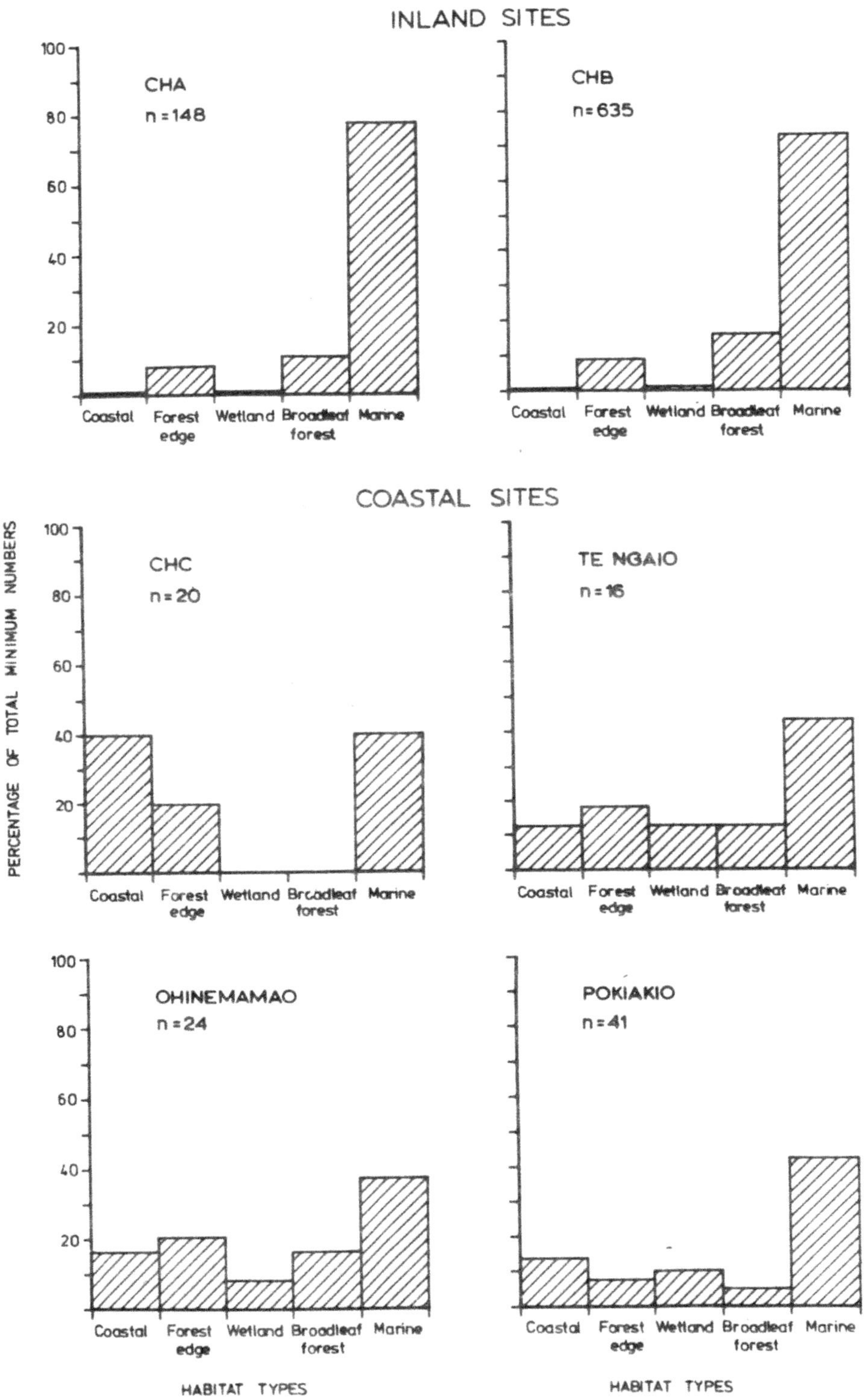

Fig. 9.3 Percentage of Birds Represented from different Habitat Types.

bones, representing deaths during the fledgling stage (Sutton and Marshall, 1977:8, Figure 6; Table 9.2). It is emphasised that these proportions are minimum figures due to the differential destruction of small, fragmentary and immature bones.

The second strategy represented is fowling in the broadleaf forest. The large, slow and conspicuous Chatham Island Pigeon would have been taken out of the low canopy of local broadleaf forest (Cockayne, 1901) with spears (vide Skinner's (1923:78 ff) section, "Implements Used in Fowling"). Broadleaf forest was mainly confined to a narrow and irregular corridor on sheltered aspects of the coastal terrace (Begg, 1977; Dodson and Kirk, 1978). The other species from the broadleaf forest (Kaka and Tui) may have been speared, netted or snared. The frequency with which the three species occur is clear evidence for selective hunting of pigeon.

The forest-edge and bogshrubland bird species represent a third discrete fowling strategy. It was, however, clearly less important than either of the two mentioned above. Snares and traps were probably used as the Polynesian dog has not been identified in a prehistoric context for the Chatham Islands (Sutton, n.d.). No detailed ethnographic description is available for rail hunting (vide Shand, 1911:6), nor have any Moriori artefacts been associated with it. The absence of dog may account, at least in part, for the relative unimportance of rails in the inland and other middens, despite the presence in the islands of 7 rail species (Olson, 1977), including an endemic genus (Diaphorapteryx).

Wetland and coastal fowling were both similarly unimportant at the inland sites. The absence of open wetlands nearby (Begg, 1977) may account for the former. The absence of the dog is not sufficient explanation as a small midden near Lake Huro (Figure 9.1) contained a total minimum number of 233 birds of which almost 63 percent are Anseriformes (Sutton, n.d.). The few coastal birds present at CHA and CHB reflect the highly specialised nature of most Moriori settlements. This point is developed below.

Evidence of distinctive fowling strategies is much less clear in the coastal middens. The apparent concentration on marine birds at all four sites includes little, if any, immature bone, suggesting the recovery of stranded adults, possibly during hunting or gathering along the shore. Diving Petrels were taken at Pokiakio (Table 9.2) but the minimum of 8 individuals represented there would be a very small proportion of those then available in the Waitangi Tuff cliffs (Hay et al., 1970: Fig. 6) above the site. Again, in the case of coastal birds, the seven Pitt Island Shags at CHC (Table 9.2) may have been taken when the chance arose or have been included in the gut contents of the many seals (n = 153) butchered at the site (Smith, 1977). Similarly, the few rails represented in the coastal sites may have been taken as they came on to the shoreline to feed on carrion.

There is, therefore, no clear evidence of specialisation in fowling at the coastal sites. Fowling was incidental to the main economic activity at these sites, whether shellfish gathering at Te Ngaio, Ohinemamao and Pokiakio (McIlwraith, 1976) or seal slaughter as at CHC (Smith, 1977). It appears to have consisted only of the opportunistic capture or recovery of live or stranded birds from a range of species.

Seasonality

Season of occupation can be assessed for both inland and coastal sites. The juxtaposition of high proportions of immature bone from Taiko, Diving Petrel and Broad-billed Prion can be used to establish seasonality at CHA and CHB. Broad-billed Prion fledglings are available in December (Fleming, 1939; Richdale, 1944). Diving Petrel fledglings are available in January (Richdale, 1943a). The breeding season of Taiko has not been established from live birds (Imber, 1976). However, its presence in CHA and CHB in the absence of immature bone of the Whitefaced Storm Petrel, Sooty Shearwater and Mottled Petrel whose fledglings are available in March, April-May and May respectively (Richdale, 1943b, 1963, 1964) suggests Taiko fledglings were available in the interval December to January and that the inland sites were occupied in that short summer season.

The seasonality for the coastal shellfish middens cannot be clearly ascertained. However, we know that they were occupied in the 16th century, and that this was the period of greatest severity of the Little Ice Age (Leach, H. M., 1976). Further, the efficiency with which intertidal shellfish can be gathered is dependent on sea and wind conditions (Anderson, 1973). Current wind data for Chatham Island indicates that any exacerbation of the critical southerly and westerly winds to the degree thought to have been associated with the Little Ice Age (Leach, H. M., 1976; Lamb, 1977), would have limited efficient and sustained shellfish gathering to intervals within the calmer months of October to February, and occasional days in other seasons.

Smith (1977:72-78) has established occupation of the CHC sites during summer, most probably in winter and possibly for other seasons of the year.

Interpretation

The seasonality data suggest strongly that shellfish gathering at areas of maximum intertidal exposure and muttonbirding at Taiko breeding colonies were both summer activities. Body parts representation for the Taiko at CHA and CHB (Sutton and Marshall, 1977:10-12; Figure 6) suggest the use at both stations of a single butchering pattern in which wing tips and lower leg were broken off and discarded, probably on the colony where the birds were taken. The meatier inner portions of wing and the thigh were kept. Heads and necks were discarded in a large proportion of the birds. This general pattern is also evident in the Broad-billed Prion and Diving Petrel remains (*ibid.*). It is consistent with one of the methods of preservation of the Sooty Shearwater described by Richdale (1948).

Meat from both shellfish gathering and fowling was evidently taken back to a centrally located village settlement for storage and then consumption in the colder winter months. This pattern of complementary exploitation of a range of resources at seasonal stations some distance from the village accounts for the very high ratio of small and shallow middens to complex "Village" sites in the Chathams (Simmons, 1964) and for the very specialised nature of the middens sites. For instance, the Pokiakio midden contained the remains of over 21,000 seellfish drawn from the largest and most economic of Chatham mollusca; paua (*Haliotis iris*), limpet (*Cellana strigilis chathamensis*) and

and Cookia sulcata (McIlwraith, 1976: Tables 7 and 8). However, there were no structures, and only a few portable artefacts present.

The central place village site type is represented in the Durham area by the Waihora mound site. This is known to be contemporary with McIlwraith's (1976) coastal middens (Sutton, 1977) and the CHC site. The contemporaneity of CHC and Waihora is based on Nitrogen determination (Smith, 1977) pending availability of C14 dates. There are positive indicators in the faunal assemblage for occupation during all seasons of the year and materials present from areas throughout the Chathams archipelago. These include bone of albatross chicks evidently from offshore breeding colonies (Dawson, 1973) and a total of over 15 different stone materials (Sutton, n.d.).

The site is central to the spatial distribution of resources discussed above. However, the nearby Fur Seal breeding colony, exploited from CHC, does appear to have been the only one on this stretch of coast at the time of occupa-tion of these sites (Smith, 1977). It was a major source of meat, fat, bone and presumably skins for the Waihora site. The seals there were the only large meat source available all year round to the Moriori. The location of the seal colony is thus seen as the principal factor effecting location of the village. A similar pattern is likely to apply throughout sea coast settlements in the Chathams. Chathams and has been reconstructed for the Foveaux Strait Coast of Southern New Zealand (Sutton and Marshall, n.d.).

Conclusion

The prehistoric Moriori were coastal hunter-gatherers occupying a medium-high latitude environment. Prehistoric horticulture was never pos-sible. The limitations of the terrestrial resources (McCartney, 1975) brought about a dependence on the marine resources of the area. The most important of these were seals (the only accessible and perennial mammal present, apart from man) and marine birds.

The seals taken were: Southern Fur Seal (Arctocephalus forsteri), Southern Elephant Seal (Mirounga leonina), Leopard seal (Hydrurga leptonyx) and the New Zealand Sea Lion (Neophoca hookeri). The total population of Fur Seals present before European sealing must remain a matter of conjecture. However, a figure of 20,000 animals for the whole archipelago seems reason-able as the basis of Wilson's (1974) data for the few remaining colonies. The other, larger seals all bred on the Subantarctic Islands (Gaskin, 1972) and visited the Chathams where they would haul out for varying periods.

The marine birds exploited were: 3 species of penguin, at least 3 Diomedea species, 4 Pterodroma species, 2 Prions, a Procellaria species, 3 Shearwaters, a Storm Petrel and the Diving Petrel (Marshall, Scarlett and Sutton, n.d.). This unique concentration of marine birds (Bourne, 1967) is explicable on biogeographical grounds (Heather, 1966). The birds present formed a year-round succession of species coming ashore to breed for 6-8 weeks in the case of the smaller ones, then departing.

The bird bone assemblages discussed above result from the very successful adaptation (Richards, 1972) of Polynesian man to the "utter limits of (their) colonisation" (Bellwood, 1978:416). A similar pattern of coastal adaptation is known to have occurred, with local modifications, on other southern high latitude coasts (vide Jones, 1971; Parkington, 1976; Stuart, 1977; Anderson, n.d.; Sutton and Marshall, n.d.)

BIBLIOGRAPHY

Anderson, A. J., 1973. Archaeology and Behaviour, M.A. thesis, Anthropology, University of Otago.

Anderson, A. J., n.d. Southern Hunters: models for the study of prehistoric communities in Southern New Zealand. Paper read to 49th A.N.Z.A.A.S. Conference, Auckland, January 1979.

Begg, A., 1977. Development of the Otonga Landblock, Southwest Coast, Chatham Island. Working Papers in Chatham Islands Archaeology 17. Anthropology Dept., University of Otago.

Bellwood, P., 1978. Man's Conquest of the Pacific. Collins, London.

Bourne, W. R. P., 1967. Subfossil Petrel Bones from the Chatham Islands. Ibis 109:1-7.

Bowdler, S., 1974. An Account of an Archaeological Reconnaissance of Hunter's Isles, Northwest Tasmania, 1973/1974. Records of Queen Victoria Museum, Launceston, 54:1122.

Cockayne, L., 1901. A Short Account of the Plant Covering of Chatham Island. Trans. N.Z. Inst. 34:243-325.

Dawson, E. W., 1973. Albatross Populations at the Chatham Islands. Notornis 20:210-230.

Dodson, J. R. and Kirk, R. M., 1978. The Influence of Man at a Quarry Site, Nairn River Valley, Chatham Island, New Zealand. J. Roy. Soc. N.Z. 8:377-384.

Dodson, J. R. and Kirk, R. M., n.d. Holocene Tectonism and Palynology of Waihora, Southwest Coast, Chatham Island. Manuscript, Geography Dept., University of Canterbury.

Fleming, C. A., 1939. The Birds of the Chatham Islands. Emu 38:380-409.

Gaskin, D. E., 1972. Whales, Dolphins and Seals. Heinemann, Auckland.

Hamel, G., 1977. Vegetation and Archaeology on Chatham Island. Working Papers in Chatham Islands Archaeology 6. Anthropology Dept., University of Otago.

Hay, R. F., Mutch, A. R. and Watters, W. A., 1970. Geology of the Chatham Islands. Bulletin of New Zealand Geological Survey n.s. 83. Government Printer, Wellington.

Heather, B. D., 1966. A Biology of Birds. Unity, Auckland.

Imber, M. J., 1976. The Taiko. Wildlife - A Review, 7:35-40.

Jones, R., 1971. Rocky Cape and the Problem of the Tasmanians. Ph.D. dissertation, Anthropology, University of Sydney.

Lamb, H. H., 1977. Climate: present, past and future. Volume 2. Metheun, London.

Leach, B. F., this volume. Maximizing Minimum Numbers: avian remains from the Washpool Midden Site.

Leach, H. M., 1976. Horticulture in Prehistoric New Zealand. Ph.D. Dissertation, Anthropology, University of Otago.

Marhsall, Y. M., Scarlett, R. J. and Sutton, D. G., n.d. Bird Species Present on the Southwest Coast of Chatham Island in the Sixteenth Century A.D. Paper submitted for publication.

McCartney, A. P., 1975. Maritime Adaptations in Cold Archipelagoes. Pp. 281-337. In W. Fitzhugh (ed.), Prehistoric Maritime Adaptations of the Circumpolar Zone. Aldine, The Hague.

McIlwraith, M. A., 1976. Archaeology, Middens and People. M.A. thesis, Anthropology, University of Otago.

Molloy, B. P. J., 1977. Palaeobotanist, Botany Division, D.S.I.R., Christchurch. Pers. comm., 1977.

Oliver, W. R. B., 1955. New Zealand Birds. Reed, Wellington.

Olson, S. L., 1977. A Synopsis of the Fossil Rallidae. Pp. 339-374 In Rails of the World, S. D. Ripley. Godine, Boston.

Parkington, J., 1976. Coastal Settlements Between the Mouths of the Berg and Olifant Rivers, Caper Province. South African Archaeological Bulletin 31:123-124 and 127-140.

Richards, R., 1972. A Tentative Population Distribution Map of the Morioris of Chatham Island, circa 1790, J. Polynes. Soc. 81(3):350-374.

Richdale, L. E., 1943a. The Kuaka or Diving Petrel (Pelecanoides urinatrix) Emu 43:24-48 and 97-107.

Richdale, L. E., 1943b. The White-faced Storm Petrel or takihi-kare-moana, Pelegodroma marina maoriana (Mathews).

Trans. Roy. Soc. N.Z. 73:97-115, 217-232 and 335-350.

Richdale, L. E., 1944. The Perara or Broad-billed Prion, Pachyptila vittata Emu 43:191-213.

Richdale, L. E., 1948. Maori and Muttonbird. Otago Daily Times and Otago Witness, Dunedin.

Richdale, L. E., 1963. Biology of the Sooty Shearwater. Proc. Zool. Soc. Lond. 141:1-117.

Richdale, L. E., 1964. Biology of the Birds of Whero Island, with Special reference to the Diving Petrel and the White-face Storm Petrel.

Trans. Zool. Soc. Lond. 31:1-87.

Shand, A., 1911. The Moriori People of the Chatham Islands. Memoir of Polynesian Society 11. Avery, New Plymouth.

Simmons, D. R., 1964. Chatham Island Archaeological Survey. N. Z. Arch. Assn. Newsletter 6:51-69.

Simpson, K., 1972. Birds in Bass Strait. Reed, Wellington.

Skinner, H. D., 1923. The Morioris of the Chatham Islands. B.P. Bishop Museum Bulletin 9(1). Bishop Museum Press, Honolulu.

Smith, I. W. G., 1977. Prehistoric Fur Seal Exploitation on the Southwest Coast of Chatham Island. Working Papers in Chatham Islands Archaeology. 7. Anthropology Dept., University of Otago.

Stuart, D. E., 1977. Seasonal Phases in Ona Subsistence, Territorial Distribution and Organisation: implications for the archaeological record. Pp. 251-281 In For theory Building in Archaeology, (ed.) L. R. Binford, Academic Press, NewYork.

Sutton, D. G., 1976. Radiocarbon Dates from the Waihora Mound Site N.Z. Arch. Assn. Newsletter 19:195-196.

Sutton, D. G., 1977. Radiocarbon Dates from Chatham Island - II. N.Z. Arch. Assn. Newsletter 20:127.

Sutton, D. G., n.d. Polynesian Coastal Hunters in the Chatham Islands. Ph.D. Dissertation, in prep.

Sutton, D. G. and Marshall, Y. M., n.d. Coastal Hunting in the Subantarctic Zone. In Press N.Z.J. Archaeology.

Wallace, R. T., 1977. The Land and Freshwater Mollusca of the Chatham Islands. Working Papers in Chatham Islands Archaeology 15. Anthropology Dept., University of Otago.

Wallace, R. T., n.d. Landsnails and Palaeobotanical Reconstruction. M.A. thesis, University of Otago.

Wilson, G. J., 1974. Distribution, Abundance and Population Characteristics of the New Zealand Fur Seal (Arctocephalus forsteri) M.Sc. thesis, Zoology, University of Canterbury.

PREHISTORIC EXPLOITATION OF CRAYFISH IN NEW ZEALAND

B. Foss Leach and Atholl J. Anderson

Introduction

If present day exploitation is anything to go by, it would be surprising if crayfishing had not been an important aspect of prehistoric fishing in New Zealand. Yet it is an aspect which has been all but ignored by archaeologists. Basic issues such as the recognition and identification of crayfish species from midden refuse, and the reconstruction of body weight from common anatomical survivals of these animals have not previously been investigated, and it is one purpose of this paper to make a start in rectifying this situation. The second general interest lies in interpreting the role of crayfish in prehistoric economies. Specifically, what effects does preferential selection for large size have on local crayfish stocks over long periods of time? An obvious place to begin the consideration of these questions is with the European observations of Maori crayfishing during the 18th and 19th centuries.

Early Historic Records of Crayfish in New Zealand

The relevant ethnographic literature is not particularly revealing of the role of crayfish in the economy at European contact. One reason is because the most detailed studies were of inland communities, and therefore references to crayfish were most commonly to the freshwater species. These were regarded as a delicacy, and were comparatively easily caught. Hamilton comments under 'crustaceans' in his work Fishing and sea foods of the Ancient Maori (1908:13) as follows:

> "The principle item in the bill of fare taken in this group was the koura—the large red crayfish (Palinurus or Jasus)—and the freshwater koura of the lakes and streams (Paranephrops). A number of traps and pots of various kinds were baited and set for these dainties, and they were also, in time of plenty, slightly smoked and dried, and strung on a string of flax".

Some quite revealing observations concerning marine crayfishing are to be found in the journals of the visitors who came to New Zealand in the 18th century with Captain Cook. Crayfish were apparently often brought to the boats as gifts and items of trade, and were clearly very welcome. Anderson remarked during February, 1777 that:

> "The indians likewise sometimes brought us crawfish equal to our largest lobsters, which are very fine" (Anderson, 1967:808).

Furneaux noted in his journal for November 4, 1772:

> "We again got in shore near Cape Pallisser and was visited by a number of the Natives in their Canoes with a great quantity of Cray fish, which we bought of them for nails and Otaheite Cloath...." (Furneaux, 1969:741).

Banks had earlier remarked on similar transactions (1963 Vol. 1:422) in November 1 1769 when they were somewhere near White Island in the Bay of Plenty. He noted that people were trading both mussels and crayfish with them. The method for catching them was vividly described as follows:

> "Dr. Solander who was today in a cove different from that I was in [in Mercury May] saw the natives catch many lobsters in a most simple manner: they walked among the rocks at low water about middle deep in water and still felt about with their feet till they felt one, on which they dived down and constantly brought them up. I do not know whether I have before mentioned these lobsters but we had them in tolerable plenty in almost every place we have been in and they are certainly the largest and best I have ever eat" (Banks, 1963 Vol. 1:434).

This was the entry for November 14, 1769 and shows well just how unneees-sary elaborate fishing equipment was for catching the animals at this period. Banks was very impressed with this item of food, and made some general comments on them in March 1770:

> "But above all luxuries we met with the lobsters or sea crawfish must not be forgot, which are possibly the same that in Lord Ansons Voyage are mentioned to be found at the Island of Juan Fernandes; they are large tho not quite so large as those at Juan Fernandes and differ from ours in England in having many more prickles on their backes, and being red when taken out of the water. Of them we bought great quantities of the natives every where to the Northward, who catch them by diving near the shore, feeling first with their feet till they find out where they lie" (Banks, 1962 Vol. 2:7).

The reference to Anson was expanded by Beaglehole in a footnote to this passage as follows:

> "Walter in his account of Anson's voyage, pp. 125-6, writes, 'we found here one delicacy in greater perfection, both as to size, flavour, and quantity, than is, perhaps, to be met with in any part of the world: this was a sea-cra-fish; they generally weighed eight or nine pounds a-piece, were of a most excellent taste, and lay in such abundance near the water's edge, that the boat-hooks often struck into them, in putting the boats to and from the shore'" (ibid.).

The species found about Juan Fernandes Island is *Jasus frontalis* (Waugh 1973: 256) and specimens are indeed sometimes larger than those of *Jasus edwardsii*, although they are surpassed by examples of the New Zealand 'packhorse crayfish' Jasus verreauxi. These latter crayfish can reach enormous size, and may have been the species referred to by Parkinson who spoke of obtaining crayfish up to 5 kg. His remarks related to Tolaga Bay after his visit there 23 to 30 October 1773!

> "This bay abounds in a variety of fifh, particularly fhell and cray-fifh; fome of the latter, which we caught, weighed eleven pounds; thefe are found in great plenty, and feemed to be the principle food of the inhabitants, at thif feafon of the year, though they have a kind of fern, the roots of which, roafted, make a good fubftitute for bread, efpecially when their koomarra is young and unfit for ufe" (Parkinson, 1773:99. Beaglehole notes that "a pencil sketch of a small specimen of the crayfish is on the back of Plate 12 in Parkinson II" in Banks, 1963 Vol. 1:422).

Best recorded several methods of catching crayfish (1929:52-59). On the subject of diving for them he noted that the Maori, unlike the pakeha, always dived feet first, sometimes to a depth of 3 fathoms (5.5 m). Apparently some Maoris claimed that women were actually more expert in the art of crayfish diving (this is also mentioned by Banks, 1963 Vol. 2:33; and the same comment has been made of Tasmanian Aboriginal women, q.v. Jones, 1977). Hoop-shaped drop nets were recorded in which either paua (Haliotis iris) or kina (Evechinus chloroticus) were used as bait in a small suspended netted bag. This is still a common non-commercial method today. One gains the impression from Best's records though, that the baited wickerwork trap was the commonest early Maori method of catching them. Again the bait was most often paua, but he also recorded the use of starfish which is unusual. Colenso noted that crayfish itself was an esteemed bait for catching groper (Taylor, 1959:6). Freshwater crayfish were taken both by hand and with a dredge net, and also by an ingenious method which involved sinking bundles of bracken fern and allowing them to remain on the bed of still water for a time. The crayfish would hide in the bundles and if these were gently raised the crustaceans could be shaken out in large numbers.

Many apocryphal stories are told of crayfishing, and these have become prodigal since the great crayfish boom in the late 1960s in the Chatham Islands. However, twentieth century fishermen have no monopoly on extravagant fishy folk stories. The great size of crayfish in the Chatham Islands is an established fact, and Baucke remarked that the now defunct Moriori of the Chatham Islands excelled in the art of crayfish diving and that:

> "a diver who appeared with three—one in each hand and one held by the antennae between the teeth, was acclaimed a tchim' tchakat' me' kye [possibly tenei tangata mahi kai (a food provider of renown)]" (Baucke, 1928:360).

What Survives Archaeologically?

Crayfish present a daunting exterior to would-be predators, but in fact this structure is not very strong, and rots quickly in reasonably moist aerobic conditions. The basic material of the exoskeleton is calcium carbonate which is precipitated from the dermal layers of the animal. It is deposited in layers interspersed with a mucoprotein secretion during a process similar to the build up of nacreous shell in molluscs. In the case of crayfish, however, the final structure is not only more flimsy but the process is partly reversible. Just before the moult, some of the calcium carbonate is extracted from the exoskeleton and stored as gastroliths in the case of the freshwater crayfish; special storage bodies, which provide the material for the rapid hardening of the new shell. Crayfish also possess small gizzard raspers which are quite durable in appearance and might be expected to survive archaeologically, although this has yet to be demonstrated. The marine species, however, unlike their freshwater counterparts and marine crabs, do not possess chelae or pincers. The ends of these are very dense, and are the principal component of crab remains which are found in midden sites (for example see Anderson, 1973:Appendix E).

All these animals have one hard part in common, and this is the pair of mandibles. Strictly speaking, these are the first of the numerous paired appendages, although clearly in this case they are highly specialised in form. They occur at the ventral anterior end of the animal immediately in front of the mouth. They appear to be more durable than other parts of the body. Modern rubbish dumps of crayfish bodies show that most of the exoskeleton decomposes rather rapidly into a granular composition, and that the only easily identifiable parts are the very ends of the second antennae, and the densest cap portion of the mandibles. The tips of the antennae are sometimes seen archaeologically, but the mandibles are more common.

In appearance, crayfish mandibles closely resemble rather weathered and unidentifiable fragments of shellfish, and this fact is doubtless the reason they have been so seldom documented by archaeologists until recently. The midden excavations in Palliser Bay suggested that archaeological survival may be higher in deposits where there are substantial quantities of ash, possibly because an alkaline environment may be more favourable to their preservation than an acidic one. Most New Zealand soils are acidic, and while shell midden does raise the pH slightly (Park, 1969:26), this may not be sufficient to make much difference to the preservation of crayfish parts. Ash, on the other hand, is very rich in potash, and might be expected to raise the pH considerably in its near vicinity. In addition, the fine texture and hygroscopic nature of ash may help to create locally dry anaerobic conditions which would be most suitable to the survival of crayfish mandibles.

Despite the good chance of preservation of crayfish mandibles the first published reference to these remains in archaeological sites was in 1955, after nearly a century of New Zealand archaeology. This was from a site now known as the Washpool Stone Wall Fort (N168/27), and the discovery was by Adkin in 1952. It is worth quoting a passage from Adkin's article in full:

> "Another item was a number, 12 specimens in all, of limy bone-like objects, the largest one inch, others down to 5/8 in. in length, one side slightly convex, the other shaped somewhat like a human ear. These have not been exactly identified, but Dr. R. A. Falla, Director Dominion Museum, tentatively regards them as 'like the grinding

mouth-parts of a large kind of crayfish or some related marine crustacean.' He further remarks: '...it seems rather strange that there should be so many of them [the limy ear-shaped objects] in the absence of any other parts [of the presumed crustacean]. Perhaps they have survived owing to the quantity of lime in their composition' " (Adkin, 1955:469).

An excavation by the present authors in 1970 in the same spot as Adkin had worked revealed another dozen of these fragments, and these were later positively identified as portions of the mandibles of Jasus edwardsii (see B. F. Leach, 1976:241).

This lack of recovery of remains of crayfish or lobsters from archaeological sites is not confined to New Zealand. After an intensive search of archaeological literature only one well documented case could be found. This was from a midden-cave site at Bonteberg, 30 km south of Capetown in South Africa, and the excavations revealed 338 mandibles (both left and right) of Jasus lalandei, which is closely similar to the common New Zealand species, Jasus edwardsii (Kensler, 1967b, notes that they are synonymous). Grindley, who studied these remains (1967), performed Van Wisselingh's test on the mandibles which showed that the polysaccharide chitin was still present, but in reduced quantity compared with modern specimens. He explained the differential survival of mandibles as being due to the higher chitinous content in the rest of the exoskeleton, and higher calcification in the mandibles.

In New Zealand, large numbers of mandibles have now been recovered in excavations in Palliser Bay, the Chatham Islands, and the Fiordland area (however, the latter have not been studied ; see Coutts, 1975:28). The minimum numbers for the Palliser and Chatham finds are given in Table 10.1. It is interesting that comparatively few were found in the extensive excavations in the Chathams, bearing in mind the vast quantities of crayfish recovered commercially during the past decade (q.v. Waugh, 1973:278). It is not known whether crayfish are prone to major fluctuations in natural abundance in specific areas over centuries, but the relative lack of mandibles in prehistoric sites of the early 16th century in the Chathams, coupled with their great abundance in later periods, might argue for this. Current hypotheses regarding environmental changes during the past millennium (H. Leach, 1976: Chapter 8) suggests that these were of insufficient scale to have wrought the effects which appear to be indicated from the Chathams excavations. Besides, the Palliser Bay crayfish finds are well documented from the 12th to 17th centuries (Anderson, 1973: Appendix E; B. F. Leach, 1976:432).

The Identification of Archaeological Remains

There are four species of crayfish in New Zealand; two freshwater and two marine. The freshwater species are variously grouped, but the commonest view is that one predominates in the North Island, and the other further south (q.v. Hopkins, 1970). The North Island species, Paranephrops planifrons (Figure 10.1A) has a somewhat fuller carapace and more slender chelae than the other species. It extends into Marlborough in its distribution. The South Island species, Paranephrops zealandicus has a much narrower carapace, but is most noted for its very hairy chelae (see Figure 10.1B). Unfortunately

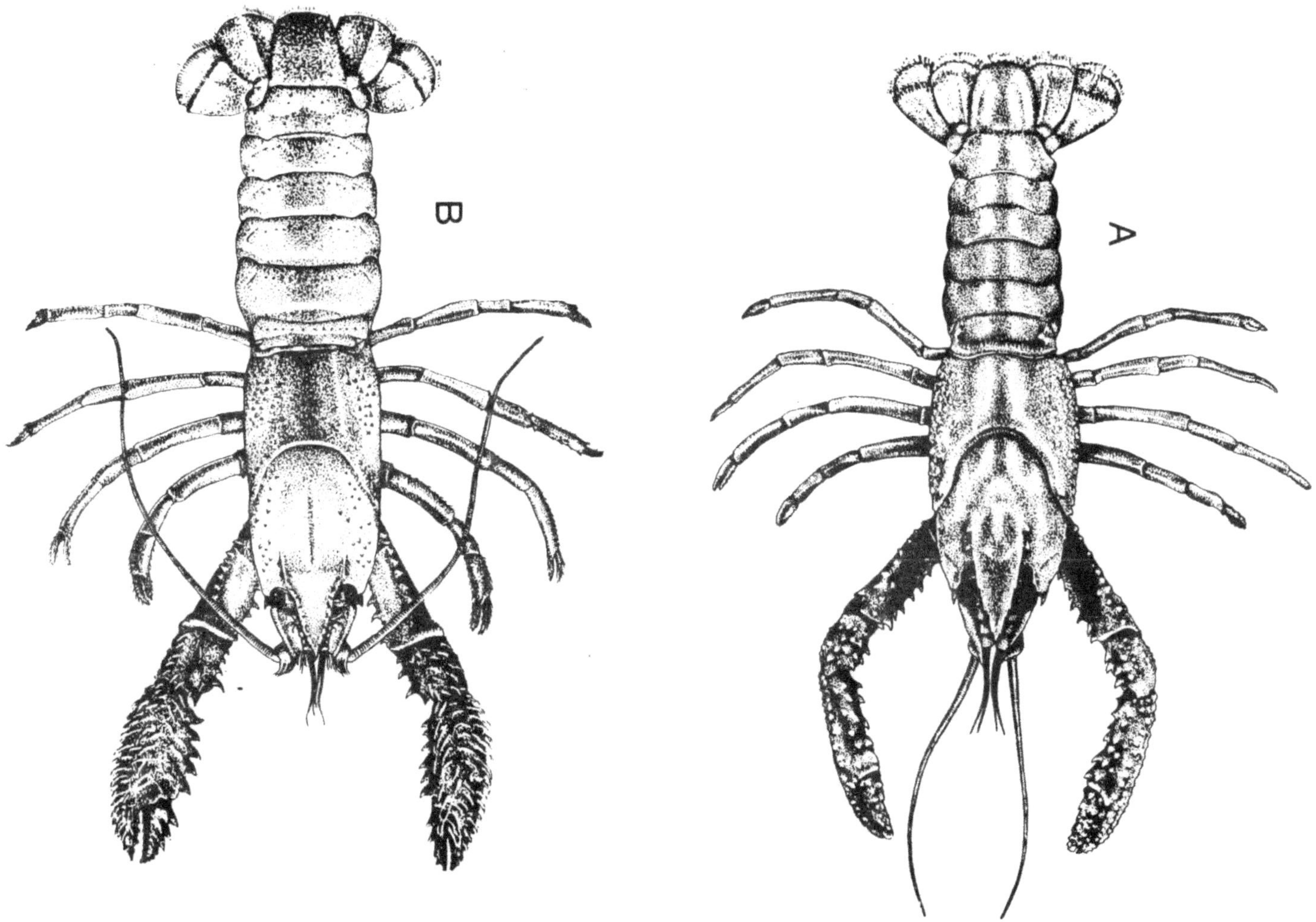

Fig. 10.1 New Zealand Freshwater Crayfish. A = <u>Paranephrops planifrons</u>; B = <u>Paranephrops zealandicus</u>

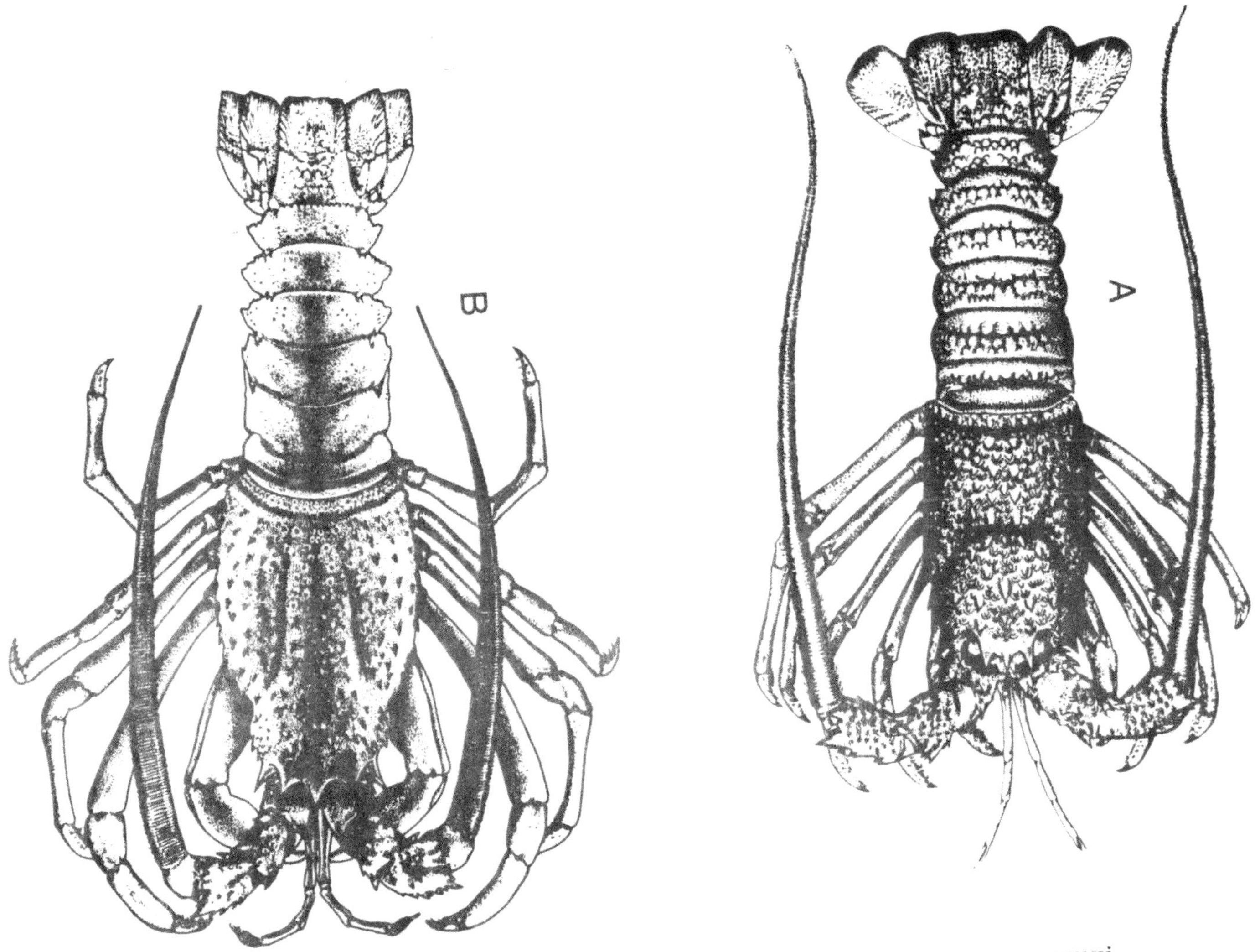

Fig. 10.2 New Zealand Marine Crayfish. A = Jasus edwardsii; B = Jasus verreauxi.

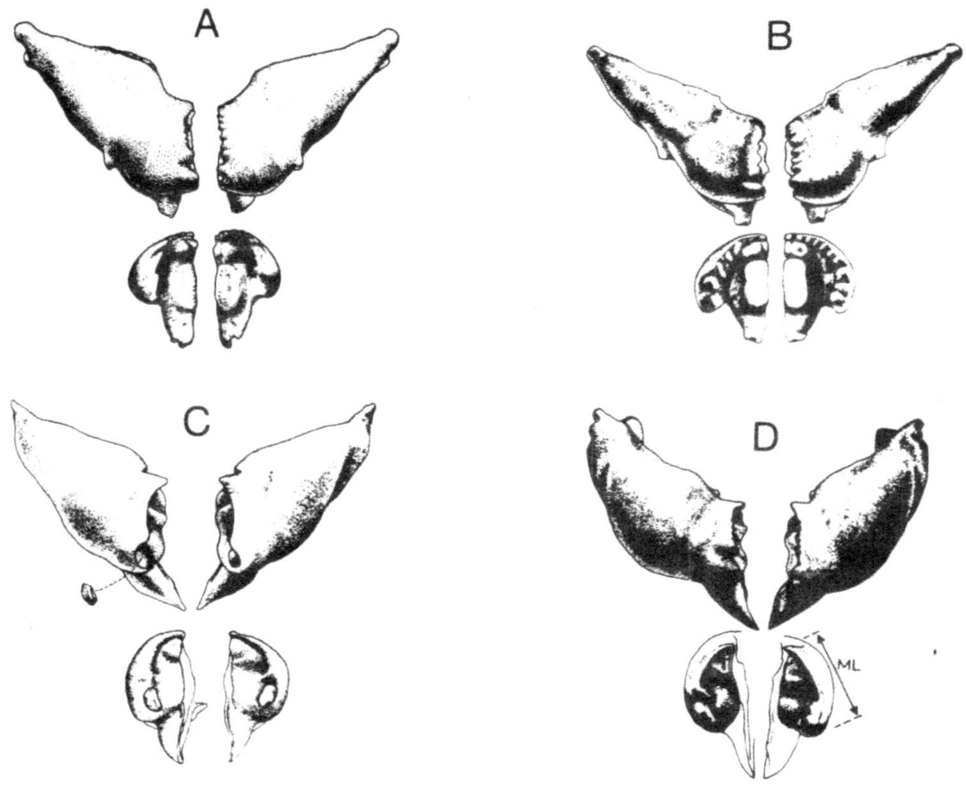

Fig. 10.3 New Zealand Crayfish Mandibles
A = Paranephrops planifrons
B = Paranephrops zealandicus
C = Jasus edwardsii
D = Jasus verreauxi
ML = Mandible Length as measured

the mandibles of these two species are very similar, and probably impossible to distinguish reliably. For large specimens, the mandibles are only about 5 mm across the occlusal surface, and although they are quite distinctive in shape, are liable to be overlooked unless midden refuse is very carefully examined. The two mandible forms are shown in Figure 10.3, and these show the same basic shape as for other crayfish, except that the outer edge is more denticulated, and there are two sensory pads along the inner edge of each mandible. Specimens of freshwater crayfish have yet to be recovered from New Zealand archaeological sites, and when they are portions of the chelae may well be found to be present also.

Marine crayfish possess mandibles of about 6 to 35 mm in size, and are readily identified during midden analysis. Again, there are two species which may be distinguished - Jasus edwardsii (Figure 10.2A), the commonest species, and Jasus verreauxi (Figure 10.2B), the packhorse crayfish, which may reach a considerably larger size. It can be seen from the two figures that Jasus verreauxi has a less sculptured and spiky tail, and also the carapace is more rounded in appearance. In addition, the foremost rostral spine is much larger than for Jasus edwardsii. The general morphology of the mandibles is similar for the two species with the exception of one important feature. At the most anterior part of the outer edge of the mandible the surface has a raised rounded cusp. In Jasus edwardsii (Figure 10.3C) this cusp is quite smooth, and many archaeological specimens show considerable attrition in this zone. However, a small sensory pad is found in the same region in specimens of Jasus verreauxi (Figure 10.3D), and this pad becomes detached during decomposition to leave a small regular shaped pit in its place. This difference would be easily detected on archaeological specimens, although a careful search of the many fragments from both Palliser Bay and the Chatham Islands failed to reveal a single example. It is quite certain, therefore, that all the remains examined from these sites are those of Jasus edwardsii.

Prehistoric Crayfish Remains from New Zealand

Grindley's study of crayfish mandibles from the Bonteberg excavation in South Africa (1967) included an assessment of the corresponding crayfish body size and weight using information from modern examples of Jasus lalandei. These data were then compared with those from modern commercial catches in the region. The commercial size-frequency distribution is severely truncated at the smaller end due to a minimum size regulation. Despite this, the prehistoric size-frequency distributions resemble those of the modern catches more closely than those of the modern natural population. Grindley argues that this indicates a pattern of relatively heavy prehistoric exploitation (1967: 99), and that the people occupying the site consumed a total of about 2000 kg of crayfish (589 kg of which would be tails). This figure should be revised to about half the stated value, since the calculation was not based on minimum numbers, but on the total count of both left and right mandibles. Moreover, the view that the size frequency distribution indicates heavy prehistoric exploitation is one which could be challenged. The modern distributional figures include very large numbers of tiny specimens, and this is typical of natural populations. However, it is likely that prehistoric people would select larger specimens, as appears to be the case with other marine resources found in archaeological sites (see Anderson, n.d.). In such cases initial and moderate exploitation would not show the dominance of tiny specimens, apparent in modern samples but in fact be similar to the distribution figures found in the site. Indeed, it might well be argued that under conditions of sustained heavy predation, the archaeological remains should eventually show a bias to ever smaller examples. The Bonteberg figures appear to show a pattern of moderate exploitation without the imposition of a 'minimum size'.

The Palliser Bay excavations produced remains of nearly 15 times as many crayfish as those from Bonteberg, and it was thought that the sample recovered would be useful to compare with some modern New Zealand figures. In addition, the excavations were of sites which covered much of the span of New Zealand prehistory, and any long term effects of human predation upon crayfish populations might be revealed from such a numerical study. To facilitate this, it was first necessary to undertake a survey of the modern crayfish in the region.

During several periods of archaeological fieldwork in Palliser Bay and the Chatham Islands opportunities arose to measure live or freshly killed Jasus edwardsii and to extract their mandibles for further study. For a number of reasons it was not always possible to weigh or sex all of the specimens, but sufficient records were obtained to enable reasonable regression formulae to be calculated. Most of the larger specimens were from the Chatham Islands and some very small examples were measured alive and returned to the water. The main aim of the study was to establish the numerical relationship between mandible length (ML) and tail length (tL), since regression formulae linking most other measurements (see Table 10.2), were already well documented in the literature. Some basic statistics from the present study are given in Table 10.3. Correlation between mandible length and tail length was high, but noticeably lower in the case of females. This is probably a reflection of the smaller sample size in this case. In Figure 10.4 a scatter plot is presented which shows the cluster of values about the regression lines predicting tail length from mandible size. The various equations obtained are given below:

$$tL_M = 9.9425 \times ML_{LM} + 4.75 \ (\pm 14.6,\ r = 0.93) \ldots\ldots 1$$

$$tL_M = 9.7057 \times ML_{RM} + 5.71 \ (\pm 15.0,\ r = 0.93) \ldots\ldots 2$$

$$tL_F = 4.8458 \times ML_{LF} + 78.38 \ (\pm 5.4,\ r = 0.77) \ldots\ldots 3$$

$$tL_F = 4.9589 \times ML_{RF} + 74.94 \ (\pm 5.2,\ r = 0.79) \ldots\ldots 4$$

$$tL_B = 9.2205 \times ML_{LB} + 17.86 \ (\pm 12.2,\ r = 0.95) \ldots\ldots 5$$

$$tL_B = 9.1206 \times ML_{RB} + 16.18 \ (\pm 12.3,\ r = 0.95) \ldots\ldots 6$$

Other Jasus edwardsii measurements have been subjected to regression analysis, and amongst these studies that of Sorensen (1970) is especially useful (see also Kensler, 1967a). Using a sample of nearly 500 crayfish, he established the relationships between total length (TL) and tail length, and also between total length and rostrum to carapace length (RCL). Unfortunately, he does not give the equation for total length from tail length. Although it is not strictly legitimate, a fair approximation may be made by transposing his formulae for the reverse relationship (Sorensen, 1970:13), viz:

$$TL_M = (tL_M - 4.98)/0.571 \; (r = 0.989).....7$$

$$TL_F = (tL_F + 0.81)/0.604 \; (r = 0.991).....8$$

$$TL_B = (tL_B - 2.31)/0.586 \; (r = 0.989).....9$$

The additional formulae for predicting rostrum to carapace length from total length are as follows:

$$RCL_M = 0.389 \times TL_M - 3.33 \; (r = 0.989).....10$$

$$RCL_F = 0.352 \times TL_F + 2.90 \; (r = 0.989).....11$$

$$RCL_B = 0.372 \times TL_B - 0.61 \; (r = 0.984).....12$$

It has been suggested several times (for example Sorensen, 1970:17) that a preferable carapace measurement is not from the rostrum, which is variable in length, and sometimes broken, but from the platform between the second antennae. Fielder (1964:78) has documented a simple conversion factor between the two measurements regardless of sex for Australian Jasus lalandei, and this is unlikely to be much different for New Zealand Jasus edwardsii. Sorensen (1970:18) gives a blanket conversion factor of 4.8 mm for Jasus edwardsii, which is equivalent to Fielder's conversion for a moderate sized specimen of TCL = 118 mm.

$$TCL = RCL \times 1.042 ....13$$

As is well known, predicting live body weight from body dimensions is usually difficult, probably more so for animals like crayfish where dehydration can be a most significant factor. Nevertheless, the study by Fielder (1964) of nearly 600 Australian Jasus lalandei resulted in fairly reliable cubic functions for this species. In the absence of similar data from New Zealand, the formulae used here are based on Fiedler's results, and are therefore subject to some scaling factor. This is not regarded as a serious problem since it is the comparative results which are important especially in the shape of size frequency distributions. The weight estimation formulae are as follows:

$$W_M = \text{antilog } (2.7389 \times \log_e (TCL_M/10) - 0.2609)....14$$

$$W_F = \text{antilog } (2.7449 \times \log_e (TCL_F/10) - 0.2760)....15$$

$$W_B = \text{antilog } (2.7419 \times \log_e (TCL_B/10) - 0.2682)....16$$

The foregoing equations allow for the estimation of all the main variables which are used in modern crayfish studies directly from the remains of mandibles in archaeological sites. Some of the calculations are obviously rather indirect; however, residuals were calculated for the 126 Jasus edwardsii collected during fieldwork and even for the weight determination these showed that the formulae are reasonably precise.

As an example of the application of these equations consider the right mandible from a male crayfish with ML = 17.44 mm. Tha tail length can be predicted using equation 2 giving a value of 175.0 mm. The total length may now be calculated using equation 7, and this gives a value of 297.7 mm. Application of equation 10 yields a value of 112.5 mm for the rostrum to carapace length, and this can be converted to the alternative total carapace length

TABLE 10.1

NEW ZEALAND CRAYFISH REMAINS

	Sites	Minimum Numbers
A	Palliser Bay	
	Black Rocks BR3 (N168/77)	481
	Black Rocks BR4 (N168/77)	947
	Black Rocks BR2 (N168/77)	101
	Washpool Midden (N168/22)	69
	Stone Wall Fort (N168/27)	3
	TOTAL:	2519
B	Chatham Islands	
	CHB (C240/680)	31
	Waihora (C240/283)	11
	Te Ngaio (C240/277)	3
	TOTAL:	45

TABLE 10.2

MODERN CRAYFISH MEASUREMENTS

TCL = Total Carapace Length: From the posterior margin of the carapace to the union of the second antennae.

TL = Total Length: From the tip of the rostral spine to the extended tip of the telson.

RCL = Rostrum to Carapace Length: From the tip of the rostral spine to the posterior margin of the carapace.

tL = Tail Length: From the extended tip of the telson to the calcified bar on the ventral side of the tail.

Unless otherwise stated all measurements given are in mm or gm, and the different subscripts used are as follows:

L = Left	M = Male	B = Both sexes pooled
R = Right	F = Female	

TABLE 10.3

SOME BASIC STATISTICS FROM MODERN CRAYFISH STUDY

		Males	Females	Both
ML_L	Mean	17.05	13.74	15.43
	SD	3.83	1.36	3.85
ML_R	Mean	17.37	14.12	15.79
	SD	3.91	1.36	3.89
tL	Mean	174.29	144.97	160.15
	SD	40.80	8.56	37.57
Sample Sizes		72	31	126

Table 10.4 Basic Statistics of Prehistoric Crayfish (Means and Standard Deviations)

Site	Age	N	ML	tL	TL	TCL	RCL	W
BR2/1	AD1750	44	19.1(3.9)	190.5(36.0)	321.1(61.4)	123.8(23.8)	118.8(22.8)	824.4(427.4)
BR4/1	AD1350	72	16.8(3.8)	169.7(35.1)	285.7(59.8)	110.1(23.2)	105.7(22.3)	608.6(368.2)
BR4/2	AD1300	93	17.7(4.6)	177.6(42.1)	299.2(71.8)	115.3(27.8)	110.7(26.7)	713.3(495.3)
BR4/3	AD1270	89	18.1(5.2)	181.1(47.5)	305.0(81.1)	117.6(31.4)	112.9(30.2)	774.1(590.0)
BR4/4	AD1250	44	17.5(4.8)	175.8(43.5)	296.0(74.3)	114.1(28.8)	109.5(27.6)	700.7(530.4)
BR3/1	AD1170	73	20.7(4.9)	205.4(44.4)	346.5(75.7)	133.7(29.3)	128.3(28.2)	1040.0(559.8)
BR3/2	AD1150	91	22.8(4.3)	224.3(39.1)	378.9(66.7)	146.2(25.8)	140.3(24.8)	1282.6(556.6)
CHB	AD1550	52	23.2(2.7)	229.5(25.1)	387.8(42.9)	149.7(16.6)	143.6(16.0)	1312.4(404.6)
Waihora	AD1550	16	23.0(3.8)	228.7(34.7)	386.3(59.3)	149.1(23.0)	143.1(22.0)	1329.2(548.0)
Te Ngaio	AD1660	3	19.7(-)	197.4(-)	332.8(-)	128.4(-)	123.2(-)	837.6(-)
Bonteberg		358	12.8(2.9)	134.6(26.2)	225.8(44.7)	86.9(17.3)	83.4(16.6)	314.3(162.4)
Gisborne	AD1976	3287	-	-	-	-	-	563.6(517.9)
Palliser	AD1972	252	15.6(3.9)	159.9(36.4)	270.1(64.0)	105.1(26.2)	100.9(25.2)	520.1(133.6)

Table 10.5 Estimates of Crayfish Meat Weight

Site	Mean Weight	Minimum Number	Minimum Weight (kg)	Percentage Site Excavated	Total Meat Weight (kg)
BR2/1	824.4	101	83.26	25	333.0
BR4/1	608.6	162	98.59		
BR4/2	713.3	543	387.32	15	4423.4
BR4/3	774.1	185	137.66		
BR4/4	700.7	57	39.94		
BR3/1	1040.0	259	269.36	14	3957.8
BR3/2	1282.6	222	284.74		
CHB	1312.4	31	40.68	80	50.8
Waihora	1329.2	11	14.62	66	22.2
Bonteberg	314.3	192	60.35	8	754.4

NB: The figures given are total body weight. Of this, tail weight is gauged to be about 30% (q.v. Grindley, 1967:99).

Table 10.6 Prehistoric Crayfish: Skewness and Kurtosis of Weight-Frequency Distributions

	Skewness			Kurtosis		
Site	W1	g1	Sig	W2	g2	Sig
BR2/1	2.7	0.9	S	0.6	3.2	NS
BR4/1	4.6	1.3	HS	6.8	6.4	HS
BR4/2	4.6	1.1	HS	2.2	4.0	PS
BR4/3	4.6	1.2	HS	2.6	4.2	S
BR4/4	3.7	1.3	HS	3.8	5.2	HS
BR3/1	2.2	0.6	PS	-1.1	2.4	NS
BR3/2	2.1	0.5	PS	-0.1	2.9	NS
CHB	2.9	0.9	S	1.9	4.0	NS
Waihora	1.8	0.9	NS	2.1	4.3	PS
Bonteberg	6.3	0.8	NS	-0.6	2.8	S
Modern	60.4	2.6	HS	60.4	17.2	HS

NB: Significance levels are as follows:

Not significant $> .05$
Possibly significant .05-.01
Significant .01-.001
Highly significant $< .001$

of 117.2 mm using formula 13. Finally, the weight of the live animal may be estimated using equation 14 giving a body weight of 652.4 gm. This is remarkably close to the actual weight of this specimen which was 650 gm. The sex of mandibles from archaeological sites cannot be determined at present, and therefore equations 6, 9, 12, 13 and 16 would have been used in this case. A weight of 601.1 gm would have been thus obtained, which is still quite close to the correct figures.

Computer programmes were written for the analysis of the archaeological finds from Palliser Bay and the Chatham Islands. In addition, the above formulae were applied to the mandible data from the Bonteberg site for comparative purposes. The results were similar but not identical to those obtained by Grindley (1967).

The basic results are set out in Table 10.4. In the case of the Bonteberg material all the data is placed in one assemblage. Grindley grouped the mandibles into two levels, but since the basis of the division was not properly explained, this procedure could not be followed here. However, of the 358 mandibles, only 12 belonged to the earlier level, and their inclusion in the present analysis should make little difference. In Table 10.5 the total meat weight figures are given for each site. The values of the proportion of each site excavated follow estimates given by Anderson (1973:118) and Grindley (1967:99).

The weight frequency histograms of the crayfish are given in Figure 10.5. The data were first standardized to give zero mean and unit variance, and then scaled so that the largest value does not exceed unity. In this way, standard histograms of 20 cells are formed, and this facilitates direct comparison of one distributional shape with another. In addition, statistics for skewness and kurtosis (q.v. Snedecor and Cochran, 1967:86 ff) were calculated (for reasons discussed below) and these appear in Table 10.6. The values for g1 are all positive, indicating positive skewness throughout. Values for g2 greater than 3.0 indicate a leptokurtic shape, and less than this figure platykurtic. The W statistics given in Table 10.6 are the normalised deviates and are used to assess the significance of the g values using the distribution of t. Overall significance of departure from normality in each case is also given in Table 10.6.

Discussion

The interpretation of differences in mean size and weight-frequency distributions between prehistoric and modern populations is not without difficulties, particularly in the case of marine invertebrates where so little background research has been carried out. Live populations of these animals frequently exhibit changeable, and sometimes multi-modal size-frequency distributions. Archaeological samples of species in which the surviving parts of young individuals are fragile can blur or bias these characteristics. Another source of error is introduced by the impossibility at present of sexing crayfish from surviving mandible fragments. Perhaps the single most important problem, however, is one which at least partly may be solvable by archaeological means. Crayfish, more than most marine invertebrates, are strongly seasonal in their movements (see Kensler, 1969). At various times of the year, usually in late summer, some large adult crayfish enter shallow water to moult and mate. Reliable seasonal evidence is clearly a matter of crucial importance in interpreting differences in mean size encountered in different archaeological sites. In the case of those at Black Rocks, all are firmly believed to have been occupied over the summer months (see Anderson, 1973:117 ff). In this case therefore, the seasonal effect can be regarded as a constant.

Even with these problems firmly in mind, there are gross differences between the archaeological samples and the modern one, and this calls for comment. Clearly, the crayfish in the modern samples considered are significantly smaller than those obtained during the prehistoric period, and the variation about the mean size is, today, much narrower. There are several possible explanations. One is that the archaeological samples are biased by differential survival of the larger mandibles. Given the excellent condition of most of the prehistoric mandibles recovered, and the fact that much smaller and less resistant objects were abundant in the sites in question (such as numerous examples of Haliotis iris of about 1 cm in length), this does not appear to be a reasonable hypothesis. Much more likely is that prolonged pressure of exploitation has driven down the mean size of the accessible population. In support of this interpretation for the Black Rocks samples the following must be noted. Firstly, the general pattern is in the right direction - a progressively lower mean size through time (see Figure 10.6). Secondly, there is a minor recovery in mean size following the well attested settlement hiatus along the eastern Palliser Bay coast for several hundred years prior to the occupation of BR 2 - the Pond Midden Site (see B.F. Leach, 1976:296, 306, 326; H. M. Leach, 1976:188; Anderson, 1973: 163). Thirdly, other species in the sites are also prone to similar fluctuations in mean size, and those for Haliotis iris are shown in Figure 10.7. With this species it can be seen that the occupation hiatus resulted in dramatic recovery in the mean size of the shellfish, and that more recent exploitation seems to have had only a small impression on local stocks. On the face of it, comparison of Figures 10.6 and 10.7 seems to show that given the same lack of predation for a period of about 300 years prior to A.D. 1700, the crayfish population has been unable to rejuvenate itself to anywhere near the same extent as Haliotis iris. In addition, modern crayfish size in Palliser Bay is well below anything seen in this region during the previous 800 years.

It will be noticed in Figure 10.7 that paua appear to recover to about twice their mean size in the earliest archaeological sample. However the sample from BR3/2 may well have come from an already exploited population, since there is no reason to suggest that this site represents the first exploitation of the Black Rocks area. Similarly crayfish may well have been larger before the exploitation phase represented by BR3/2.

Finally, it is clear from Table 6 that there have been dramatic changes in the shape of the weight-frequency distributions over the period concerned. This is illustrated in Figure 10.8 where it will be seen that the crayfish from the earliest sites (BR3/1 and BR3/2), and presumably representing the most natural populations, exhibit a platykurtic shape, and a minimum of positive skewness. As time goes by, sustained exploitation by the people responsible for the BR4 site resulted in a rise in positive skewness, and pronounced leptokurtosis. The modern catches are at the extreme end of this trend, suggesting

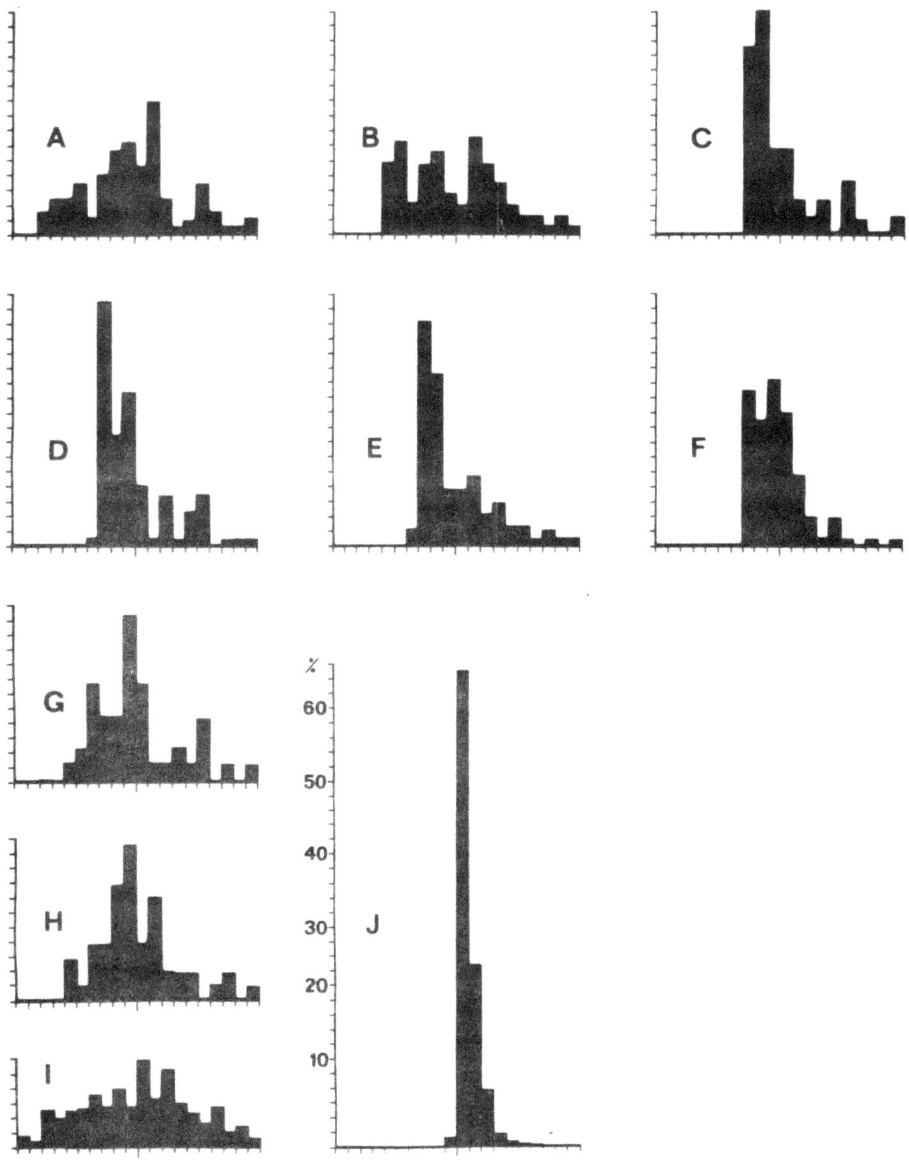

Fig. 10.5 Crayfish Weight-Frequency Distributions
A = BR3/2
B = BR3/1
C = BR4/4
D = BR4/3
E = BR4/2
F = BR4/1
G = BR2/1
H = CHB
I = Bonteberg
J = Modern Gisborne Sample, N = 3287

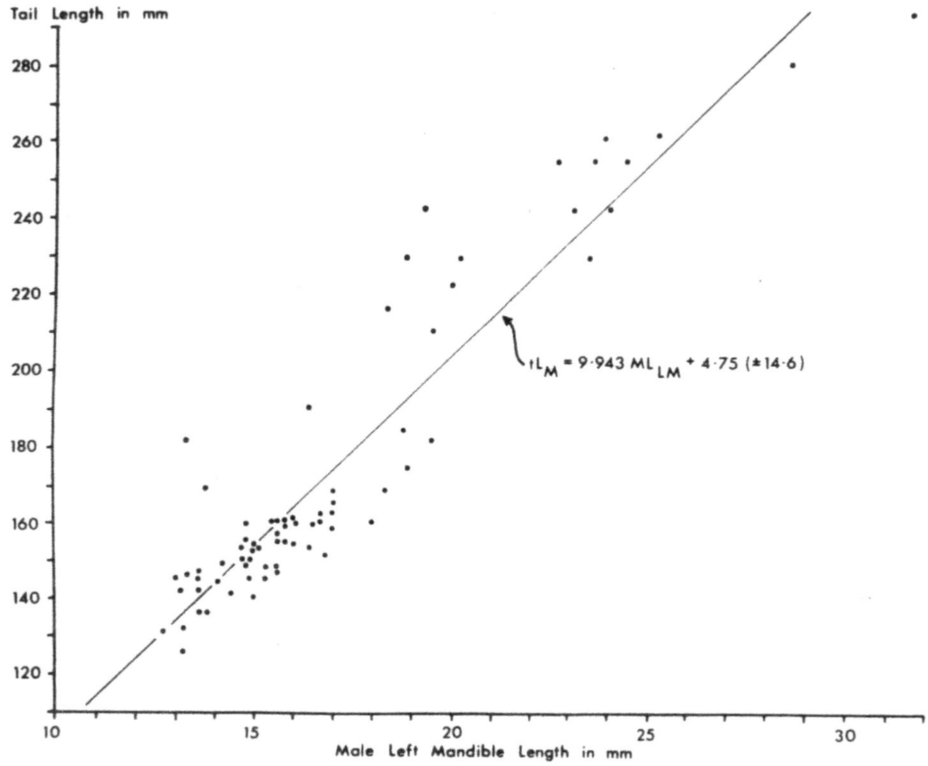

Fig. 10.4 Correlation between Mandible Length and Tail Length.

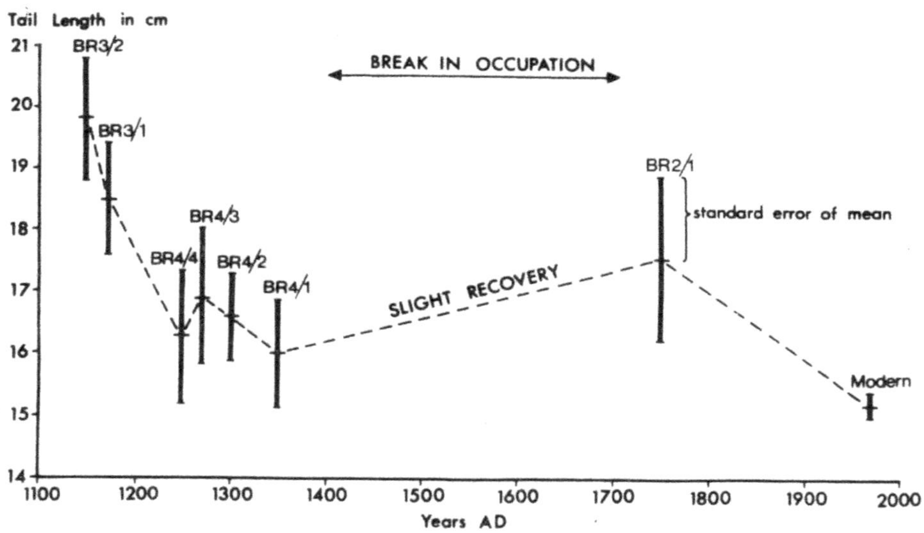

Fig. 10.6 Time Trends in Crayfish Size.

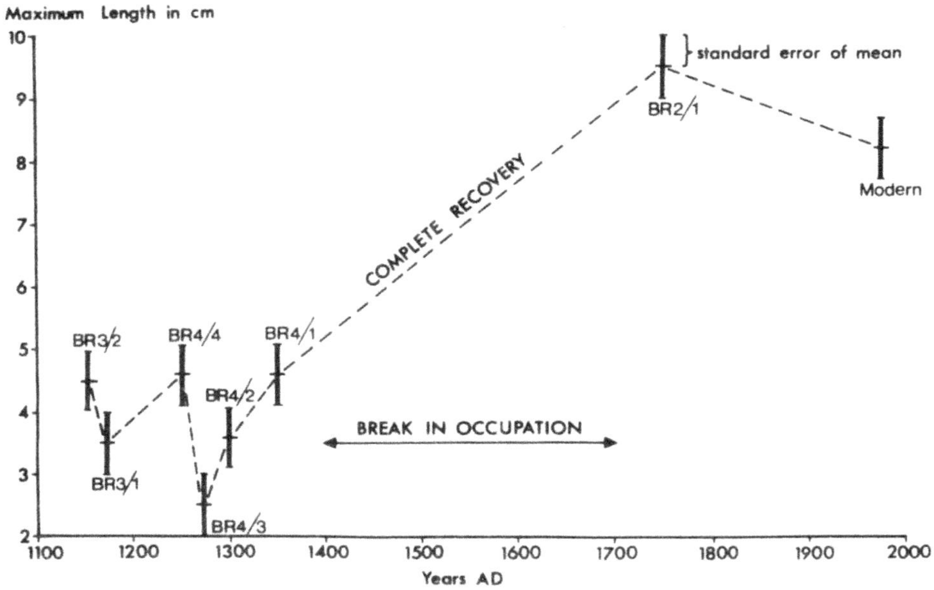

Fig. 10.7 Time Trends in the Size of Haliotis iris.

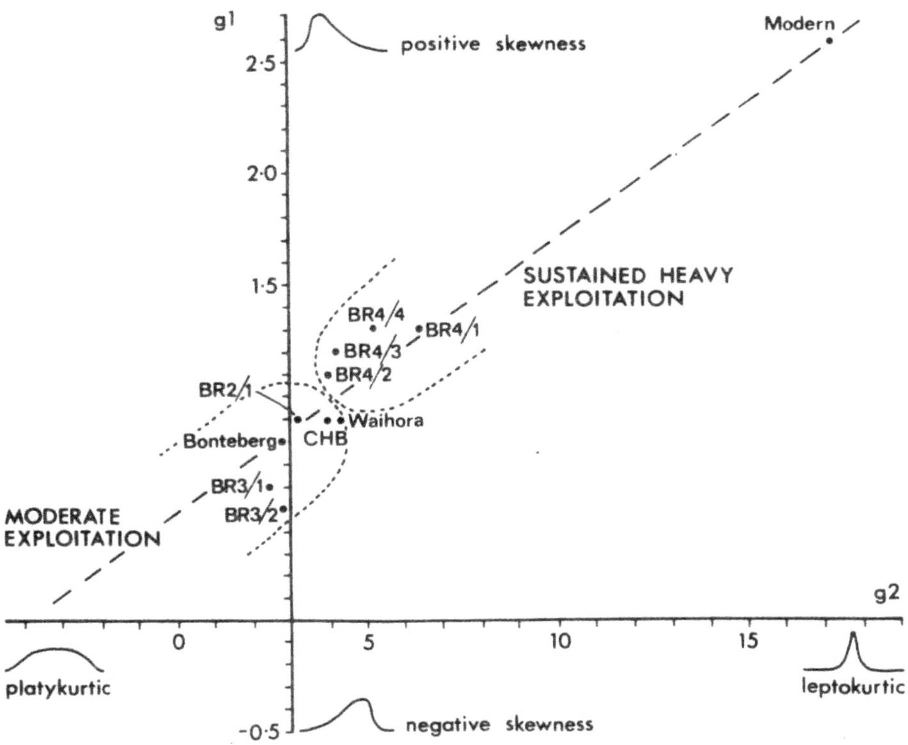

Fig. 10.8 Departure from normality for Crayfish.

considerable departure from a pristine condition. It is interesting that the hypothesised break in occupation resulted in a return to the earlier population shapes (represented by BR2/1 in Figure 10.8), even though the mean size hardly recovered in the same interval (evident in Figure 10.6). It will also be noted that both samples from the Chatham Islands archaeological sites fit within the shapes apparently more characteristic of relatively unexploited populations. Although the Chatham sites are fairly late by New Zealand standards, there is little direct evidence of habitation much earlier than this in the region examined in the Chathams so far.

Conclusions

This study has been instructive from a number of viewpoints. It is clear that crayfish as a source of food were just as important in antiquity in New Zealand as they were when the first Europeans came to these shores. Equally, it has been demonstrated that reasonable estimates may be made of the characteristics of prehistoric catches of these animals - their overall size, and size-frequency shapes. Comparing the figures obtained for several archaeological sites and from modern catches suggests that the effects of previous exploitation may be identified from these characteristics. It would appear that while a long period of relaxed pressure does result in a return to the shape structure of earlier catches, mean size is much slower to respond to lack of predation. On the whole, it seems unlikely that the modern New Zealand fisherman will ever gain access to the kind of crayfish stocks which were available to prehistoric man a millennium ago.

Acknowledgements

The authors wish to thank John McKoy and John Annala of the Fisheries Research Division, for valuable comments and for supplying modern catch information for the Gisborne district, and also Murray Webb (Anthropology Department) for his illustrations.

BIBLIOGRAPHY

Adkin, G. L., 1955. Archaeological Evidence of Former Native Occupation of Eastern Palliser Bay. J. Polynes. Soc. 64:450-480.

Anderson, A. J., 1973. Archaeology and Behaviour. M.A. Thesis, Anthropology, University of Otago.

Anderson, A. J., n.d. Prehistoric Collecting on the Rocky Shore , m.s. Anthropology Department.

Anderson, W., 1967. A Journal of a Voyage made in his Majestys Sloop Resolution. pp. 723-986 In: Beaglehole, J. C. (Ed.) The Journals of Captain James Cook, Volume 3 (2). Hakluyt Society.

Banks, J., 1963. The Endeavour Journal of Joseph Banks. Beaglehole, J. C. (Ed.). Angus and Robertson.

Baucke, W., 1928. The Life and Customs of the Moriori. pp. 357-384 In: Skinner, H. D. and W. Baucke, The Morioris. Memoirs of the Bernice P. Bishop Museum Volume 9 (5).

Best, E., 1929. Fishing Methods and Devices of the Maori. Dom. Mus. Bull. 12.

Coutts, P. J. F., 1975. Marine Fishing in Archaeological Perspective: Techniques for Determining Fishing Strategies. pp. 265-306 In: Casteel, R. W. and G. I. Quimby (Eds.). Maritime Adaptations of the Pacific. Mouton, The Hague.

Fielder, D. R., 1964. The Spiny Lobster, Jasus lalandei (H, Milne-Edwards), in South Australia. Parts I & II. Australian Journal of Marine and Freshwater Research 15: 77-92, 133-144.

Furneaux, T., 1969. Furneaux's Narative. pp. 729-745 In: Beaglehole, J. C. (Ed.). The Journals of Captain James Cook, Volume 2.

Grindley, J. R., 1967. The Cape Lobster Jasus lalandii from the Bonteberg Excavation. South African Archaeological Bulletin 22:94-102.

Hamilton, A., 1908. Fishing and Sea Foods of the Ancient Maori. Dom. Mus. Bull. 2.

Hopkins, C. L., 1970. Systematics of the New Zealand Freshwater Crayfish Paranephrops (Crustacea: Decapoda: Parastacidae). New Zealand Journal of Marine and Freshwater Research 4(3):278-291.

Jones, R., 1977. Why Did the Tasmanians Stop Eating Fish? In: Gould, R. (Ed.). Explorations in Ethnoarchaeology. University of New Mexico Press.

Kensler, C. B., 1967a. Size at first Maturity in Females of the Spiny Lobster Jasus verreauxi (Crustacea: Decapoda: Palinurus) New Zealand Journal of Marine and Freshwater Research 1:327-340.

Kensler, C. B., 1967b. The distribution of spiny lobsters in New Zealand waters (Crustacea: Decapoda: Palinuridae). New Zealand Journal of Marine and Freshwater Research 1:412-420.

Kensler, C. B., 1969. Commercial Landings of the spiny lobster Jasus edwardsii (Hutton) at the Chatham Islands (Crustacea: Decapoda: Palinuridae). New Zealand Journal of Marine and Freshwater Research 3:506-517.

Leach, B. F., 1976. Prehistoric Communities in Palliser Bay, New Zealand. Ph.D. thesis, Anthropology Department, University of Otago.

Leach, H. M., 1976. Horticulture in Prehistoric New Zealand. Ph.D. thesis, Anthropology Department, University of Otago.

Park, G. S., 1969. Chemical Analysis in Archaeology: its contribution to the location of material for faecal analysis. Unpublished Manuscript, Anthropology Department, University of Otago.

Parkinson, S., 1773. Journal of a Voyage to the South Seas in his Majesty's ship, the Endeavour. Australiana Facsimile Editions Number A34, 1972.

Snedecor, G. W. and Cochran, W. G., 1967. Statistical Methods. Iowa State University Press.

Sorensen, J. H. 1970. New Zealand Rock Lobster Jasus edwardsii Carapace and Tail Measurements. New Zealand Marine Department, Fisheries Technical Report Number 53.

Taylor, N. M., 1959. Early Travellers in New Zealand. Clarendon Press, Oxford.

Waugh, D., 1973. Fish and Fisheries. pp. 251-284 In: Williams, G. R. (Ed.). The Natural History of New Zealand. A. H. and A. W. Reed.

THE ROLE OF THE DOG IN THE ECONOMY
OF THE NEW ZEALAND MAORI

Jan Allo Bay-Petersen

Introduction

Most interpretations of New Zealand prehistory have, in the past, been based on changes in artefact styles, or, more recently, in settlement patterns. Only in the last few years has use been made of the knowledge that food debris in middens reflects just as much culture-determined human choice as the artefacts made by man, and that New Zealand sites are capable of providing a much more sophisticated economic analysis than a simple division into 'hunter-gatherer' or 'agricultural'. The study of midden material provides a description, not only of foods eaten and their relative importance, but also indirectly of the society which organised its economic activity to obtain these foods, and of the cultural basis of that society.

The Polynesian dog of New Zealand held a unique place in the Maori economy; not only was it the only domesticated animal of the Maori, it was also the only land mammal of any size to be found in New Zealand during the prehistoric period. In the absence of any ungulate species in prehistoric New Zealand it was of particular importance as a food animal; moreover, since it lived in close association with man throughout the prehistoric period, one could expect changes in its distribution, physical form and diet to yield valuable evidence on wider changes in the Maori economy as a whole.

The Maori dog is now extinct in New Zealand as a separate identifiable breed, but is well represented in both Archaic and Classic Maori sites. This discussion is based on a study of over 2000 skeletal remains, representing a minimum of 338 individual dogs, from 90 archaeological sites ranging from the far north to the extreme south of New Zealand. There are also a number of ethnographic descriptions, of which those written in the period immediately following European contact are particularly valuable. The later the record, the more likely it is to be describing dogs with an admixture of introduced European breeds. Similarly the role of the dog in Maori society is likely to have been influenced by the social and economic impact of European culture. Maori culture underwent an abrupt and profound change following European contact. It is hard to distinguish, in later accounts, between descriptions of traditional Maori culture, and those of a culture, which, while still recognisably different from a European way of life, had been modified as a result of culture contact in ways which the observer himself often did not recognise.

Forster described the dogs of the Pacific Islands as

> "... short, and their size varies from that of a lap-dog to the largest spaniel. Their head is broad, the snout pointed, the eyes very small,

the ears upright and the hair rather long, lank, hard and of different colours, but most commonly white or brown". (G. Forster, 1777: 378).

The elder Forster wrote a very similar description.

"... the dogs of the South Sea Isles are of a singular race, they mostly resemble the common cur, but have a prodigious large head, remarkably little eyes, prick ears, long hair, and a short bushy tail". (J. R. Forster, 1778, :189).

Crozet (1891), writing of the Polynesian dog of New Zealand, wrote:

"The dogs are a sort of domesticated fox, quite black or white, very low on the legs, straight ears, thick tail, long back, full jaws, but more pointed than that of the fox, and uttering the same cry: they do not bark like our dogs".

Anderson, who sailed with Cook on his third voyage, also described the New Zealand native dog as "a sort of fox dog" (Anderson 1784:809): while Cook, at Tolaga Bay near Gisborne, dismissed the Maori dog briefly as "very small and ugly".

All these descriptions of the New Zealand dog were written in the eighteenth century, at the time of the first European contact, and before the introduction of European dogs. Not until the nineteenth century do we read descriptions of a yellowish or reddish brown native dog in New Zealand. Dieffenbach, for instance, writes that the Maori dog "rather resembles the jackal; its colour is reddish brown, its ears long and straight". (Dieffenbach 1843). Colenso (1878) also described the Maori dog as yellowish or yellowish brown. However by the time he was writing it was almost certainly extinct. It is of interest as well to note that the majority of dogskin cloaks in New Zealand museums are black and/or white; these cloaks have no European innovations, such as sheep's wool, and are probably of pre-European or early post-European manufacture.

In summary, the ethnographic evidence would seem to indicate that the Polynesian dog of New Zealand was of small stature, brownish black, white or piebald in colour, with short legs and a relatively long body, its jaws broad at the molars but tapering sharply to a pointed snout, and with small eyes and large erect ears.

Skeletal Characteristics

Of more interest to the archaeologist are the skeletal characteristics by which Maori dog remains can be distinguished from those of introduced European breeds, or European crosses. Details of these diagnostic features are given in Appendix A: those of the cranium are a prominent sagittal crest, a narrow skull with a very long muzzle in relation to the length of the brain case, well-spaced teeth, occasional supernumerary alveoli, and the nasal bone typically ending level with the posterior borders of the maxillae. In the body skeleton there is the long narrow glenoid fossa of the scapula, and short legs in relation to body length.

None of these characteristics is enough in isolation to distinguish a Maori from a European dog, since each feature (except possibly the position of the posterior end of the nasal bone and the supernumerary alveoli) is also found in various European breeds. They are, however, most unlikely to occur together in any but a Polynesian dog, the accuracy of the identification increasing with the number of observable features.

Although the physical appearance of the Maori dog seems to have been similar to that of the dog of tropical Polynesia, there are some interesting differences in skull and jaw formation, in particular the presence of a marked sagittal crest in the Maori dog skull. The presence of a prominent sagittal crest, although characteristic of the Maori dog, is not unique to it, being found in any breed of dog with massive jaws and powerful jaw musculature. However it is not present in the Polynesian dogs of the Pacific Islands, which mainly ate a soft vegetable diet.

The skull of the Tahitian dog is described by Luomala (1967), as "slender and delicate in appearance". Wood-Jones, (1931), discussing skulls of the Hawaiian native dog, states that, consequent to a diet of soft vegetable matter, the skulls became shorter, rounder, and devoid of sagittal crests. The muzzle of the Hawaiian dog is described as short and rounded, and the palate as short and broad, markedly different to the long, powerful jaw of the Maori dog.

No dog skulls of the tropical Polynesian type have been found in New Zealand, although presumably they exist since the dog was introduced into New Zealand by Polynesians from this area. It is difficult to estimate the length of time it took for the different environment and diet of New Zealand to produce changes in skull formation, but presumably at least the first generation of Polynesian dogs in New Zealand would have had the rounded skull and broad palate typical of tropical Polynesia. The presence of such dog skulls in a New Zealand site would be a strong indication that the site represented one of the first New Zealand settlements.

Although the morphological features of the Maori dog material studied showed considerable uniformity, there seems to have been a marked size difference among adult individuals. This is clearly shown in Figure 11.1, which gives the range of cranial, long bone and mandible lengths in the prehistoric dog population.

It was of interest to establish whether this heterogeneity was found within a single site, or whether sites contained a relatively uniform population with wide inter-site variation, and if the latter case, whether size variation had any regional or chronological significance.

Sites were placed in crude chronological order, according to their earliest date (where this was known), and according to their cultural associations (Archaic or Classic). Measurements from each site, and the mean measurement (shown by a solid bar) were plotted on a graph. Since intact crania and long-bones were too few to give a meaningful average length for any one site, the condylo-symphisis length of the mandible was used. The results are shown in Figure 11.2.

It is evident that the range of variation within each site is wide (over 3.5 cm., or 30% of the mean length, at Kaikais Beach for example), while the difference in average lengths in different sites is less pronounced and shows no

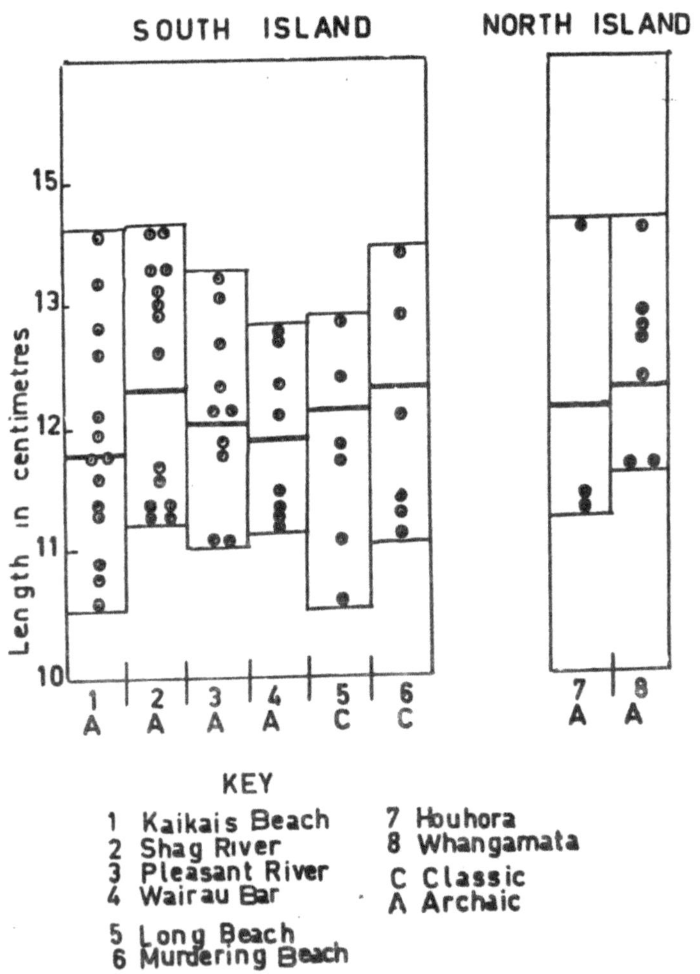

Fig. 11.2 Condylo-symphsis length of mandible

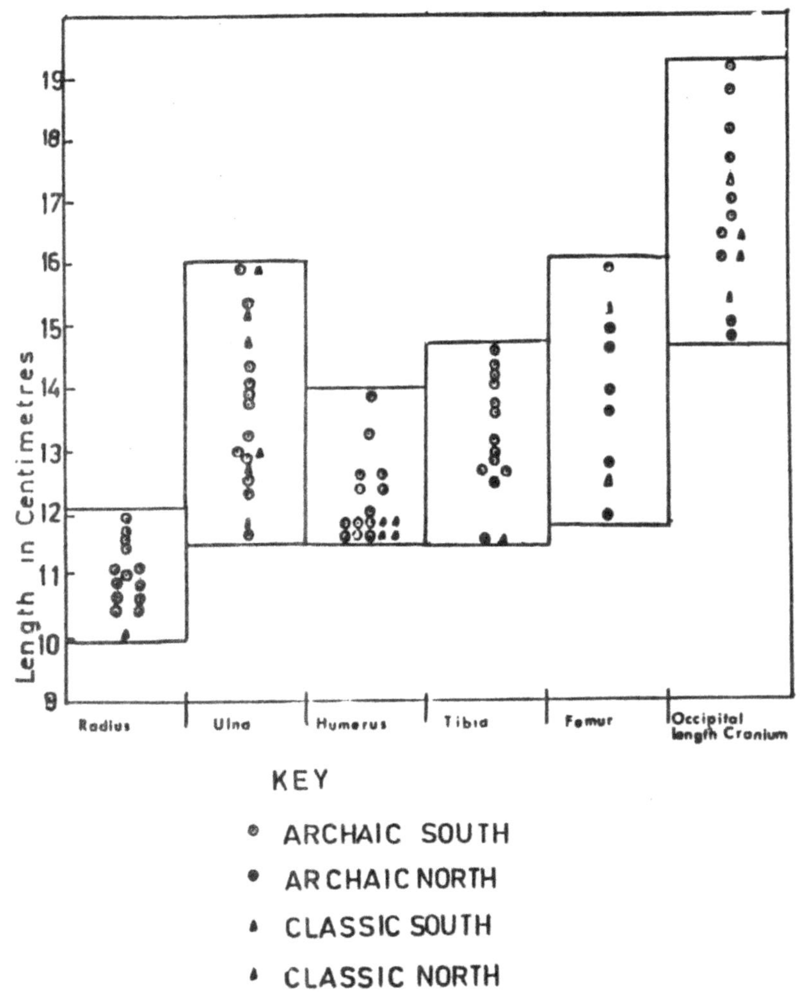

Fig. 11.1 Maximum length of Adult Skeletal Remains

obvious time/space distribution. The variation in size within a site may be explained partly by sexual dimorphism (male dogs are generally larger than females of the same breed), and partly by differences in age. Although all mandibles recorded on the graph possessed adult dentition, this only means that they were over seven months old, and some additional growth probably occurred after this stage.

There is no evidence in the skeletal measurements of the two separate breeds of Maori dog sometimes described in (late) ethnographic literature (e.g. Best 1913, Skinner 1914). Although a few exceptionally large dogs are present in some sites (e.g. Kaikais Beach), it is unlikely that the small number of dogs represented could constitute a separate breeding population; they are probably only the outer limits of the normal size variation found in any population.

The lack of any differentiation in body size and skeletal characteristics between dogs of the Archaic and Classic Maori period, implying that all Maori dogs are descended from a common stock, has interesting archaeological implications. New Zealand prehistory, at present, essentially consists of two polarised periods of Maori culture, which differ consistently in artefact and ornament styles. The connection between the two has not yet been defined. but presumably if the culture of the later period arose from a second migration to New Zealand of Polynesians bearing proto-Classic Maori culture, they would also bring with them dogs which should show some variations from those descended from the dogs originally introduced to New Zealand. The fact that there is no indication of such differences would seem to support the concept of an internal development within New Zealand of Classic Maori culture, suggested also by the number of artefact types common to both Archaic and Classic Maori culture (barbed one-piece fish hooks, bird spears, tattooing and round-sectioned working chisels, barracouta points, bone awls etc.), and the fact that Classic Maori culture seems to have no parallels elsewhere in Polynesia.

Although there seems to be no significant genetic change over space and time, there were interesting differences in phenotype between dogs of the North and South Island, notably with regard to dentition. Periodontal disease, an infection of the soft tissues of the mouth and subsequently of the alveolar bone, usually caused by attrition or irritation, was found in six jaws from Archaic sites and in four from Classic sites in the South Island, but was absent from all North Island material, including the large sample of ninety-four mandibles from Whangamata and Houhora. Another abnormality of the dentition, tooth loss before death with subsequent ossification of the alveolus (in most cases the result of trauma) was found in eight dog jaws from the South Island, while only one example was found in the North Island (see Allo 1971). Although the incidence of such pathological abnormalities was low, their almost complete absence in dogs from the North Island may indicate some difference in diet, in particular a less resistant diet with a higher proportion of vegetable foods.

This is also indicated by the difference in tooth wear between dogs of the North and South Islands. The degree of wear of the mandibular dentition was recorded for the Maori dog skeletal material according to a five point scale (see Appendix B). The results are shown in Table 11.1, and it is evident that the pattern of extreme tooth wear is found only among dogs of the South Island.

Table 11.1 Tooth wear in the Maori Dog

Site	Rate of Tooth Wear									
	.5	1	1.5	2	2.5	3	3.5	4	4.5	5
<u>Archaic South</u>										
Heaphy River	x									
Kaikais Beach	xxx	xxx	xx	xxx	xxx			x	x	x
Marfells Beach	xxx			xxx						
	xx	xxx	x	xx	x	x				
Pleasant River		xxx								
	x	xx	x	xx		xx		x	xxx	x
Tairua			xx	x						
Wairau Bar		xxx		xxx						
	x	xxx		xxx						
		xxx	xxx			xx		x		
		xxx		x						
<u>Classic South</u>										
Long Beach										xxx
	x	xxx	x	x	x		x			x
Murdering Beach		xx		x	x	xx	x		x	xx
<u>Archaic North</u>										
Houhora	xx	xx		x	x	xxx	x	xx		
Whangamata	xx	x	x	x	x	xx				
<u>Classic North</u>										
Houhora		x		xx	x		x	x		

If the average age at death of the dogs from the two areas was the same, these results would indicate conclusively that dogs of the South Island ate a more resistant diet, but since tooth wear is a product of age as well as diet, the higher frequency of heavily worn teeth in South Island sites may indicate an older population. On the other hand, it seems unlikely that there were no aged dogs in the North Island, and it seems probable that the different tooth wear patterns are at least partly related to diet.

These indications of dietary differences between dogs of the north and South Islands are of considerable archaeological interest in view of the unresolved controversy concerning the presence of agriculture during the Archaic period. There is no direct evidence of the prehistoric Maori cultigens; since these are all root crops such evidence is unlikely ever to be found, and the presence of prehistoric agriculture must be based on indirect evidence. Duff (1956), who first defined the Archaic period in New Zealand prehistory, believed that Classic Maori culture was an outside intrusion, and that not until this event was agriculture introduced into New Zealand. He bases this claim largely upon traditional evidence. This view was brought into some question by the discovery of an alleged agricultural soil in the Bay of Islands, dated at 800 A.D., and a (possible) agricultural storage pit in an early Archaic midden at Sarahs Gully in the Coromandel peninsula (Groube 1967 :21, Golson 1959). The division of New Zealand prehistory into agricultural (Neolithic) and non-agricultural (Mesolithic) may therefore be less chronological than geographical; Polynesian root crops cannot be grown in most of the South Island.

The Maori dog presumably shared the diet of its master, and the fact that there are probably diet-related differences in dentition between the dogs of the North and South Island, but no such differences between Archaic and Classic Maori material in the North Island, may imply that some dependance on vegetable foods was not confined to the Classic period, but was a feature also of the Maori economy in the North Island during the Archaic period.

The Dog as a Food Animal

Where bones on an archaeological site represent food remains they generally bear clear indications of the fact, both in postmortem damage and in their distribution. They are often burnt, broken, or carry the marks of butchering implements, and are scattered round the site, especially in cooking and eating areas. There is extensive evidence that throughout New Zealand prehistory the dog was used as a food animal.

Out of a minimum total of 335 dogs from New Zealand sites, only two were found articulated and nearly complete. The bones of the rest were widely scattered over the site, indicating that they were dismembered for eating. The removal of the posterior part of the cranium, presumably for the extraction of the brain, was a common feature on all sites, being found in 169 out of 204 crania (83%). The skull of the dog is very strong and unlikely to split cleanly in half across the parietals by accident, especially since the breakage did not follow a cranial suture. On some skulls there was even a clear indication of a heavy blow, probably made with a stone, on the supraoccipital or right parietal. These marks of percussion were always on the right side of the cranium, indicating the skull was held steady by the snout in the left hand in order to give a right handed blow. The relatively few cuts found on the bones (apart from those of recent origin) would indicate that the dog was cooked whole rather than dismembered immediately after slaughtering. A few bones do bear shallow cuts, however, almost certainly made by stone knives. These cuts, nearly always on the long bones, may have been made in butchering the animal, or as a preliminary stage in manufacturing a bone artefact.

Carbonised bones were relatively common in some sites, but completely absent in others. Since the Maori in both the Archaic and Classic periods used the haangi, or earth oven, where the food is cooked by hot stones and never comes into direct contact with fire, the burning of bones is probably not a result of cooking, but is accidental.

The archaeological evidence that most Maori dogs were used for food is well supported by ethnographic evidence. There are several descriptions of the dog as a food animal. Crozet, for instance, who gave one of the first descriptions of the Maori dog, wrote in 1772: "It appears that the savages only raise them [i.e. dogs] for food". (Crozet, 1891) . The elder Forster, in 1778, commented that "[Maori dogs] are kept by the natives chiefly for the sake of their flesh, of which they are very fond, preferring it to pork".

Although we know that the dog was killed for food, it is difficult to evaluate its economic importance in terms of the amount of dog meat eaten in relation to other foods. Owing to the lack of time control over the skeletal material, and the relatively small sample that excavated sites represent, it is impossible to estimate, from archaeological evidence, the dog population in New Zealand at any one time, but the distribution of skeletal material would seem to indicate that dogs were relatively more common during the Archaic period. The reduction in the number of dogs in Classic sites is difficult to establish quantitatively, since few sites were totally excavated, but it seems significant that there has been no pre-European Classic site excavated in the North Island with more than two or three dogs; and there have been numerous carefully excavated Classic Maori sites where all skeletal material has been kept and which lack dog bones completely; e.g. Station Bay (Motutapu), Hamlin's Hill (Auckland Isthmus), Bald Hill (South Auckland). It must be remembered, however, that there are also Archaic sites with little dog bone, mostly in the North Island; the site of Tairua, on the Coromandel Peninsula, for instance, has only two dogs. On the other hand we have North Island Archaic sites with plentiful dog material, such as Whangamata and Houhora.

It seems that the keeping of dogs was a common, although not universal feature in Archaic times, but rare during the North Island Classic period. There was apparently a general decline in numbers of dogs with the adoption of Classic Maori culture in the North Island. Yet it would be unwise to advance from this a simple explanation that the dog was relatively unimportant in Classic Maori culture. The available ethnographic evidence would seem to indicate that Classic Maori culture placed a high value upon dogs. The elder Forster (1778) wrote "The New Zealanders continually living on fish are glad when they can get a dog or bird to eat, which with them always is reckoned a dainty". Similarly Anderson, on Cook's third voyage, wrote, at Queen Charlotte Sound: "They also breed considerable numbers of the dogs mentioned before for food, but these cannot be considered as a principal article of diet". (Cook 1784:809)

Cruise, who lived for nearly a year in New Zealand in the early nineteenth century, describes the "... dogs, which themselves are sometimes eaten, and are considered a great delicacy". (Cruise, 1823:185).

The most comprehensive description of the dog as a food animal in the Classic Maori period (though perhaps unreliable because of the late date of the observations), is given by Best (1902:47)

> "The domestic dog (kuri) was not numerous enough to form an important item in the native bill-of-fare, but its flesh was highly esteemed. This dish, however, only appeared on important occasions, as at a feast, or when prepared for a distinguished visitor.
>
> The hind quarters of the kuri are said to have been the best eating".

This account indicates that the dog was highly valued and limited in quantity, and that dog meat seems to have had an almost ritual importance in Classic Maori culture. As with the other main non-marine source of protein, forest birds, it was treated with ceremony and eaten only on important occasions.

Colenso writes (1878:151):

> "The flesh of the dog was not only deemed a dainty but it was also a tapu (or sacred) dish. A dog was always killed for the priest to eat on performing certain tapu or religious ceremonies over the children of chiefs, and on other great and formal occasions; also as food for the tohunga-toa-moko, or tattooer, when operating on chiefs".

Augustus Earle (1966), the painter, who travelled through New Zealand in the early nineteenth century, gives an interesting description of a feast at Pakanae on the Hokianga. Dog meat had been chosen for this feast, in spite of the availability of fish foods and European meat animals: there were also pigs in the village.[1]

Considering these accounts, it would seem that Classic Maori society suffered from a shortage of dogs; presumably dogs would have been preferred as a regular food resource rather than as an occasional delicacy if their numbers had permitted. Concurrently the Classic Maori period also saw the growth of large fortified sites, indicating the emergence of warfare; the appearance of cannibalism; and a very marked increase in the exploitation of shell-fish (See, e.g. Lockerbie 1959, Groube 1967). All these factors may indicate a growing population in a situation of some economic scarcity; a situation which was exacerbated by the decline in moa during the Classic period. The decrease in the number of dogs during the Classic period may be a related factor, in that the dog population in a settlement would be regulated largely by the food surplus available to feed them. Competition for scarce resources may have thus reduced the food available to feed a dependant dog population.

Although the resource exploitation at a single site cannot be assumed to represent a general economic pattern, the site of Whangamata Beach in the North Island, one of the very few New Zealand sites with both Archaic and Classic Maori remains, offers an interesting example of changes in dog exploitation over time. The site lies at the head of a large estuary, which opens out onto a long ocean beach: at the present day both types of littoral have very dense shellfish populations. The site contains two distinct cultural levels separated by over a metre of sterile sand, the later (Classic Maori) level consisting of a thick midden layer. Three main types of edible fauna were found on the site: dogs, shellfish, and man (burnt and shattered human bones were probably evidence of cannibalism). The total midden excavated was retained for analysis and the faunal material translated into the minimum number of individuals and meat weights they represented. (Allo, 1971, 1972). Results are shown in Figure 11.3.

Since only twenty-four square metres were excavated in a site of unknown extent, any results must of course be tentative. However, it would appear that dog meat in the Archaic period constituted the major article of diet: shellfish were very rare although presumably the same nearby shellfish beds existed as today. In the Classic period dogs declined abruptly in importance, and the major source of protein seems to have been shellfish.

It is interesting to compare with Whangamata the Classic Maori site of Galatea Bay (Ponui Island). Shellfish were by far the most important food found in the midden (representing an estimated 4224 kg of meat), with fish second in importance (c. 136 kg of meat). A single dog was found on the site, representing only 5.6 kg of meat, (See Shawcross, 1967).

The Dog in Maori Technology

The dog also played an important role in Maori technology. The working of long bones of dogs into bone needles and points was a feature in Maori culture throughout the prehistoric period, and long bones, particularly the radius, are typically under-represented on sites (see Allo 1970:170-175). One of the two articulated prehistoric dog skeletons, that from Wairau Bar, was intact except for the tibiae and femora, which had been removed, presumably for the manufacture of artefacts. The manufacture of fish-hooks from the mandibular bone and canines seems to have been characteristic of the Classic rather than the Archaic period (see Allo *ibid*).

Dogs also provided skins for highly prized dogskin cloaks. The value of these cloaks seems to have been ceremonial rather than functional. All the clothing worn by the prehistoric Maori could be woven out of the native flax (*Phormium tenax*), and the use of dogskin did not incorporate any functional advance (except possibly as protection against spear thrusts); the dogskin cloak consisted basically of decorative strips of dogskin laid over a flax cloak base. The dogskin cloak had a significance, however, far beyond that of utilitarian efficiency, since it served as an indicator of social rank. Only chiefs were permitted to wear dogskin cloaks. Mead (1969:24) writes that "The cloak of dogskin is a concrete symbol of chieftainship, a badge of office".

An immense amount of labour was invested in each cloak; firstly a closely woven flax cloak was made, with *taaniko* embroidered borders along the side edges, and sometimes along the neck border. Subsequent methods of manufacture then depended on the type of cloak to be made, but usually long narrow strips of dogskin, about one inch wide, were sown to the cloak in a vertical position with a fine flax thread and bone needle. The hairy side, of course, faced outwards.

Most ethnographic descriptions portray the Maori dog as a food animal, indicating that its role as a provider of skins for cloaks was a subsidiary one.

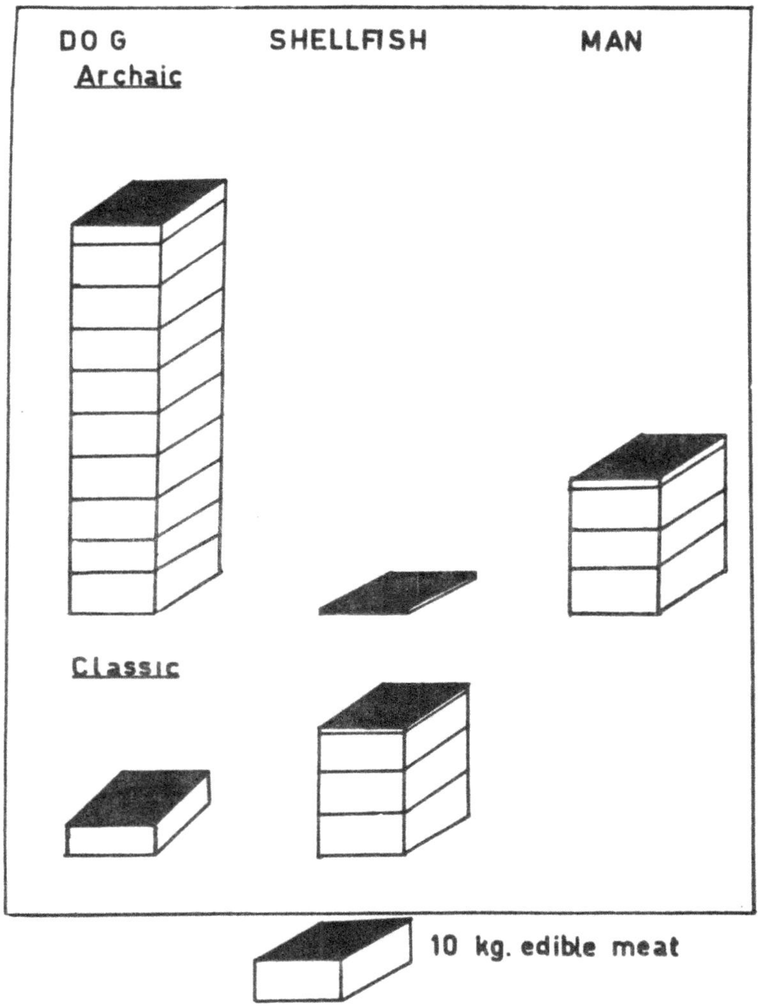

Fig. 11.3 Calculated meat weights of foods at Whangamata

Brunner (1848:281), who travelled through the South Island in the first half of the nineteenth century, stated that "... they [the Maori] never kill a dog unless for its skin". By this time, however the Maori dog was probably extinct, and European dogs were not used for food; moreover dogskin cloaks were being made extensively as trade goods.

Whether the Maori regarded the dog primarily as a source of food or of valuable clothing is not known, but in any case the dogskin cloak was of sufficient traditional importance to impart a certain ceremonial status to the dog, quite apart from its role as a highly regarded food animal. Evans-Pritchard (1940) has suggested that the ceremonial importance of a food resource among primitive people can be directly related to its economic importance. In terms of its calorific contribution to the Maori economy, the dog seems to have been of major importance only during the Archaic period. However the economic importance of a particular food resource cannot be measured simply in quantitative terms. As the only domesticated animal, the dog was the only source of animal protein whose distribution was governed by human movement; other animal resources could only be exploited by human movement to them. Furthermore, apart from the rat and the bat, it was the only mammal living in prehistoric New Zealand, and after the extinction of the moa was the only land animal of any size available for exploitation.

APPENDIX A. SKELETAL CHARACTERISTICS OF THE MAORI DOG.[2]

1. The cranium has a very prominant sagittal crest, a feature characteristically associated with a powerful bite and heavy jaw musculature. This was an almost universal feature in the Maori dog skull, and was found in 106 out of the 110 crania (96%). Only four adult crania were found with a reduced sagittal crest, giving the back of the skull a rounded juvenile appearance. Two of these, from Matatoki and Hawkes Bay, were surface finds, while a third comes from a site, Oruarangi, which contained early European dogs, or of mixed Maori-European breed.

2. The glenoid fossa, where the scapula articulates with the humerus, is relatively long and narrow in the Maori dog, and tends to be broader and more circular in most European breeds.

3. The limb bones are proportionally short compared to head and body size: this was particularly marked in the radius, ulna, and to a lesser extent, the tibia (see Allo 1970:69-76).

4. The nasal bone of the Maori dog characteristically ends in a relatively anterior position, level with the posterior borders of the premaxillas. In contrast the nasals of all observed modern breeds, both brachycephalic and dolichocephalic, end some distance behind the maxillas.

5. Present in the dentition of some Maori dogs are supernumerary alveoli, representing either additional teeth or the double roots of a normally single-rooted tooth. These alveoli are usually found behind the third molar of the mandible, although sometimes they are also present, in both mandible and cranium, between the first premolar and the canine. They are not a normal feature of Maori dog dentition—they are found in less than 1% of Archaic South Island mandibles—but where present in a site they are usually fairly common. They do not seem to be found in European dogs, even in breeds with a long jaw and well-spaced teeth (see e.g. Crawford 1937:216).

6. The teeth of the Maori dog are well-spaced in a long tooth row: tooth crowding was very rare, and in every case involved simply the lack of a diastema between premolars and never the overlapping of teeth. The tooth crowding index proposed by Degerbøl (1961), in which the combined lengths of the premolars are compared to the distance from the posterior edge of the alveolus of the canine to the anterior border of the alveolus of the carnassial, gave uniformly low results, indicating a very well-spaced tooth row. Results ranged from 75 to 81, similar to the indices of free-ranging wolves.

7. The Maori dog seems to have had a long, narrow skull, slightly broader than that of the modern grey-hound. The cephalic index, measuring the relation of head length to head width, of the few completely intact skulls ranged from 51 to 62: this can be compared to dolichocephalic, mesati-

cephalic and brachycephalic breeds such as the grey hound, fox terrier and pug with cephalic indices of 50, 70 and 90 respectively.

8. The palate of the Maori dog has a very characteristic shape: it is wide at the molars and narrows abruptly at the third molar into a narrow snout This feature was also observed by Luomala (1967) in the prehistoric dog of Tahiti. She describes the marked constriction of the muzzle just in front of the fourth premolar or carnassial.

9. The snout of the Maori dog was very long compared to the length of the braincase. This is shown by the high cranio-facial index of the Maori dog cranium, measuring the relation of the distance between the nuchal crest and the fronto-nasal suture to the length of the nasal bone. The cranio-facial index of the Maori dog ranged from 10:6.5 to 10:8.5 (median 10:8). The cranio-facial index tends to increase with length of head, but that of a dolichocephalic breed such as the grey-hound is only 10:7, while that of the pug is 10:3. The high cranio-facial index of the Maori dog is even more remarkable considering the characteristic anterior position of the posterior end of the nasal bone.

APPENDIX B. TOOTH WEAR STAGES IN THE MAORI DOG

1. Some abrasion on the inner surface of the carnassial, but this very slight. The incisors still have the 'fleur-de-lys' shape.

2. The cusps on the carnassials and first and second molars show signs of wear. The 'fleur-de-lys' shape of the incisors begins to disappear.

3. The carnassials are very worn, and there is partial disappearance of the lingual cusps on the upper first molar. The canine shows signs of wear, and the 'fleur-de-lys' shape disappears from the incisors. In the European dog this stage occurs at c. two years.

4. The cusps are worn flat on the second molar, and on the third molar in the lower jaw, also the posterior cusps of the lower carnassial. There is extensive wear on the crowns of the incisors and premolars.

5. The incisors are worn almost flat to the original gum surface, while wear on the canine produces a spatulate flat top to the crown. The carnassials and molars are worn to a smooth concave surface, and the large anterior cusps of the premolars are worn completely away.

NOTES

1. The partiality of the Maori to dog meat does not seem to have extended to the European dog, which was said by the Maori to be "perfectly unpalatable". (See Dieffenbach, 1843; also Thomson, 1859:155; Wullerstorf-Urbair 1861-2:116).

2. For detailed measurements of the skeletal material see Allo 1970.

BIBLIOGRAPHY

Allo (Bay-Petersen), J. L., 1970. The Maori Dog: A Study of the Polynesian Dog of New Zealand. Unpublished M.A. Thesis, University of Auckland.
1971. The Dentition of the Maori Dog of New Zealand. Records of the Auckland Institute and Museum 8:29-45.
1972. The Whangamata Wharf site (N49/2): Excavation of a Coromandel Coastal Midden. Records of the Auckland Institute and Museum 9.

Anderson. See Cook 1784.

Best, E., 1902. Food Products of Tuhoe-land. Transactions of the New Zealand Institute 35,:47-8.
1913. Letter to Mr. Drummond. Lyttleton Times, 11th. January.

Brunner, T., 1848. Journal of an Expedition to explore the interior of the

Middle Island of New Zealand. Nelson (N.Z.).

Colenso, W., 1878. Notes, Chiefly historical, on the ancient dogs of the New Zealanders. Transactions of the New Zealand Institute 10 pp. 135-55.

Cook, Captain James. 1773. An Account of a Voyage round the World. J. Hawkesworth, Account of Voyages, Vol. II, London.
1784. A Voyage to the Pacific Ocean. Vol. I, London.

Crawford, E. M., 1937. Preliminary Study of Dentition in Dogs. Veterinary Medicine 32:212-7.

Crozet, J., 1891. Voyage to Tasmania, New Zealand, the Ladrone Islands and the Philippines in the years 1771-1772. London.

Cruise, R., 1823. Journal of a 10 months' residence in New Zealand. Longman London.

Degerbøl, M., 1961. On a find of a Pre-Boreal domestic dog (Canis familiaris L.) from Star Carr, Yorkshire, with remarks on other Mesolithic dogs. Proceedings of the Prehistoric Society 27:35-55.

Dieffenbach, E., 1843. Travels in New Zealand. John Murray, London.

Duff, R., 1956. The Moahunter Period of Maori Culture. 2nd. Ed., Wellington.

Earle, A., 1966. Narrative of a Residence in New Zealand. Oxford.

Evans-Pritchard, E. E., 1940. The Nuer: A description of the modes of livelihood and Political Institutions of a Nilotic Peoples. Oxford.

Forster, G., 1777. A Voyage round the World in H.M.S. 'Resolution' during the years 1772-1775 (inclusive). London.

Forster, J. R., 1778. Observations made during a Voyage round the World on physical Geography, natural history and ethic philosophy. London.

Golson, J., 1959. Culture change in Prehistoric New Zealand. In Anthropology in the South Seas (Ed.) Freeman J. D. and W. R. Geddes, Avery, New Plymouth.

Groube, L. M., 1967. Models in Prehistory: a consideration of the New Zealand evidence. Archaeology and Physical Anthropology in Oceania 2:1-27.

Lockerbie, L., 1959. From Moa-hunter to Classic Maori in Southern New Zealand. In Anthropology in the South Seas, (Ed.) Freeman J. D. and W.R. Geddes, Avery, New Plymouth.

Luomala, K., 1967. Comparison of Moorean and Old World Dogs. Unpublished M.S., University of Hawaii.

Mead, S. M., 1969. Traditional Maori Clothing: a study of technological and functional change. A. H. & A. W. Reed, Wellington.

Shawcross, W., 1967. An investigation of prehistoric diet and Economy at Galatea Bay, New Zealand. Proceedings of the Prehistoric Society 33: 107-31.

Skinner, W. H., 1914. The Ancient Maori Dog. Journal of the Polynesian Society 23:173-6.

Thomson, A. S., 1859. The Story of New Zealand. 2 Vols., John Murray, London.

Wood-Jones, F., 1931. The Cranial Characters of the Hawaiian Dog. Journal of Mammalogy 12 no.1:39-41.

Wullerstorf-Urbair, B. von., 1861-63. Reise der Oesterreichischen Fregatte 'Novara' um die Erde in der Jahren 1857, 1858, 1859 unter der Befehlen des Commodore B. von Wüllerstorf-Urbair. 2 vols. Vienna.

ARCHAIC MIDDENS OF THE COROMANDEL REGION: A REVIEW

Janet Davidson

The Coromandel Peninsula has long been known as an area productive of Archaic artefacts and moa bones in human association. An early find was made by Cormack in a midden at Opito in the 1850s (Duff 1956:250). Duff, in his definitive review of moa-hunter culture, described the Coromandel area as "a virtual 'island' of discovered artifacts of the Moa-hunter type" (ibid.). Golson drew on excavated material from Sarah's Gully and Opito in his definition of the North Island Archaic (Golson 1959a:44-45). Since then, a number of other sites containing moa bones and/or Archaic artefacts have been investigated. The identification of faunal remains, particularly bird bones, has played a major part in the interpretation of most of these sites. It therefore seems appropriate to offer a review of them in this volume as an acknowledgement of the vital contribution to their interpretation made by Ron Scarlett.

THE SITES (Fig. 12.1)

Investigations of Archaic middens in the Coromandel region have ranged from highly controlled excavations to surface collecting and fossicking. This review is largely restricted to sites which were investigated by excavation rather than surface collection or chance finds, and to sites from which moa bones and/or artefacts known to be associated with moa bones elsewhere in the country have been recovered. Two sites which do not meet the former criterion have been included because they add important information not otherwise available.

Harataonga Bay Western Midden (N30/5)

A small eroding midden on Great Barrier Island, fully described by Law (1972).

Port Jackson (N35-36/88)

A sand-dune site formerly deeply buried; exposed by deflation in recent years. Local residents have numerous artefacts which are almost certainly from this site. The only evidence considered here, however, is a collection of faunal material not previously published, obtained from the remains of an *in situ* deposit by Mr. A. Fransham and myself.

Sarah's Midden (N40/13)

A small stratified midden at the base of Sarah's Gully Pa. Excavated by Birks in 1960, referred to by Green (1963).

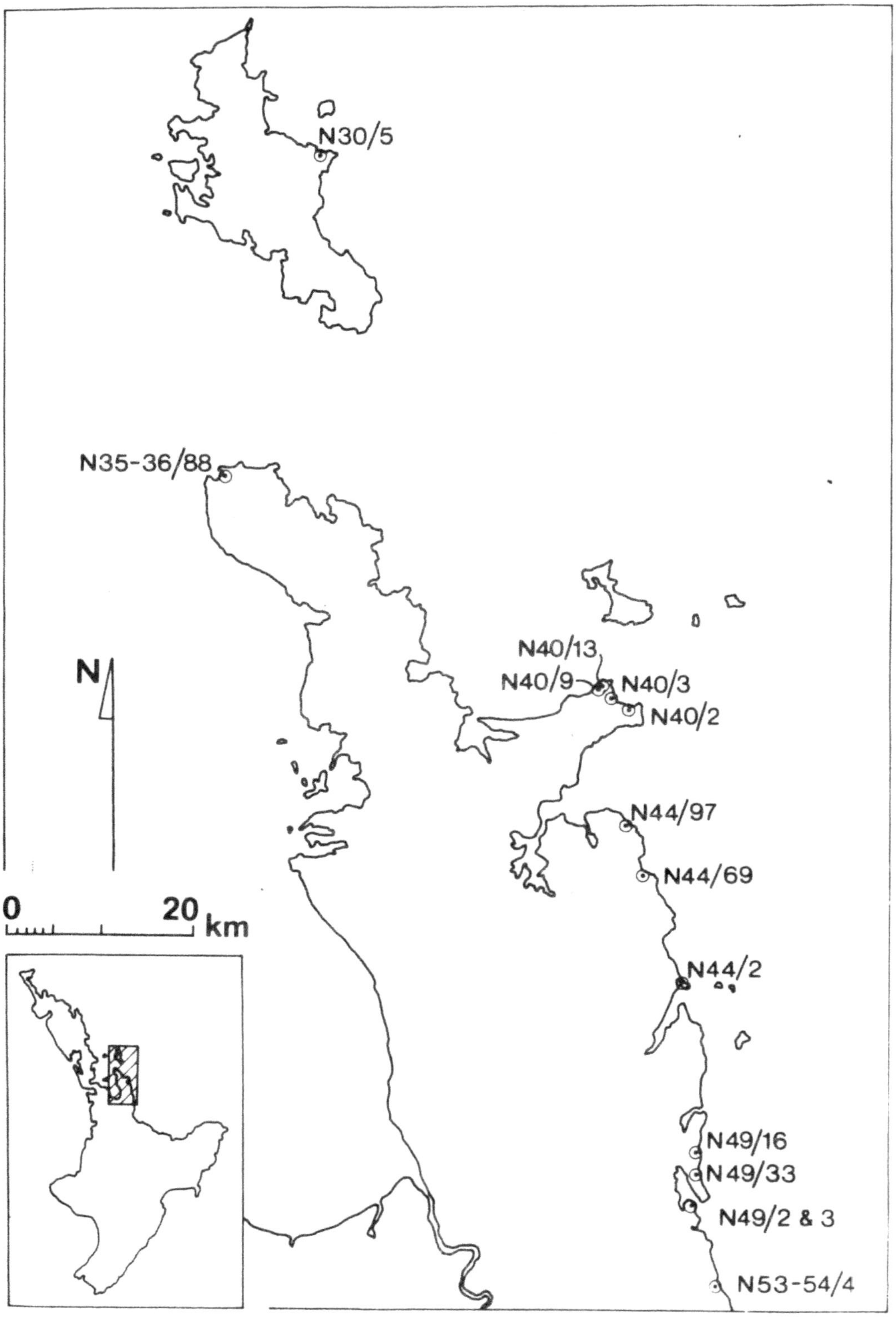

Fig. 12.1 The Coromandel Peninsula and Great Barrier Island, showing sites mentioned in the text.

Sarah's Gully (N40/9)

Excavated by Golson. Published references include Golson (1959a, 1959b) Golson and Gathercole (1962) Green (1963) Scarlett (1969, 1974).

N40/3 (Opito Beach; Skipper's Midden)

Excavated by Golson. Later investigations by Trower and Jolly were probably in this site, now destroyed by erosion. References include Golson (1959a, 1959b), Golson and Gathercole (1962) Trower (1962) Green (1963) Scarlett (1969, 1974).

N40/2 (Parker's Midden; Pohutukawa flaking floor)

A series of small excavations over a long period of time. Earlier surface collection by Parker was probably from the same site. References include Jolly and Green (1962) Green (1963) Murdoch and Jolly (1967) Jolly and Murdoch (1973) Scarlett (1974).

Hahei (N44/97)

Remains of burial with two necklaces of bone reels and central whale tooth pendant, salvaged by Mr. J. D. Osborne during subdivision. Limited salvage excavation by Edson and Brown (1977).

Hot Water Beach (N44/69)

Salvage excavation in car park by Leahy (1974).

Tairua (N44/2)

First excavated by Green and Smart and subsequently by Green and others. References include Smart and Green (1962) Green (1967) Jones (1973) Scarlett (1974) Rowland (1975).

Whangamata Wharf (N49/2)

Excavated and published by Allo (1972); further surface collections of faunal material from eroding section by Jolly.

Cabana Flats, Whangamata (N49/2)

Salvage excavation by Jolly and East in private garden following disturbance during construction of goldfish pond. Material on deposit in Auckland Museum.

Whangamata Road (49/2)

Small salvage excavation in road directly opposite Cabana Flats, probably part of the same site. Referred to by Shawcross (1964).

Whitipirorua (N49/16)

A series of small excavations in an extensive sand dune site by Jolly and various assistants. Material in Auckland Museum. Site now largely destroyed by subdivision.

Wheritoa (N53-54/4)

A large site at the north end of the beach investigated by Jolly and others, notably Crosby, who also reviewed artefacts from the same site in private collections. Destroyed by subdivision. References include Crosby (1963, 1977) Foreman and Jolly (1965). Distinct from site at south end of beach from which moas cited by Scarlett (1974) were probably collected.

The Settings

All are sand dune sites, mostly close to stream or river mouths. The cluster of three Whangamata sites and Wheritoa are on the northern tips of sandspits, with open ocean beaches to the east and estuaries to the west. Tairua is on the inner side of a sandspit linking the rocky outcrop of Paku, at the entrance to Tairua harbour, with the land to the north. The remaining sites are not on major estuaries. Tairua is unique in apparently not having ready access to fresh water.

All the sites are close to others of quite different type, including pa, open ridge habitation or storage sites, and concentrated shell middens which appear to be significantly different from the middens under review here.

Internal Structuring

Most investigations have been small samples of large and/or eroded sites and little information on internal patterning has been obtained. An important exception is Tairua, a small settlement whose lay-out revealed several different activities within an excavated area of 125 m^2 (Jones 1973). At Sarah's Gully, also, important data on site lay-out were obtained. Although the details have not been published, it is known that the extensive midden strip contained different concentrations of artefacts and faunal material in different areas. Burials were found in one area and on the slopes and knoll behind the midden, numerous postholes and several small pits, some with drains, were found.

Stratigraphy

At two sites, Tairua and Whangamata Wharf, two strikingly different cultural layers were separated by sterile sand. In each case the lower layer contained moa bones and Archaic artefacts, whereas the upper layer consistec predominantly of lenses of concentrated shell midden; of nineteenth century age at Whangamata and seventeenth or eighteenth century age at Tairua. At Sarah's Midden, also, a layer containing moa bones was stratified beneath one with European material.

At Hot Water Beach, on the other hand, three principal cultural layers contained essentially similar material, although with some changes through time. The youngest of these layers (3b) postdates A.D. 1500 and contains little that is recognisably Archaic in Golson's terms.

Of the sites, Sarah's Gully and N40/3 are known to have been stratified, the first with up to ten layers and the second with at least three. They may have reflected trends similar to Hot Water Beach, but details are not availabl The two lenses at Harataonga almost certainly represent a single occupation.

There is no information to indicate whether there was more than one occupation at the other sites, although at some of them (Wheritoa, N40/2, Whitipirorua, Cabana Flats) there appears to have been more than one layer.

Dating

Those sites for which dates are available can be broadly bracketed between A.D. 1250 and 1500, with the possible exception of Tairua. Harataonga, N40/3 and Hot Water Beach, layer 5, are associated with Loisels pumice in ways which suggest their approximate contemporaneity with it. N40/3 also has a radiocarbon date and there are comparable dates for Sarah's Gully. Layer 4 at Hot Water Beach is dated by radiocarbon to approximately A.D. 1500. Although Tairua was originally regarded as an earlier site, the general similarity of its content to the other dated sites suggests it may be broadly contemporary with them. This view finds some support from a fourteenth century radiocarbon date on shell from the lower layer.

SUBSISTENCE

The Coromandel Peninsula is an area where Polynesian cultivation has been practised. The possibility that horticulture was contemporary with the Archaic middens was raised by Golson (1959a:45) on the basis of his Sarah's Gully evidence, but remained unresolved and indeed disputed for many years. Radiocarbon dates for early storage structures at Sarah's Gully Pa and Skipper's Ridge, Opito, are now supported by growing evidence of early horticulture in other parts of the North Island. Accordingly, the Archaic middens of the Coromandel can now be regarded as representing one or more components of a settlement pattern which also included cultivation and storage sites.

The interpretation of the Archaic middens in relation to the wider subsistence base of the communities concerned has never been adequately considered. Detailed information on food remains is available only for the fully published sites. However, some information exists about faunal remains, particularly birds, in some of the other sites also.

Moas and other birds

The presence of moas and other extinct birds (or birds extinct in the North Island) is shown in Table 12.1. Only small numbers of individuals of any species seem to be involved. Nonetheless, it can be seen that a considerable variety of now extinct avifauna was available to the inhabitants of the Coromandel Peninsula during the period in question.

Euryapteryx moa, probably *E. geranoides*, was definitely present at Harataonga; the material was interpreted as industrial but nonetheless suggests contemporaneity of man and moa in the wider region. The natural presence of moas on Great Barrier remains uncertain. The moa at Whangamata Wharf, on the other hand, included a juvenile, and was probably used for food. At Hot Water Beach, moa bone was present in all three layers, but positive identification to genus or species was possible only for layer 5. It is doubtful whether moas were used for food in layers 4 and 3b. *Pachyornis mappini* was definitely present in the site, but identified only from a disturbed context. In addition to the moas shown for Wheritoa, which were possibly industrial rather than food remains, another site at the south end of the beach yielded examples of Dinornis novaezealandiae, D. struthoides, Pachyornis mappini and possibly Euryapteryx curtus.

One of the principal arguments in favour of Tairua being an early site is the number of moa species used as food. Apart from Pachyornis septentrionalis and the tentatively identified D. gazella, however, the moas present at Tairua are found in at least two other sites, while one of the most widely represented species, E. exilis, is not recorded from Tairua. Similarly, Tairua has only one other extinct species of bird, compared with four at both Port Jackson and Hot Water Beach. It is by no means certain that the number of moa species at Tairua is due to its greater age, rather than to a slightly different focus of hunting activity or a different season of occupation (cf. Rowland 1975).

A pattern of generalised hunting without marked concentration on particular species is, generally, reflected in other bird remains also. In Table 12.2, presence/absence of bird species according to habitat zone, are shown together with minimum numbers where available. The largest number of species is found at Tairua and Hot Water Beach, layer 5. At Hot Water Beach, there is a steady decline in bird species from layer 5 to layer 3b. The number of species represented in layer 4 at Hot Water Beach is similar to Harataonga and the Opito and Sarah's Gully sites. The deposits with relatively few bird species are also of interest. Wheritoa was a very small sample, but Hot Water Beach, layer 3b, reflects a decline in most kinds of food, not merely birds. Whangamata Wharf was an unusually specialised deposit primarily concerned with butchering dogs; fish, birds and shellfish are all rare compared with other sites. At Port Jackson, however, the relatively small number of bird species is at least partly due to the poor representation of sea birds. The small sample and a possible collection bias towards larger bones may explain the predominance of moa, but it can hardly account for the proportions of sea and land birds. The kaka was by far the most common bird in the sample with considerably more than 50 individuals represented. The part of the deposit sampled, at least, seems to reflect land-oriented hunting.

Unfortunately, details of minimum numbers of birds are available from only four sites. There is little evidence, for the most part, of concentration on only one species, although the kaka tends to predominate among bush birds. Only Port Jackson exhibits heavy exploitation of kaka compared with all other birds. Another interesting exception is provided by N40/2. Here there was a minimum of 26 individuals of the flesh-footed shearwater, whereas other birds are represented by one or at the most two individuals. This is one of very few instances yet reported of midden bones reflecting specialised birding. Even so, the bones could be the result of one day's work; it is noteworthy that all these bones were found in one square. The flesh-footed shearwater still breeds on the Mercury Islands and possibly other islands off the Coromandel Coast.

Table 12.1 Occurrence of Moas and other Extinct Birds

	Haratoanga	Pt. Jackson	Sarah's Gully	N40/3	N40/2	HWB/5	HWB/4	HWB/3b	Tairua	Whangamata Wh.	Wheritoa
Euryapteryx curtus	-	X	-	-	X	?	-	-	-	-	-
Euryapteryx exilis	-	X	-	X	X	X	?	?	-	X[1]	X
Euryapteryx geranoides	?	X	-	-	X	-	-	-	X	-	?
Euryapteryx sp.	X	-	-	-	X	X	?	?	-	X	-
Pachyornis septentrionalis	-	-	-	-	-	-	-	-	X	-	-
Pachyornis mappini	-	-	-	-	-	?	?	?	X	-	-
Dinornis gazella	-	-	-	-	-	-	-	-	?	-	?
Dinornis struthoides	-	-	X	X	-	-	-	-	X	-	-
Dinornis novaezealandiae	-	X	-	-	-	-	-	-	X	-	-
D. giganteus/hercules	-	X	X	X	-	-	-	-	X	-	-
Dinornis sp.	-	-	-	-	X	X	?	?	X	-	-
Palaeocorax moriorum extinct crow	-	X	X	-	-	-	-	-	X	X[1]	-
Notornis mantelli takahe	-	X	-	?	-	-	-	-	-	-	-
Nesophalaris chathamensis extinct coot	-	X	-	-	-	X	-	-	-	-	-
Circus eylesi subsp. extinct hawk	-	-	-	-	-	-	-	X	-	-	-
Heteralocha acutirostris huia	-	X	-	-	-	X	-	-	-	-	-
Gallirallus minor extinct little woodhen	-	-	-	-	-	X	-	-	-	-	-
Apteryx oweni little grey kiwi	-	-	-	X	X	-	-	-	-	-	X
Strigops habroptilus kakapo	-	-	-	-	X	?	?	?	?	-	-
Pelecanus conspicillatus subsp. pelecan	-	-	-	-	-	-	-	-	-	X[1]	-

1. 1976 collection by R. G. W. Jolly

Table 12.2 Occurrence of Other Birds

Ocean/sea/coast	Haratoanga	Pt. Jackson	Sarah's Gully	N40/3	N40/2	HWB/5	HWB/4	HWB/3b	Tairua	Whangamata Wh.	Wheritoa
?<u>Eudyptes</u> sp. large penguin	-	X	-	-	-	-	-	-	-	-	-
<u>Eudyptula minor iredalei</u> little blue penguin	4	X	X	X	1	5	3	1	X	1	X
<u>Diomedea</u> sp. albatross	1	-	X	X	1	1	1	-	X	-	-
<u>Thalassarche</u> sp. mollymawk	1	-	X	X	X	1	?	?	X	-	-
<u>Puffinus griseus</u> sooty shearwater	-	-	X	X	?	1	-	-	-	-	-
<u>Puffinus gavia</u> fluttering shearwater	2	-	-	X	1	1	-	1	-	-	-
<u>Puffinus assimilis</u> little shearwater	-	-	-	-	-	1	-	-	-	-	-
<u>Puffinus carneipes</u> flesh-footed shearwater	-	-	-	X	26	-	-	-	-	-	X
<u>Puffinus</u> sp. shearwater, undeterm.	-	-	X	X	-	-	-	-	-	-	-
<u>Pterodroma macroptera</u> grey-faced petrel	5	-	X	-	X	-	-	-	X	-	-
<u>Pachyptila turtur</u> fairy prion	?1	-	X	X	-	-	-	-	-	-	-
<u>Pachyptila vittata</u> Broad-billed prion	-	-	-	-	-	1	-	-	-	-	-
<u>Pachyptila</u> sp. prion, undetermined	-	-	X	-	-	-	-	-	X	-	-
petrel, undetermined	-	X	X	X	-	2	1	1	X	X	-
<u>Macronectes giganteus</u> giant petrel	-	-	-	-	-	-	-	-	?	-	-
<u>Pelagodroma marina</u> white-faced storm petrel	-	-	-	X	-	-	-	-	-	-	-
<u>Pelecanoides urinatrix</u> diving petrel	-	-	X	X	-	1	-	1	-	-	-

Table 12.2 (contd.)

Ocean/sea/coast	Harataonga	Pt. Jackson	Sarah's Gully	N40/3	N40/2	HWB/5	HWB/4	HWB/3b	Tairua	Whangamata Wh.	Wheritoa
Phalacrocorax melanoleucos little shag	1	X	-	-	1	1	-	-	X	-	-
Phalacrocorax varius pied shag	-	?	-	-	-	-	-	-	X	-	-
Phalacrocorax carbo black shag	1	?	-	-	1	1	-	-	-	-	-
Stictocarbo punctatus spotted shag	?2	?	-	-	-	1	-	-	X	-	-
shag, undetermined	1	X	X	X	1	-	-	-	X	3	-
Egretta alba white heron	-	-	-	-	-	-	-	-	-	1	-
Haematopus sp. oyster catcher	-	-	-	-	-	-	-	-	X	-	-
Himantopus sp. stilt	-	-	X	-	-	-	-	-	-	-	-
Larus sp. gull	1	X	X	X	-	1	-	-	X	-	-
Hydroprogne caspia caspian tern	-	X	-	-	1	-	-	-	?	-	-
Chlidonias hybrida black-fronted tern	-	-	-	-	-	-	-	-	X	-	-
Forest											
Apteryx australis kiwi -	-	-	-	-	-	1	1	1	-	1	-
Hemiphaga novaeseelandiae pigeon	1	X	-	X	1	2	1	-	X	-	-
Nestor meridionalis kaka	3	X	X	X	1	10	5	3	X	-	X
Cyanoramphus sp. parakeet	-	X	-	-	1	2	1	-	X	-	-
Ninox novaeseelandiae morepork	-	-	-	-	-	1	1	-	-	-	-
Prosthemadera novaeseelandiae tui	1	-	X	-	1	6	1	1	X	1	-

Table 12.2 (contd.)

Forest	Harataonga	Pt. Jackson	Sarah's Gully	N40/3	N40/2	HWB/5	HWB/4	HWB/3b	Tairua	Whangamata Wh.	Wheritoa
<u>Callaeas cinerea</u>											
kokako	-	X	X	-	1	4	2	-	X	-	-
<u>Turnagra capensis</u>											
thrush	-	-	-	-	-	1	-	-	-	-	-
Lake/river/swamp											
<u>Podiceps cristatus</u>											
southern crested grebe	-	-	-	-	-	-	-	-	X	-	-
<u>Anas superciliosa</u>											
grey duck	-	-	X	X	-	-	-	-	X	-	-
<u>Tadorna variegata</u>											
paradise duck	-	-	X	-	-	-	-	-	X	-	-
<u>Anas rhynchotis</u>											
shoveller	-	-	-	X	-	-	-	-	-	-	-
<u>Anas aucklandica chlorotis</u>											
brown teal	?1	-	-	-	-	-	-	-	-	-	-
<u>Hymenolaimus malacorhynchos</u>											
blue duck	-	X	-	-	-	-	-	-	-	-	-
<u>Anas</u> sp.											
duck, undetermined	-	-	X	X	-	X	-	1	X	-	-
duck, undetermined	-	X	-	-	-	-	-	-	-	1	-
Open country											
<u>Falco novaeseelandiae</u>											
falcon	-	-	-	-	-	1	1	-	X	-	-
<u>Gallirallus australis</u>											
weka	-	-	X	X	-	3	1	-	X	-	-
<u>Anthus novaeseelandiae</u>											
pipit	-	-	X	-	-	-	-	-	-	-	

Mammals

Seals are reported from all sites for which any faunal evidence at all is available. The fur seal is present in all sites, and the elephant seal is also reported from N40/3, Tairua and probably Wheritoa. The importance of seal in the diet at Whangamata Wharf has been demonstrated by Allo. Similar importance is likely at most if not all other sites, reflecting the heavy reliance on marine products in the economy.

Dogs are known from all sites. At Whangamata Wharf, a minimum number of 17 dogs is in striking contrast to the small numbers of birds and fish. On present evidence, no other site has such a large number of dogs, although several individuals are present in each site for which data are available. Immature dogs are represented in all these sites.

Cetacean remains have been less fully identified. Fragmentary whale remains are listed from Hot Water Beach, layer 5, Whangamata Wharf and N40/3, and undetermined cetacean from N49/2, N30/2 and Tairua. The extent to which any of these represent food is uncertain.

The rat is not reported from Sarah's Gully, Harataonga or Whangamata Wharf (where it is present in the upper layer). It is found in small quantities in other sites, declinging through time at Hot Water Beach from 10 individuals in layer 5 to 1 in layer 3b.

Human bones are recorded only from Whangamata, where scattered charred fragments in the midden were interpreted as evidence of cannibalism.

Fish

In view of the importance of fishing as reflected by fishing gear in most of these sites, the identification of fish remains constitutes an important field for future enquiry. At Hot Water Beach, 82 fish of at least 15 species were found in layer 5, declining to 33 fish of 6 species in layer 4 and 18 of 6 species in layer 3b. Snapper was much the most numerous in all layers. At Harataonga, six species were identified, snapper greatly predominating. At Whangamata Wharf, snapper comprised three of the total six fish of four species. Snapper was most numerous of seven species identified at Tairua. Snapper appears to be present in all sites for which any information is available. Fish bone generally and snapper in particular was very common at Wheritoa. At Port Jackson, snapper and kahawai were almost equally represented in the small area sampled, with several other species present in smaller numbers.

Tuatara

The tuatara, which is still found on some of the islands off the Coromandel coast, is present at Port Jackson, Hot Water Beach, layer 5, Sarah's Gully and the Opito sites, but has not been reported from Tairua, Whangamata and the upper layers at Hot Water Beach.

Shellfish

An apparent shift from rocky shore to estuarine and sandy beach species was demonstrated at Tairua. Subsequent work, however, has revealed a more complex picture, in which rocky shore shellfish continued to be represented in small numbers as part of the concentrated, predominantly sandy beach, shell midden dumps (Rowland 1975). At Whangamata Wharf, the great quantity of shell in the upper midden contrasted with the unimportance of shell in the lower midden; the same species were present in both deposits, however. At Hot Water Beach, a similar range of both rocky shore and sandy/mudflat species was present in all layers. There was a marked change from the dominance of rocky shore species (particularly Lunella smaragda) in layer 5 to a dominance of sandy/mudflat species (Paphies australe and Amphibola crenata) in layer 4, but in layer 3b the proportions were again similar to layer 5. At Harataonga, the midden was dominated by rocky shore species, but this was also true of the much later Eastern Midden. Clearly, local conditions are important in evaluating shellfish exploitation.

Discussion

For most sites, little or nothing is yet known about the relative contribution of different foods to the diet; season of occupation; or whether the sites reflect preservation of food to be eaten elsewhere. Law has suggested that Harataonga was certainly occupied during spring, but that there is nothing to show whether it was also occupied for the rest of the year. He found no clear evidence of food preservation. Similar conclusions were reached by Leahy for Hot Water Beach. Rowland has suggested that Tairua was probably occupied in autumn-winter, although other seasons could not be definitely excluded. By contrast, year-round occupation with no particular seasonal emphasis was indicated for the upper shell midden lenses (Bed 6) at the same site.

Although general trends such as the decline and extinction of moa and an increase in exploitation of estuarine shellfish are still evident, small individual variations in food remains from site to site show that the picture is more complex than appeared from Tairua alone. It is impossible to seriate sites or estimate age on the basis of number of extinct bird species or proportions of rocky shore to estuarine shellfish and these trends seem unimportant compared with continuity in overall economic activity in the sites under review. Most sites reflect generalised exploitation of a range of food resources, but with a strong marine emphasis. Specialisation is the exception rather than the rule. The sites with some evidence of specialisation (N40/2 with its shearwaters, Port Jackson with its kaka, and Whangamata Wharf with its dogs) also contain evidence of the use of the moa for food in small quantities.

MATERIAL CULTURE

Although no one site has yielded a full inventory of Archaic artefacts, the range of material from all the sites together includes a large proportion of demonstrably Archaic forms known from elsewhere in New Zealand. In considering the small numbers of most artefact types it is important to remember that the scale of most excavations has been minute compared with operations at Wairau Bar, for example. Many of the sites themselves are small, but others, such as Wheritoa and Whitipirorua were of considerable extent. The problem of site delimitation is also acute; are the scattered remains at Whangamata, now separated by roads and wharf facilities, part of one large site with distinct activity areas, or are they three separate sites?

Fishing Gear (Fig. 12.2)

Trolling gear is represented in small amounts at most sites. Stone shanks are rare and the triangular shank, so popular in the Cook Strait area, for instance, is hardly represented in the excavated material. The flattish-sectioned shank with dorso-ventral perforation has been found in several sites. The pearl shell specimen from Tairua is well known (Green 1967). It is matched by a ahell specimen from Whitipirorua, which although much smaller, duplicates the shield shape of some Murihiku examples, and a small stone shank, also from Whitipirorua. A bone shank with dorso-ventral perforation was found in Hot Water Beach, layer 5; a similar example was found in Sarah's Midden. A piece of worked shell from Hot Water Beach, layer 4, is probably an unfinished lure shank. Bilaterally perforated shanks are represented only by a stone specimen of trapezoidal section from Wheritoa, and a substantial bone shank, of rounded triangular section, from the Cabana Flats. The latter was associated with three bone shanks and a stone shank, all of the grooved type. There is also a grooved shank from Whitipirorua Site 3, and one from Harataonga. Moa bone blanks from N40/2 and Whitipirorua may have been intended either for dorso-ventral or grooved line attachment. Private collections from Wheritoa contain both dorso-ventral and grooved shanks.

The importance of shell lure shanks in New Zealand has been underemphasised, partly because they are not found at all in some well studied areas like Murihiku. They are often powdery and fragile when found and appear unfunctional. The Whitipirorua example, however, which is particularly well preserved, is a very practical item. Although the Tairua specimen in pearl shell may have survived as an heirloom ornament, it was clearly originally functional, and demonstrates the Polynesian origins of the dorso-ventral form (Green 1967). Dorso-ventral shells shanks have been found at Houhora, and Hjarnø has shown that shell shanks were numerous at Wairau (1967:45)

In contrast to the variety of lure shanks, only uni-perforate points with neither proximal nor distal projections have been reported. Examples include one from Sarah's Gully, N40/3, Cabana Flats and Hot Water Beach, layer 4; a possible unfinished point from Hot Water Beach, layer 5, and a shell example from Harataonga. Private collections contain several from Wheritoa.

Bait hooks are almost all unbarbed one-piece bone specimens, with numerous examples from Harataonga, Sarah's Gully, N40/3, Hot Water Beach, Cabana Flats, Whitipirorua and Wheritoa. A few examples also come from N40/2, Tairua and Whangamata Wharf. Both incurved and straight varieties are found and there is some variation in size, head form and presence or absence of bait notch. The majority of hooks from all sites are of moa bone, but other bone and ivory are also recorded.

Two-piece bone hooks are virtually unknown. Re-use of broken one-piece hooks by lashing point and shank legs together is reported from Wheritoa. An Oruarangi point was found at N40/2 but was not certainly associated with the main deposit there. Barbed hooks are known from other Coromandel sites, but the only barbed hook reported from any of the sites under discussion is a small shank-barbed shell hook from Hot Water Beach, layer 3b.

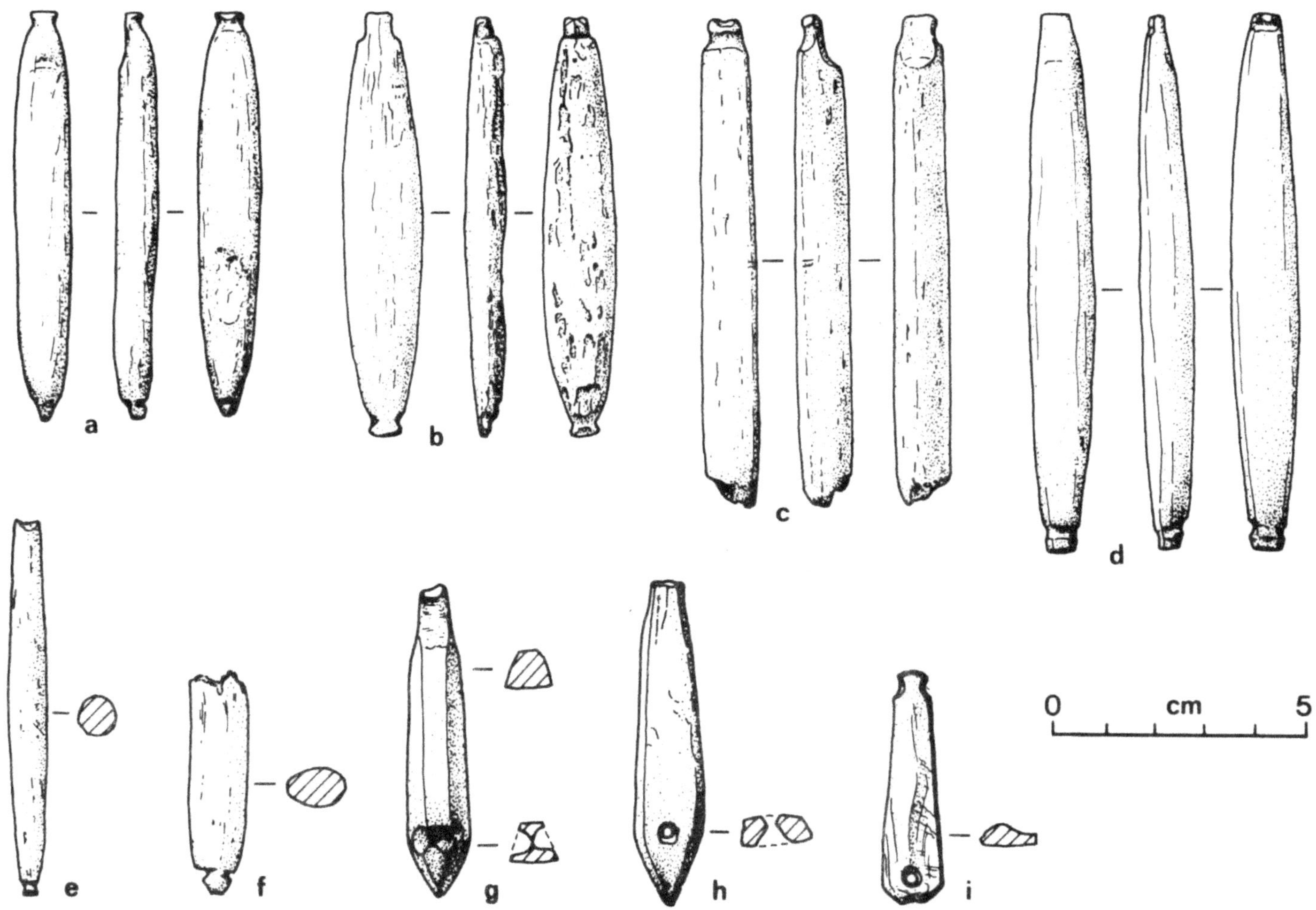

Fig. 12.2 Trolling lure shanks. a-e, Cabana Flats, Whangamata; a-c and e, bone, d, stone. f, Whitipirorua Site 3, bone. g, Wheritoa, stone. h-i, Whitipirorua, h, stone, i, shell.

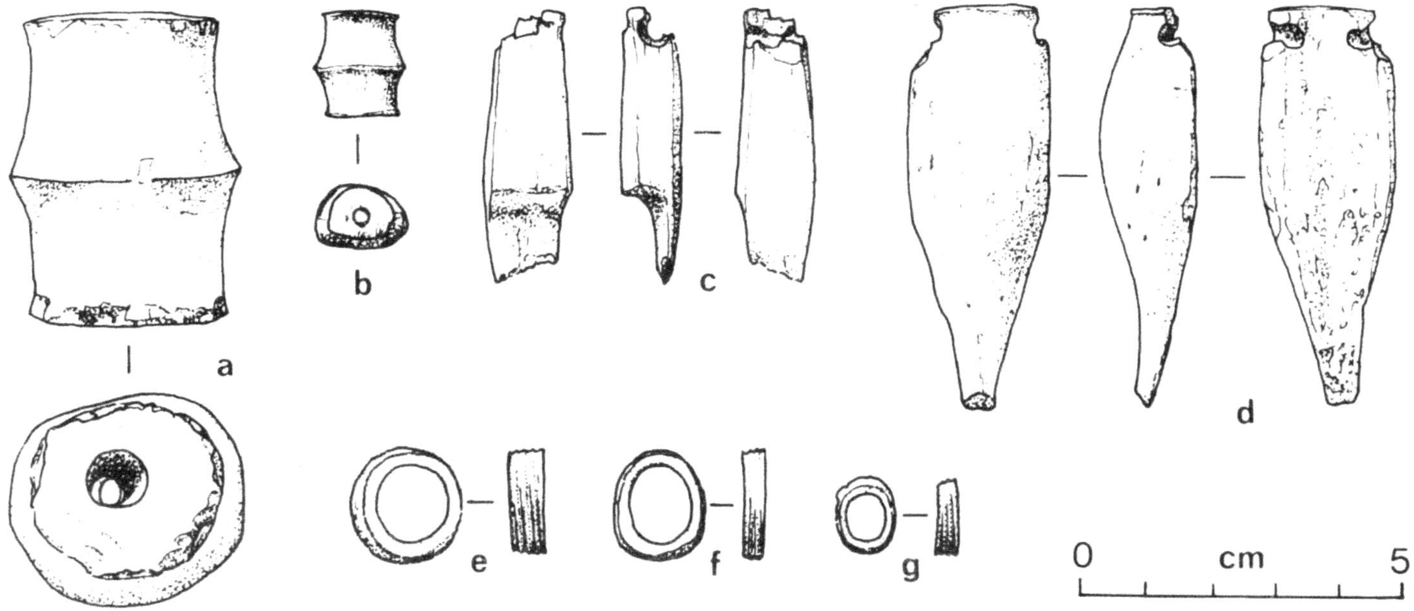

Fig. 12.3 Personal ornaments. a, Hot Water Beach, stone reel. b. Whitipirorua, bone reel. e, Wheritoa, bone pendant unit. d, Cabana Flats, Whangamata, bone pendant unit. e–g, Whitipirorua, fossil dentalium segments.

The position of shell fishhooks in the Coromandel region is obscure. Both unbarbed one-piece and unbarbed two-piece examples (with notches for lashing) were found at N40/2 and have generally been assumed to be "late Archaic". On the faunal evidence, however, there is no reason to assume that this site is any later than other sites reviewed here. Shell hooks may have been found and not published, or found and not recognized at some other sites. Shell two-piece hooks are also known from Wheritoa and Whitipirorua, and shell one-piece hooks are represented in private collections from Wheritoa

Although the position of shell hooks remains to be clarified, the other evidence suggests that both the unbarbed bone one-piece hook and the Polynesian trolling lure were sufficiently appropriate to Coromandel conditions to continue in use until at least A.D. 1500. There is no sign of the barracouta point or the barbed point, both well established elsewhere long before A.D. 1500. Stone sinkers are rare, and have so far been reported only from N40/2.

Ornaments (Fig. 12.3)

The recent discovery at Hahei of burials with necklaces of reels and central whale tooth pendant raises hopes that other complete necklaces may yet be found. The Hahei necklaces consist of ivory reels with a central pendant both probably made from elephant seal teeth (I. Smith, pers. comm.).

Reels have long been known from the Coromandel, including Opito (Duff 1956:93, 100). Other examples from recent excavations include a large serpentine reel and small bone and fossil dentalium reels from Hot Water Beach, layer 5. An ivory reel and fossil dentalium segments are known from Whitipirorua. Dentalium nanum has been found at N40/3, Sarah's Gully, N40/2, Hot Water Beach and Whitipirorua. There is also the dentalium "workshop" discovered and described by Scarlett (1958).

Imitation whale tooth pendants of both round and rectangular section are known; a rounded example from Cabana Flats are rectangular specimens from Wheritoa and N40/3. Bird bone beads have been found at Whitipirorua, Hot Water Beach, N40/3 and Sarah's Gully. Imitation Carcharodon shark teeth are reported from N40/3; similar examples were found at Hahei. A perforated shark tooth was found at Whitipirorua, and an unperforated example, possibly intended as a pendant, at Hot Water Beach, layer 4. Perforated porpoise teeth are known from Wheritoa and Whitipriorua. Perforated shells which may be ornaments have been reported from N40/2, Hot Water Beach and Harataonga.

The only example of what has been considered an exclusively Classic Maori ornament is the possible bone comb fragment from Hot Water Beach layer 5, although some forms, such as Dentalium nanum and bird bone beads are found throughout New Zealand prehistory.

It is clear that a full range of Archaic ornaments was in use on the Coromandel Peninsula. Hahei has shown that the custom of burying people with their personal ornaments was also known there.

Miscellaneous Bone Artefacts

Tattooing chisels are not common. One small perforated example is known from Harataonga and another from Hot Water Beach, layer 4. Similar examples are also reported from Wheritoa. These are similar to examples from Mount Wellington and Motutapu (N38/37) and in marked contrast to the larger specimens from Houhora. An unusual specimen found recently at Whitipirorua differs from other known Coromandel examples in being broader, and being apparently made on a thin fishhook tab.

Bird spear points are also rare, being reported only from Harataonga and Wheritoa.

Awls are not numerous, although examples are known from Tairua and Harataonga. Needles, however, are fairly common, with small examples from Cabana Flats, N40/2, Tairua, Harataonga, Hot Water Beach and Port Jackson. Large thick bone items, which may be thatching needles, cloak pins or pendants are known from Cabana Flats and Hot Water Beach.

A moa bone burnisher or chisel from Harataonga, similar to specimens from as far afield as Houhora, Taranaki and Paremata is not so far matched from the other sites, except for a doubtful example from Tairua.

Adzes

The typology of adzes from the Coromandel Peninsula is intimately related to the use of the Tahanga adze factory, and beyond the scope of this paper. Evidence of Tahanga basalt working is found in many, though not all, of the sites reviewed (Best 1975, Moore 1976). Broken and unfinished adze rough-outs are numerous in some sites, such as N40/2, and a few specimens are found in almost all sites. Both quadrangular and triangular sections are well represented. The most common complete adzes recovered, however, are small untanged quadrangular adzes approximating to Duff's (1956) Type 2A, and indeterminate small adzes made on flakes.

The distinctive Archaic types found include Duff's type 1A (N40/3 and surface finds at Harataonga and Whitipirorua); 2A (most sites including Tairua); 3 (mostly untanged, from Whangamata Wharf, Hot Water Beach and Wheritoa, but 3A reported from N40/3); 4A (Sarah's Gully, N40/2, N40/3 and Wheritoa); and 5 (Hot Water Beach, N40/2 and Wheritoa). Greenstone is limited to a chisel found by Shawcross at Whangamata and several gouges from Wheritoa in private collections.

Other Stone Tools

The wide range of other stone tools present in at least some of these sites includes drill points, files, grindstones, hammers, saws, used pumice and used flakes as well as waste flakes from core tool production. Jones (1973) concluded that most of the flakes at Tairua resulted from adze manufacture or were the result of maintenance activities for tools in other materials. With the exception of Port Jackson and Hahei, about which too little is known (although surface collections from Hahei suggest similar activities to Hot Water Beach), and Whangamata Wharf (which seems to have been a specialised butchering or skin dressing area at which only flakes suitable for these activities were prepared), most sites contain evidence suggesting a wide range of industrial activity. Fishhook manufacture was practised at most sites, as was stone working, including manufacture of adzes and drill points; wood working probably also took place.

CONCLUSION

Since the first controlled excavations on the Coromandel Peninsula more than 20 years ago, evidence has accumulated from many sites. Extensive and generalised hunting and collecting with relatively little specialisation is reflected by the midden remains. Moas were hunted but appear to have been unimportant in the diet compared with seals and probably dogs and fish. Considerable industrial activity is evident is most sites, and the range of stone resources used, like the extensive midden remains indicates reasonable familiarity with the local environment. The similarities in midden content and material culture suggest that all sites reviewed may fall within the period between approximately A.D. 1250 and 1500. In particular, Tairua compares closely with the group of sites thought to date to the fourteenth century. Earlier material may be mixed in the Whitip.rorua or Wheritoa assemblages, or concealed in the little known Hahei and Port Jackson sites, but the best documented sites do not belong to a Settlement Phase. What these sites reflect is a successful local adaptation, associated with recognisably Archaic material culture. Durable fishing gear is a small selection of the Archaic range which was evidently highly suited to Coromandel conditions. Adze technology is based on the Tahanga quarry; other stone technology is geared to other local resources. On the evidence of Hot Water Beach, the same type of exploitation of local food resources and similar material culture continued with only minor changes until A.D. 1500 and quite possibly later.

Two main interpretations are possible for most of these sites. Either they were seasonal camps, or they were specialised activity areas within larger settlements (remembering that in Polynesian ethnography a "settlement" may be considerably larger than what most archaeologists would recognise as a "site"). The best evidence for a whole constellation of site components in close proximity comes from Sarah's Gully, where midden, working area, burial ground, and cultivation and storage components could all be contemporary. Similarly, N40/3 has sometimes been correlated with nearby Skipper's Ridge. Spring and early summer occupation has been suggested for Harataonga and Hot Water Beach, but the lack of evidence of food preservation and the difficulty of showing that the sites were <u>not</u> occupied in winter stand in the way of their positive identification as seasonal camps. On the other hand, Tairua has recently been identified as a winter camp (Rowland 1975). Against this variety of interpretation may be set the fact that all these sites are far more like each other than they are like any other sites. They should therefore be susceptible to one interpretation. Perhaps fuller publication of more of them and renewed enquiry into seasonality and dating, will result in more satisfactory interpretation.

BIBLIOGRAPHY

Allo, Jan, 1972. The Whangamata Wharf Site (N49/2): excavations on a Coromandel coastal midden. Rec. Auckland Inst. Mus. 9:61-79.

Best, S. B., 1975. Adzes, rocks and men. Unpublished MA research essay, Anthropology Department, University of Auckland.

Crosby, Eleanor, 1963. Preliminary report on Whiritoa. N.Z. Arch. Assn. Newsletter. 6:46-48.

Crosby, Eleanor, 1977. Wheritoa: a post-settlement dune midden on the Coromandel Peninsula. Microfiche. Oceanic Prehistory Records, 2.

Duff, Roger, 1956. The Moa-hunter period of Maori culture. Government Printer, Wellington. 2nd edition.

Edson, S. and Brown, D., 1977. Salvage excavation of an Archaic burial context, N44/97, Hahei. Rec. Auckland Inst. Mus. 14:25-36.

Foreman, J. M. and Jolly, R. G. W., 1965. Report on Whiritoa Beach site, Coromandel Peninsula. N.Z. Arch. Assn. Newsletter 8:149-151.

Golson, Jack, 1959a. Culture change in prehistoric New Zealand. In J.D. Freeman and W.R. Geddes (eds.) Anthropology in the South Seas. Avery, New Plymouth.

Golson, Jack, 1959b. Excavations on the Coromandel Peninsula. N.Z. Arch. Assn. Newsletter 2 :13-18.

Golson, Jack and Gathercole, P. W., 1962. The last decade in New Zealand Archaeology. Part 1. Antiquity 36:168-174.

Green, R. C., 1963. Summaries of sites at Opito, Sarah's Gully and Great Mercury Island. N.Z. Arch. Assn. Newsletter 6:57-69.

Green, R.C., 1967. Sources of New Zealand's East Polynesian Culture: the evidence of a pearl shell lure shank. Archaeology and Physical Anthropology in Oceania 2:81-90.

Hjarnø, Jan, 1967. Maori fish-hooks in Southern New Zealand. Rec. Otago Mus. Anthropology 3.

Jolly, R. G. W. and Green, R. C., 1962. Beach middens on the Coromandel coast. N.Z. Arch. Assn. Newsletter 5:41-43.

Jolly, R. G. W. and Murdoch, Pat, 1973. Further excavations at site N40/2, Opito Bay. N.Z. Arch. Assn. Newsletter 16:66-72.

Jones, Kevin, 1973. Excavations at Tairua (N44/2) 1958-64: a synthesis. N.Z. Arch. Assn. Newsletter 16:143-150.

Law, R. G., 1972. Archaeology at Harataonga Bay, Great Barrier Island. Rec. Auck. Inst. Mus. 9:81-123.

Leahy, Anne, 1974. Excavations at Hot Water Beach (N44/69), Coromandel Peninsula. Rec. Auck. Inst. Mus. 11:23-76.

Moore, P. R., 1976. The Tahanga Basalt: an important resource in North Island prehistory. Rec. Auck. Inst. Mus. 13:77-93.

Murdoch, C. J. and Jolly, R. G. W., 1967. An excavation at Opito Bay. N.Z. Arch. Assn. Newsletter. 10:157-166.

Rowland, M. J., 1975. Tairua and the offshore islands in early New Zealand prehistory. Unpublished MA thesis, University of Auckland.

Scarlett, R. J., 1958. A Dentalium shell necklace 'workshop'. J. Polynes. Soc. 67:75.

Scarlett, R. J., 1969. Avifauna and man. Paper presented to N.Z. Arch. Assn. conference in Dunedin, May 1969. Mimeographed. (See this volume).

Scarlett, R. J., 1974. Moa and man in New Zealand. Notornis 21:1-12.

Shawcross, Wilfred, 1964. Stone flake industries in New Zealand. J. Polynes. Soc. 73:7-25.

Smart, C. D. and Green, R. C., 1962. A stratified dune site at Tairua, Coromandel. Dom. Mus. Rec. Ethnology 1 (7).

Trower, D., 1962. Opito Beach: two sites. N.Z. Arch. Assn. Newsletter 5:43-46.

TAI RUA: A MOA-HUNTER SITE IN NORTH OTAGO

Michael M. Trotter

The late 1950s were exciting years in New Zealand archaeology. Following pioneer work by Les Lockerbie in Otago, American field methods were introduced by Robert Bell and the sophisticated techniques of the English school were brought over by Jack Golson. The New Zealand Archaeological Association was formed and the New Zealand Historic Places Trust contracted its first site recording and archaeological salvage project; Roger Duff published a second edition of his Moa-hunter Period of Maori Culture, incorporating some minor revisions necessitated by new discoveries; museum expeditions still looked for signs of live moas in Fiordland, and radiocarbon dating—"the answer to every archaeologist's prayer"—had arrived.

All of this stimulated research that today, only twenty years later, seems naive; yet because the widely accepted version of New Zealand prehistory was so simple and straightforward at that time, it was very important to those who did not entirely agree with it to obtain evidence to support their alternative ideas. It was of vital importance to followers of the Otago school, for instance, to be able to prove that more than one species of South Island moa had survived into the human period; disciples of H. D. Skinner knew that the barb was an ancient feature of Polynesian culture and obtained and published evidence to prove it indisputably; and if remains of *Cnemiornis* or *Aptornis* could be found in primary association with such cultural material, then that was the sort of bonus that made the work so much more exciting. For a few years we almost experienced a revival of that flamboyant period of last century when Haast, Hector and Hutton were making momentous discoveries and repeatedly proving and disproving numerous hypotheses to each other's dissatisfaction.

It was in this climate that the investigation of the Tai Rua site commenced in 1956, as a purely salvage operation of occupational material in a ploughed paddock. Soon the discovery of an underlying, undisturbed deposit, plus the nature of the material being revealed, was responsible for a change in objective to a more research-oriented investigation. One of the principal factors influencing this change was the discovery of barbed fish-hooks in association with moa bones—there being some controversy at that time as to the place of the barb in the South Island cultural sequence.

Locality

The adopted name for the site, Tai Rua, is one which was recorded (by Stevenson 1947:81) as applying to an adjacent headland, now called Bridge Point, and it has been found convenient to use it in referring to the nearby archaeological site. A disadvantage of using this name is possible confusion with a North Island archaeological site called Tairua (first excavated in 1958-59).

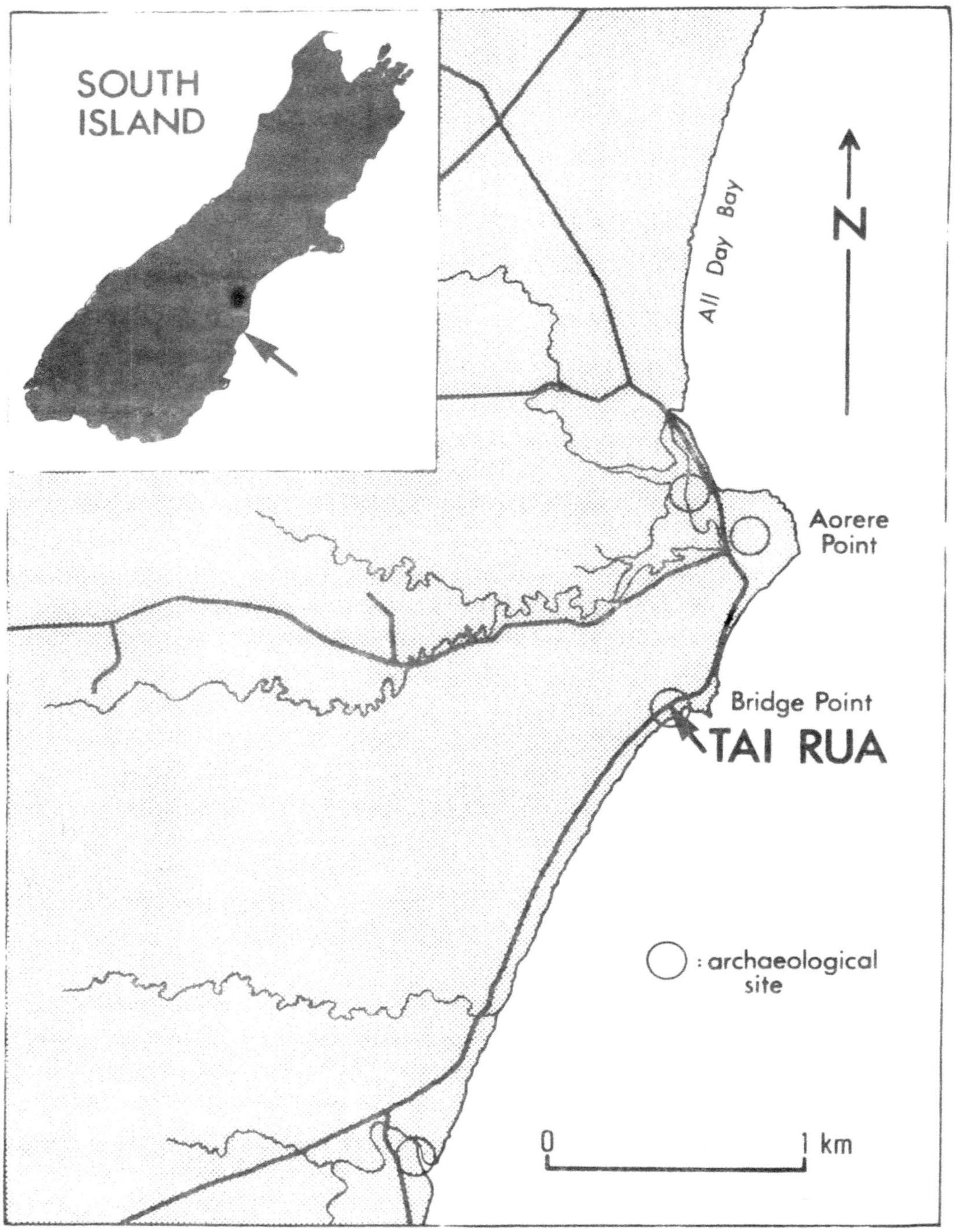

Fig. 13.1 Location map of the Tai Rua Moa-hunter site, South Island, New Zealand.

This, however, is outweighed by its common usage since the beginning of the investigations and in several publications (Trotter 1959, 1965a, 1965b, 1966, 1970; Otago Anthropological Society 1960; Gathercole 1961; Hjarnø 1967, etc.). Tai Rua is recorded as site number S136/1 with the New Zealand Archaeological Association.

The site is situated at the northern end of a narrow strip of coastal flat land immediately behind the present beach north of the Waianakarua River in North Otago (South Island, New Zealand), Fig. 13.1. This flat is about 150 metres wide and ita loamy top-soil thins out towards the beach giving way to sand dunes above high tide mark. High ground to the north provides some slight shelter from prevailing north-easterly winds but the site is exposed to the main rain bearing winds from the south. While bush or forest on the flat at the time of occupation could have given some shelter, the virtual absence of small forest-bird remains in the site suggests that bush, if present, was sparse. The predominant vegetation at present is pasture grass, and all suitable land in the area is used for mixed farming. In very wet weather a normally dry water course on the coastal flat carries surface water into a swamp which lies between the site and the hill that forms the northern boundary of the flat. This swamp is at present retained by the foundations of a road which parallels the coast (and runs across the site), but originally the water probably drained directly into the sea.

Methods and History of Excavations

With the help of a team of members of the North Otago Scientific and Historical Society (occasionally augmented by up to ten pupils from Waitaki Boys' High School), salvage work at the site was commenced by sieving the cultivated topsoil and recording the location of objects found according to a grid of ten foot (approximately three metre) squares.

When it became evident that there was a considerable amount of undisturbed occupational deposit beneath the cultivated level, a different method was adopted. A grid of five-foot squares was established, based on a brass peg set in concrete and with coordinate lines running parallel to and perpendicular to the surveyed road boundary. Excavation was done layer by layer with trowels (Fig. 13.2). (Because those assisting with the excavation varied widely in their experience and skill, many sieved the material after trowelling.) All artefacts and bones were retained, and samples were taken of shells, burnt stones, charcoal and soil. A record was kept of the square and layer in which each item was found, and the exact location and associated material of important items were noted. At this stage work was restricted to week-ends, not more than once a month, and it was deliberate policy to leave a major portion of the site untouched.

As the excavations progressed it became apparent that some prehistoric activities had been confined to certain parts of the site, and in order to sample the complete range of evidence, permission was obtained from the Waitaki County Council to extend the investigation to both sides of the road in 1958 and 1959.

In January 1960 I relinquished my direction of the excavations to Peter Gathercole, then of the University of Otago and the Otago Museum, who continued

Fig. 13.2 Excavations in progress, Area D, Tai Rua, looking south (1962)

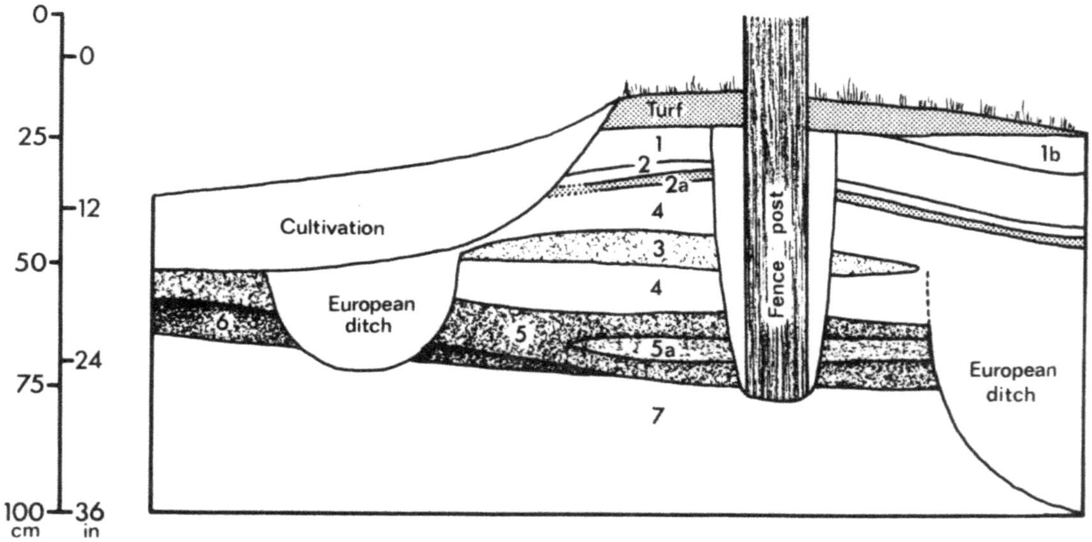

Fig. 13.3 Stratigraphy at fence line—see text for explanation.
The main occupational deposit comprises layers 5, 5a and 6.

the work in close consultation with me as an extension of my investigations. This was done largely to provide training for students and others, and excavations were made on four occasions during the next two years (January and April 1960, April 1961 and January 1962).

In February 1968 Gathercole briefly reopened an area on the seaward side of the road where he had previously found a puzzling wooden structure, while the North Otago Scientific and Historical Society collected some pollen samples for me.

Finally, in 1973, I inserted a dosimeter probe in the site to record environmental radiation for a thermoluminescent dating experiment.

Stratigraphy

Although the stratigraphy varied considerably over the site, there was basically one main occupational deposit which was divided into two, three or even four layers in some places. On either side of the road some re-deposited occupational material, apparently derived from the main deposit as a result of European disturbance, occurred at a higher level. Near the surface were several layers that were either of European origin or connected with European settlement, and ditches, post-holes, rabbit burrows and cultivation disturbances intruded into the occupational layers in some places.

Strata revealed by excavation near the road boundary fence-line in 1958 were as follows (see Figure 13.3):

Surface:	turf	Present day
Layer 1:	sterile sandy humus	Historic
Layer 1b:	scattered road gravel	
Layer 2:	clean wind-deposited sand	
Layer 2a:	old turf line	
Layer 3:	re-deposited occupational material	
Layer 4:	sterile stained sand	Prehistoric
Layer 5:	midden etc. in sandy matrix	
Layer 5a:	lens of burnt midden	
Surface at time of occupation		
Layer 6:	black stained sand	
Layer 7:	sterile beach sand	Pre-human

It would appear that since the time of occupation, wind-deposited sand gradually built up over the occupational material, probably supporting some light growth which would account for its organic stain (layer 4). On either side of the fence, and running parallel to it, were found two European-made ditches, dug through the occupational layers into the sterile beach sand. Their purpose is unknown; drainage would not be necessary in such dry sandy soil. Their proximity to, and alignment with, the road strongly suggests some connection with it. Spoil from the ditches formed layer 3, a fairly homogeneous deposit of stained sand containing occupational material. The spoil had apparently been placed in a long mound between the ditches (which were five to six feet apart). Some of it had soon moved back (probably by wind) into the easternmost ditch, which was larger and more roughly dug than the other. Stained sand, without any occupational material, also filled the smaller (westernmost) ditch, and later some of the layer 3 material shifted back over it, leaving only a very low mound of spoil between the ditches. These soil movements and infilling of the ditches are probably the effects of strong winds, first from the west and then from the east, shortly after the ditches were dug. There is no evidence that they were dug at the same time, but it is likely that they were approximately contemporaneous; that they are of European origin is deduced from their alignment and from pieces of iron wire in one of them. Their purpose could doubtless have been determined by further excavation, but only by removing the road boundary fence, which was not warranted. Local residents suggested that the ditches may have been remains of European defensive earthworks made at the time of the first World War. Turf (layer 2a) then grew on the levelled-out spoil, and was later covered with a thin deposit of clean sand (2), probably wind-blown from the beach. Above this were deposited layers of sandy humus (layer 1) and road rubble (1b), on which grew the present-day turf.

Over most of the site the stratigraphy was much less complex than this. Further into the paddock, north-west of the fence-line, was a strip where the occupational deposit had been completely disturbed by cultivation. In places there was only one occupational layer, or where it was divided, the layers did not necessarily correspond to any of the individual layers 5, 5a and 6. The composition of the occupational deposit, too, varied considerably over the site, although in general it could be described as scattered burnt stones, artifacts, shells, bones, and charcoal, in a dark stained sandy matrix.

Useful horizon markers were the clean wind-blown sand of layer 2, and small orange coloured quartz beach pebbles in the matrix of layer 5. The pebbles were assumed by me to be a natural beach deposit, although the Gathercole party considered them to be moa gizzard stones. They appeared to have lain on or very near the ground surface at the time of occupation, and were often associated with beach shells and definite beach stones.

Occasional seal bones, apparently of natural origin, were found in the sterile beach sand underlying the occupation deposit. On the inland margin of the site this sand had an increased mud content, or pinched out altogether so that the occupational material lay upon an old developed soil.

Layout of Site

It can be seen from Figure 13.4 that the excavated squares were mostly concentrated in three main groups, divided by the present roadway and by a strip of ground in which the occupational deposit was sufficiently near the surface to be destroyed by cultivation. Initially, when the excavations were of a purely salvage nature, the area concentrated on was that part of the central group on the inland (north-west) side of the road fence, Area C, as it was here that the most occupational material had been exposed by cultivation. A series of isolated five-foot squares was later dug to gauge the extent and basic characteristics of the site; further intermediate and often adjoining squares were then excavated in those areas shown to warrant fuller investigation.

For convenience the site can be divided into five areas (labelled A to E on Figure 13.4) and general characteristics of the occupational deposits in these

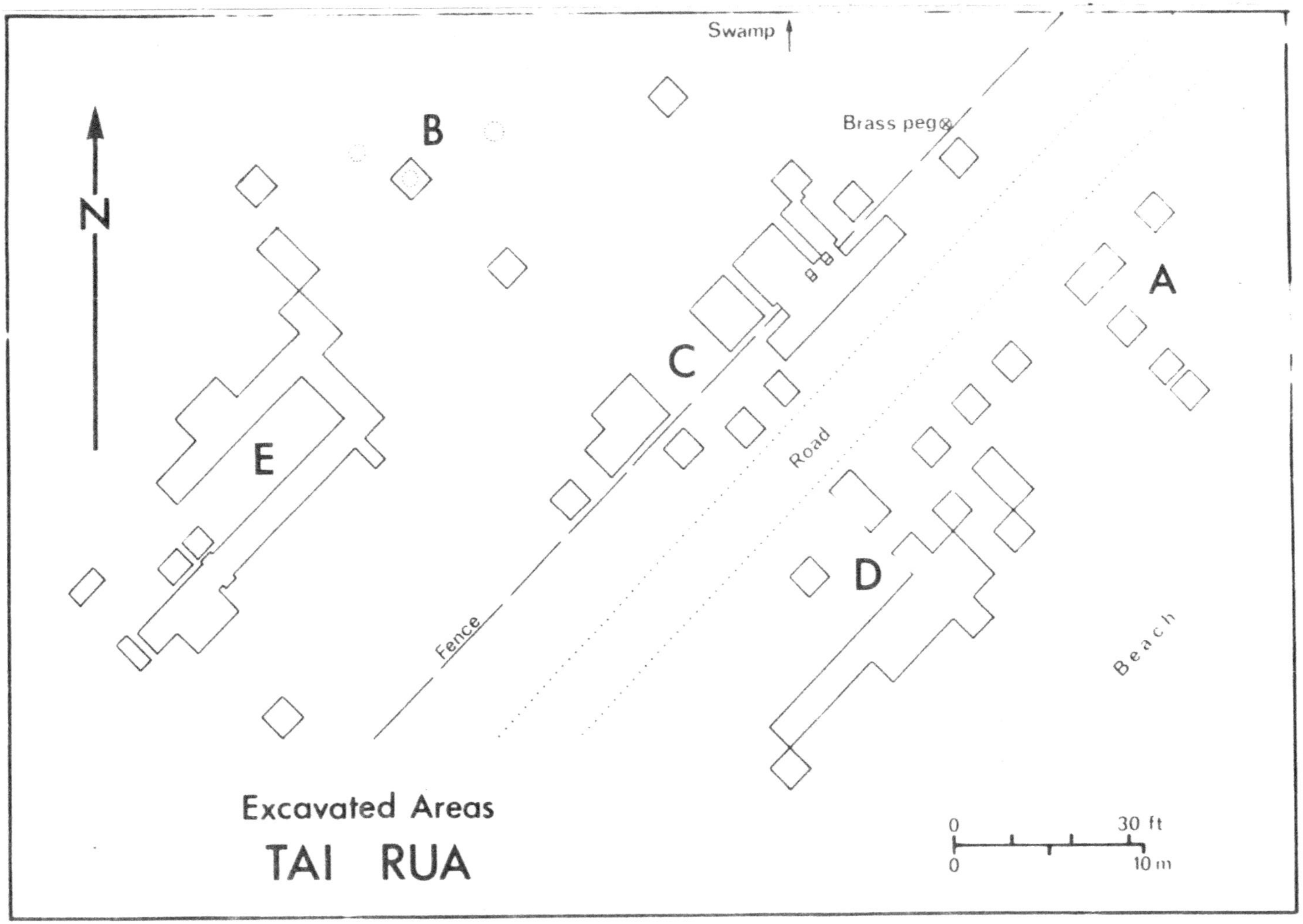

Fig. 13.4 Plan of Tai Rua, showing areas excavated by Michael Trotter and Peter Gathercole, 1958-1962.

areas can be summarised thus:

A. A buried soil blackened by charcoal and decayed organic material; faunal remains and artefacts sparse.

B. Very sparse occupational material except in three fireplace (cooking) hollows filled with charcoal and burnt stones. The ground surface at the time of occupation was higher here than in adjacent areas C and E, and as a consequence less overlying material had since been deposited over it.

C. Several layers of material from a single period of prehistoric occupation, partly disturbed in places by European activities. The description of the stratigraphy given previously specifically applies to this area. Shell midden was most prominent here.

D. Most notable for its post holes and small fireplace hollows projecting downwards from the occupational layers into the underlying clean sand. Features of this nature were very rare elsewhere in the site.

E. Concentrations of moa bones, with some tracheal rings, gizzard stones, and eggshell. The vertebrae of two complete necks and of some sections of necks were in articulation. Some shell midden.

This distribution of features and faunal remains immediately suggests that basically different activities were carried out in different parts of the site. For example, even though it was not possible to distinguish any discrete patterns in the fireplace and post hole distribution in Area D, it is highly probable that many, if not all, represented the remains of shelters or sleeping quarters. Area C was largely a midden or dumping area for food remains; the principal activity in B appears to have been cooking, while the butchering of moa carcasses was carried out in Area E. To the north-west, Area A was the outskirts of the living area.

This hypothetical pattern is not obviously supported by the distribution of artefacts. One might, for example, have expected to find a relatively high proportion of flakes and blades in the butchering area E, but in fact the most numerous artefacts there were fish-hooks. Similarly, manufacturing could have been carried out in the habitation area D, and certainly most of the drill points were found here, yet only four drilled tabs of moa bone were obtained from here as compared with 22 elsewhere. (This can, of course, be easily explained if the drill points were mainly for use on wood—which has not survived—yet they are generally associated with fish-hook manufacture in southern Moa-hunter sites.) Area D had as many adze portions as the other areas combined (no whole adzes were found at Tai Rua), but grindstones and files were mostly in Area E.

Notwithstanding the artefact distribution, however, I believe that the pattern outlined above is basically correct, and that the apparently anomalous association of certain classes of artefact with particular activities may be due to more than one activity being carried out in the one area or to some as yet unrecognized connection.

Nowhere in the site was found any indication of a large cooking pit such as is often present (in my experience, e.g. Trotter 1975a) in sites of this nature for cooking moas and seals. Such a pit could, of course, be still present in the fairly large part of the site that remains unexcavated.

Artefacts

This report is not the place to describe in great detail all of the portable artefacts found at Tai Rua. Rather, I propose to limit the analysis and description to a minimum required to be able to say something about the culture of the inhabitants, and to enable broad comparisons to be made with assemblages from other sites. Any intensive or comprehensive studies should in any case, I feel, be made on the artefacts themselves, not merely on written descriptions or on drawings.

For some years now, portable artefact analysis has been widely relegated to a rather inferior position in New Zealand archaeology. There have been several reasons for this. Firstly, there has been, I think, a natural reaction against the amassing, by both public and private collectors, of particular classes of artefacts for their intrinsic rather than their informative value. In the course of obtaining these collections, vast amounts of archaeological data have been destroyed, albeit often in ignorance. Allied with this reaction has been the determination to make full use of other aspects of archaeological information, hence emphasis has been placed upon such things as midden analysis, post-holes and pits. As well, however, a fairly important reason for the relative absence of artefact studies seems to be that many of the sites that have been investigated in recent years have produced a dearth of portable artefacts.

At Tai Rua the character of the site as an archaeological unit was determined largely by the artefacts. Post-hole patterns have not been amenable to interpretation (possibly because we were looking for a regularity that was not there); there were no "pits" or prehistoric earthworks, nor any burials.

For the purposes of this discussion, a portable artefact can be defined as a piece of some natural material (usually stone, bone or shell) that has been shaped by man to produce an object of some particular use or that has been modified in the process of making such an object. Hence tools, implements and fishing gear are artefacts. So too are ornaments or pieces of waste material discarded during a manufacturing process. But bones broken in the course of preparing or obtaining food, or stones broken by heating or cooking fires, are not included here. This definition leaves at least two classes of objects in a limbo. Many pieces of broken moa leg bones are of a suitable size and shape for use as fish-hook "tabs" or "blanks" but they may have been produced accidentally when breaking open the bones to extract marrow. As well, pieces of red ochre may have been deliberately prepared by heating for use as pigment, or they may be just pieces of natural material in which the iron content has been naturally or artificially oxidised to produce a red colour. Neither of these classes are included in the artefact totals.

Careful analysis of the distribution of different classes of artefact throughout the site has not provided any significant information on the division of the encampment into separate activity areas—apart from that already mentioned above. There was, however, a general correlation between the density of artefact distribution and faunal remains, particularly bones.

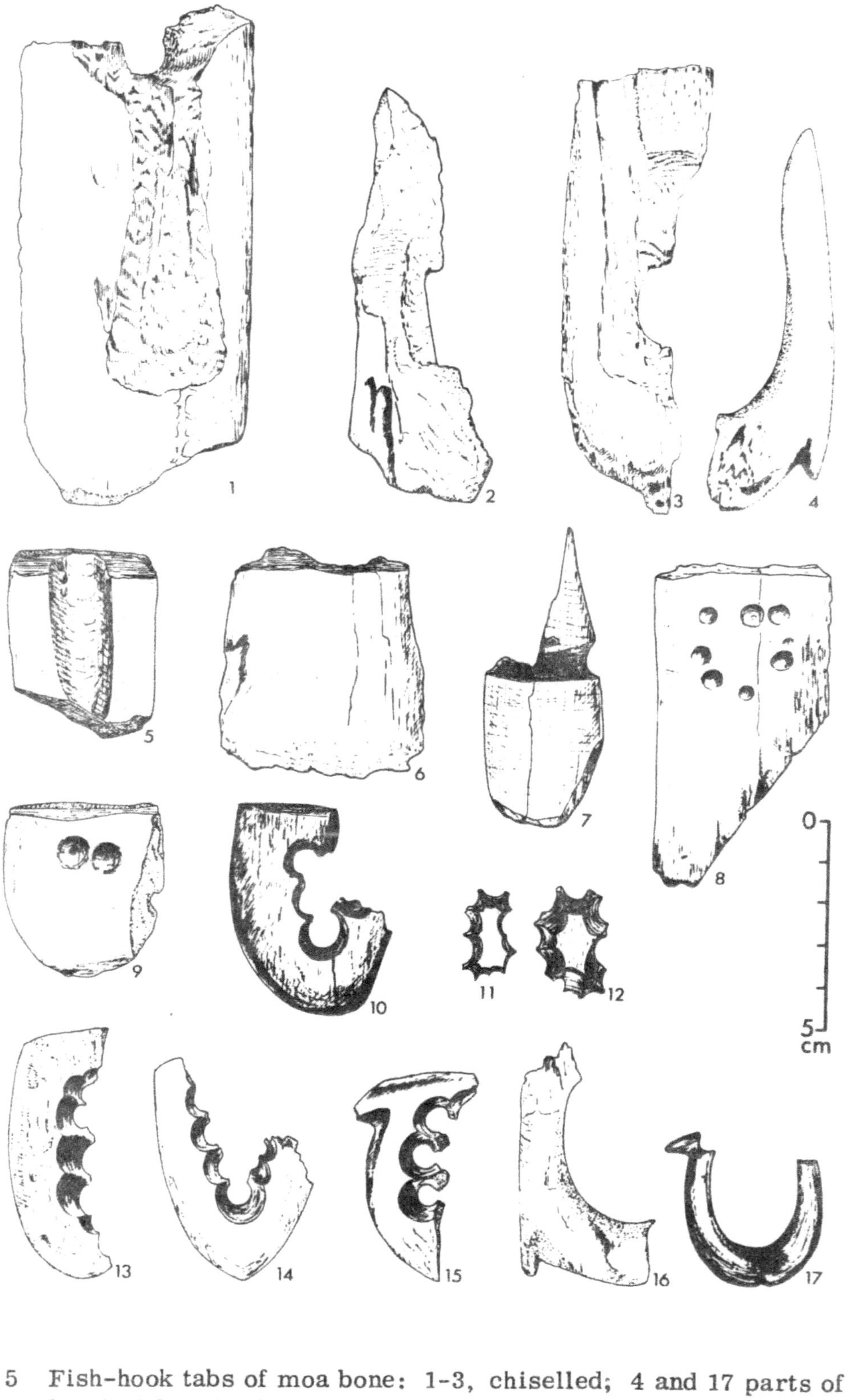

Fig. 13.5 Fish-hook tabs of moa bone: 1-3, chiselled; 4 and 17 parts of finished fish-hooks; 5, tab shaped by attrition cutting, breaking and grinding; 6, cut and chiselled; 7, cut to shape; 8, 9, partly drilled; 10, 13-15, shaped and drilled; 11, 12, drilled out centres; 16, part of a nearly finished fish-hook. Only 1, 4 and 5 are unbroken.

For convenience the Tai Rua artefacts may be grouped in seven classes according to the material they are made from and their apparent use—see Table 13.1. In order of numerical importance these groups are:

1. flake tools, cores and waste flakes of siliceous stone
2. unfinished and completed fish-hooks and other artefacts made from bones and teeth
3. cutting, chopping and hammering tools, plus sinkers of greywacke and similar material
4. ornaments (necklace units) of shell
5. flaked pieces of argillite, andesite and greenstone produced in the manufacture or reshaping of adzes
6. files and grindstones of abrasive material—sandstone and schist
7. pieces of clay that have been moulded to shape and baked.

I propose to discuss firstly groups 2 and 4 which are artefacts of biological materials—bones, shells and teeth—and then the remainder approximately in the order given above.

Bone and Tooth Artefacts

There were 263 artefacts made of bone or ivory and four from dog teeth. Most of the bone was moa bone which appears to have been obtained from the same species as were represented in the midden remains. Of this total number, 250 were either definitely or probably parts of fish-hooks or "tabs" of bone that were in the process of being made into fish-hooks.

Most of the latter were merely flat tabs of bone that had been reduced to approximate size by sawing or chiselling from either side and then breaking across the weakened line; some had also been ground (e.g. Figure 13.5:5-7). Further shaping had been carried out by chiselling or drilling on almost a third of them (Figure 13.5:1-3, 8-10, 13-15). Numbers of tabs were as follows (with one figured example for each group in parenthesis):

Unshaped tabs (Figure 13.5:6)	106	67.0%
Partly chiselled to shape (Figure 13.5:1)	26	16.5%
Drilled tabs (Figure 13.5:9)	23	14.5%
Drilled centres (Figure 13.5:12)	3	1.9%

Presumably the sawing was done with sharp edged flakes of siliceous rocks, and the drilling to remove the centre (of one-piece hooks) was done with the aid of the flaked, pointed artefacts of chalcedony and orthoquartzite that are generally considered to be drill points (e.g. Figure 13.9). The chiselling of moa bone in the manufacture of fish-hooks was probably done with these tools as well; no other chisels were found on the site, and some of the marks on the bone indicate a blade width of about four millimetres which corresponds with the average "drill point" width. A microscope search for wear marks (cf. Semenov 1964) was inconclusive, due, at least partly, to the degree of polish on the worn surfaces. Eighty-two abrasive tools of sandstone and schist from the site were suitable for use in the final shaping of fish-hooks from the drilled or chiselled tabs (e.g. Figure 13.9).

Table 13.1 Portable Artefacts

	Number	% No.	Weight (in g)
Siliceous rock			
Chalcedony flakes	397	24.45	6160
Orthoquartzite flakes	259	15.95	4484
Silicified tuffaceous material	22	1.35	318
Porcellanite flakes	53	3.26	968
Jasperoid flakes	21	1.29	291
Vitreous porcellanite	5	.30	11
Obsidian flakes	14	.88	35
Drill points	43	2.65	510
		50.13	
Greywacke etc.			
"Teshoa"	104	6.40	12660
"Choppers"	28	1.72	7432
"Hammers"	25	1.54	7144
Sinkers	11	.68	4448
		10.34	
Adze materials			
Argillite etc.	98	6.03	2087
Greenstone	3	.18	12
		6.21	
Abrasive stone			
Sandstone	49	3.02	2456
Schist	33	2.03	1721
		5.05	
Bone and tooth			
Fish-hooks	112	6.90	390
Tabs	196	12.07	2868
Others	8	.49	63
		19.46	
Shell			
Fossil	63	3.88	66
Recent	72	4.43	18
		8.31	
Clay			
Moulded to shape	8	.49	103
Total	1624		

As might be expected from a Moa-hunter site, the majority of fish-hooks were of the one-piece type, but they were of several varieties. Most were small and U-shaped with inturned points and were without barbs, the classic Moa-hunter type that was classified by Skinner (1942) as 1A and by Hjarnø (1967) as D (e.g. Figure 13.5:17, 13.6:1). However, at least one of these was more circular (Figure 13.6:10), two fragments were much thicker than usual (e.g. Figure 13.6:7), and one was heavily ornamented with notches (Figure 13.6:22). This latter was the only ornamented specimen from Tai Rua, and while it was recorded as coming from layer 2 during Gathercole's excavation in January 1960, there is good reason to believe that it must have been derived from layer 3 (the redeposited material). Nevertheless, it cannot definitely be associated with the rest of the assemblage from the site.

Three slightly larger one-piece hooks had external basal barbs (Figure 13.5:4, 16; Figure 13.6:23). Fourteen fragments appeared to be from a still larger variety of one-piece hooks that may have been up to 15 centimetres long (Figure 13.6:13, 14, 16-21). Two barbed point fragments may have been of this variety although I earlier considered them to be from two-piece hooks (Trotter 1965a:350-51, Fig. 9-10).

Next most common were points for trolling hooks, the kind that are generally called barracouta hooks, which comprise Skinner's type 5 and Hjarnø's type A.1. In use these would have been fixed to heavy straight wooden shanks (Figure 13.7:12-15). One of the points was made from dog jaw bone, up to five were whale ivory and the rest were made of moa bone. Ten unbroken points were between 5.5 and 6.5 centimetres long.

Three varieties of points for two-piece fish-hooks were present—long straight unbarbed points, small curved unbarbed points, and large points having an external barb. These varieties are illustrated in Figure 13.7:1-4, 5-8 and 9-11 respectively and comprise Hjarnø's types C.1 and C.3. Four of the small curved points were made from dog canine teeth. A single shank of a two-piece hook may be the re-used broken shank of a one-piece hook (Figure 13.6:15).

Fish-hook numbers were as follows:

One-piece,	small (4.0-5.5 cm)	46	40.35%
	intermediate (6-10 cm)	3	2.63%
	large (10-15 cm)	14	12.28%
Two-piece,	straight points	7	6.14%
	curved points	6	5.26%
	barbed points	3	2.63%
	indeterminate points	2	1.75%
	shank	1	0.88%
Trolling hook points		32	28.07%

The great majority of fish-hooks and tabs had probably been discarded because of breakage; less than 13% of them were unbroken. One would expect bone fish-hooks to be highly prone to breakage while in use but in fact there were almost as many point portions as there were shank portions of one-piece hooks, which indicates that they were probably broken on the site (possibly

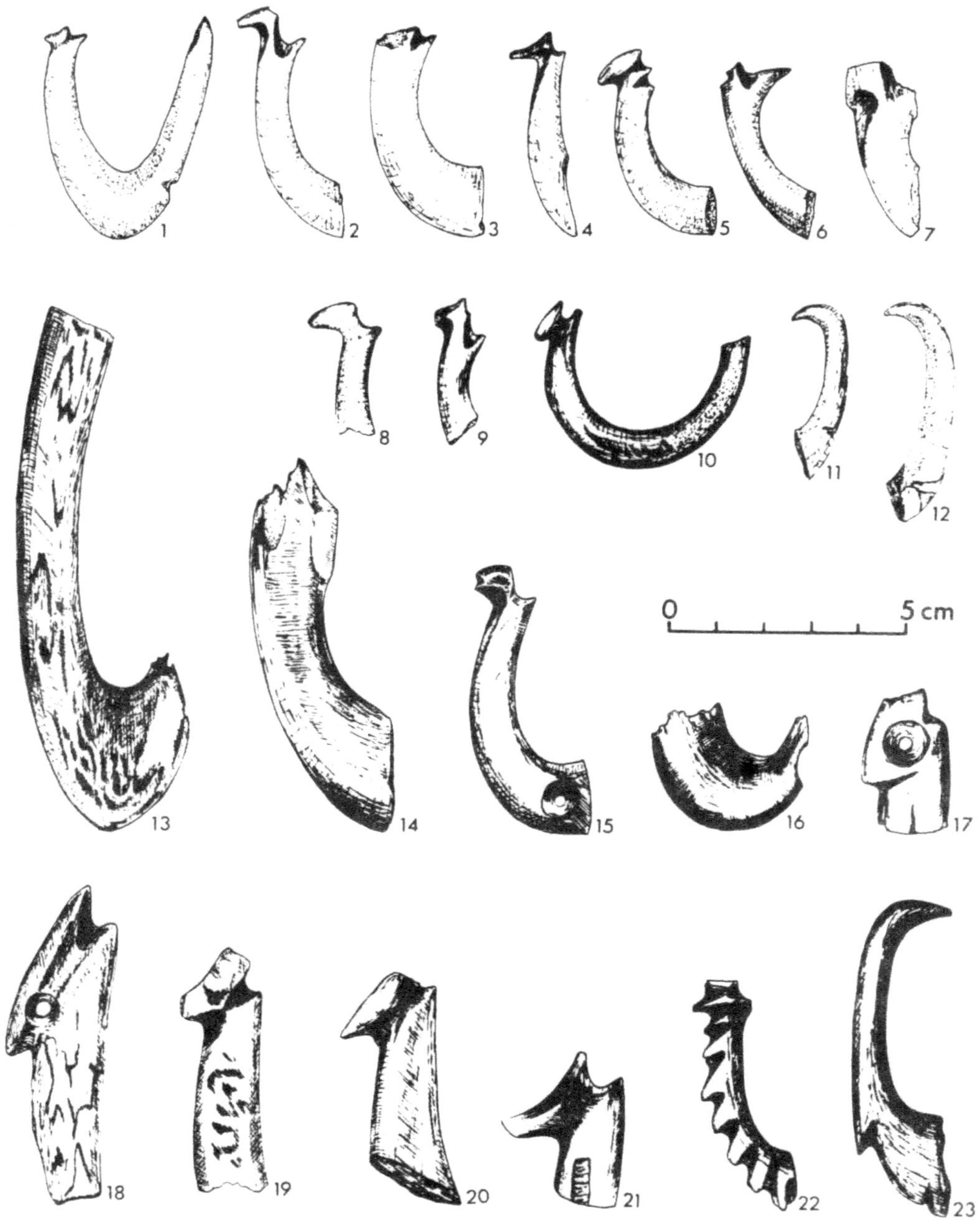

Fig. 13.6 Fish-hooks of moa bone: 1-12, shanks and points of small one-piece hooks; 13, 14, 16-21, parts of large one-piece hooks; 15, shank of two-piece hook; 22, ornamented shank of one-piece hook; 23, point leg of medium sized one-piece hook.

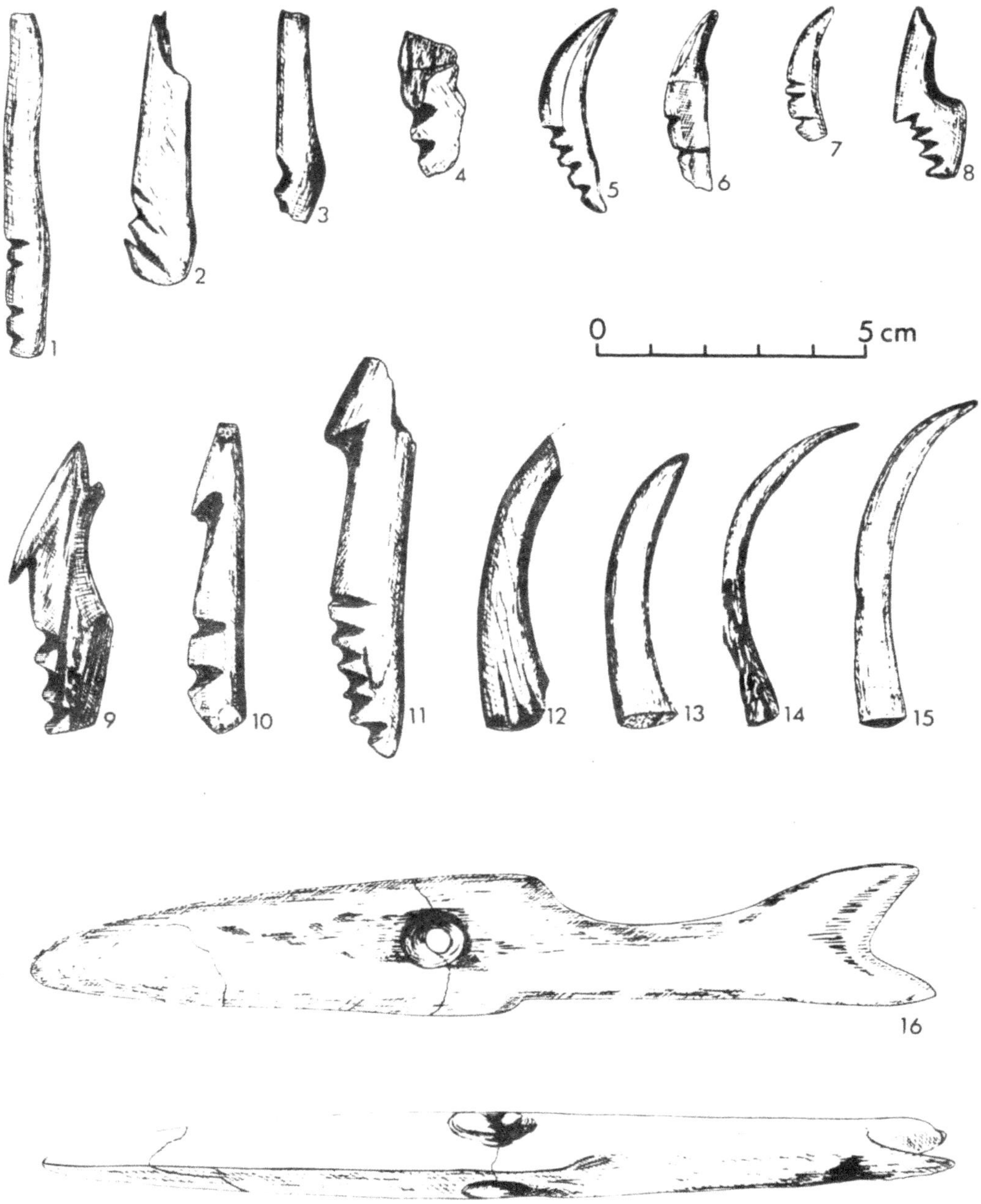

Fig. 13.7 1-4, basal ends of long straight unbarbed points for two-piece fish-hooks; 5-8, small curved points for two-piece hooks; 9-11, externally barbed points for two-piece hooks; 12-15, points for trolling hooks; 16, side and top views of harpoon point. Numbers 5-7 are dog-teeth, 16 is whale bone, the rest are moa bone.

while extracting the hooks from caught fish). Amongst the trolling hooks, seven had the tip of the point missing and three were basal portions, but twelve had the base missing (the remainder were unbroken). Again this suggests a high percentage of breakage on the site. Only three, one complete and two with the base missing, were heavily scored with fish tooth marks, indicating use.

Thirty-six per cent of the fish-hooks and a lesser proportion of the tabs were heat-stained dark grey or black, whereas only half that amount of non-artefactual moa bone, for example, had been burnt. The proportion of burnt dog bone was similar to that of the moa but hardly any of the seal, fish or small bird bones showed any sign of having been burnt. There is thus a strong, but not irrefutable, suggestion that many bone artefacts were deliberately put into fires.

A seventeen centimetre long harpoon point was found in three pieces (each burnt) in a five-foot square in Area E (Figure 13.7). This particular square contained a surprising number of artefacts, including nine hook tabs, a broken hook, drill points, and flakes of chalcedony, orthoquartzite and greywacke, as well as the harpoon point. Moa bones were also very numerous here—both body and leg bones of Euryapteryx and Pachyornis—with some shells, fish-bones and fragments of charcoal. The harpoon point is made of whale bone and was presumably designed for catching moas or seals.

Other artefacts made of bone consisted of five pieces of moa bone, eleven to twelve centimetres long, that had each been roughly fashioned to a circular sectioned point at one end, one broken piece of bird bone that may have been a needle, one "awl" of bird bone, and eight other worked fragments that were not sufficiently complete to suggest a likely function. Besides these, a small cut section of bird limb bone and a tattooing chisel were recovered from the site but have since been lost. The chisel was approximately 12 millimetres wide and 25 millimetres long to its broken proximal end; part of a single hole could be seen at one side of the break. One dog canine tooth was drilled near the proximal end, presumably for use as a pendant.

Shell Artefacts

A collection of 63 shells of the small dark top-shell, Zediloma sp., each pierced in the outer whorl, 35 cut section of fossil Dentalium sp., and a pierced flat valve of an oyster shell, Ostrea sinvata, were found together in Area D (see Figure 13.8). There can be little doubt that these had been prepared for a necklace or similar type of personal ornament. Strung together they would measure about 108 centimetres, more or less depending on how they were arranged. Besides these, five pierced Zediloma sp. shells, 28 sections of fossil Dentalium sp. and three sections of the extant Dentalium nanum were found separately, most but not all in the same general area.

The fossil Dentalium (probably Dentalium solidum) could have been obtained from limestone outcrops two or three kilometres north of Tai Rua. Zediloma sp. and Dentalium nanum could have been found on the rocks and beach respectively, adjacent to the site.

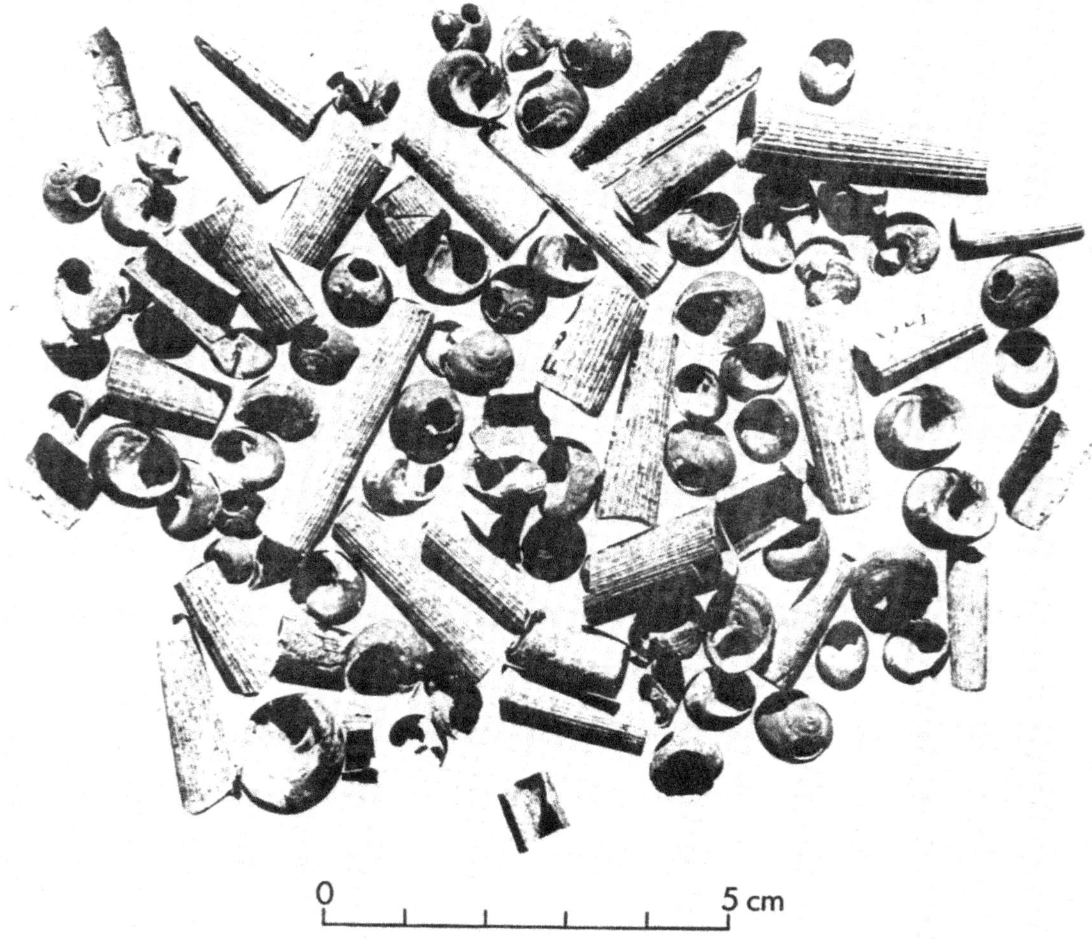

Fig. 13.8 Part of an assemblage of necklace units, found together, of sections of fossil Dentalium and pierced Zediloma shells.

Stone Artefacts

Almost half the total number of artefacts from Tai Rua were flakes, cores and waste pieces of various siliceous materials. No study has been made of them and while some have retouch flaking and some have been chipped by use, it is fairly apparent that there are very few carefully made tools among them. This could be due to lack of skill, unsuitability of available material, or possibly the absence of any need here for the skilfully produced flake tools that occur on some early sites, particularly in Murihiku (i.e. in southern New Zealand). An exception is the class of drill or chisel points mentioned above and which are included here because they are made of the same materials and produced the same way, i.e. by flaking, as the rest in this category.

There does not appear to be any universal agreement amongst petrologists —let alone archaeologists—as to the use of particular names for varieties of siliceous material, but here again the categories given below are largely for convenience. Numbers and weights of flakes, cores and chips according to materials, plus the drill/chisel points, are as follows:

	Number	% Number	Weight (g)
Chalcedony	397	48.77	6160
Orthoquartzite	259	31.82	4484
Silicified tuffaceous material	22	2.70	318
Porcellanite	53	6.51	968
Jasperoid	21	2.58	291
Vitreous porcellanite	5	.61	11
Obsidian	14	1.72	35
Drill/chisel points of orthoquartzite and chalcedony	43	5.28	510

Possible sources of the principal materials are given below.

The greywacke spawls that Haast named teshoa (see Trotter 1971:142-3, Fig. 8h) were reasonably common, but did not occur with such frequency as on the coastal Canterbury sites at Rakaia and Wakanui. Each teshoa has been knocked from a natural water-worn cobble of greywacke by a single blow. The teshoa was used for cutting and chopping soft materials, probably meat or wood, and is quite distinct in use, and hence in wear marks, from the similarly formed greywacke spawl that was used in later times as an attrition saw for cutting stone.

The twenty-five hammers found at Tai Rua are mostly water rounded stones of greywacke or other suitable material and have a median weight of 240 grams. Intermediate between hammers and teshoa, as far as the shape of the working edge is concerned, are what I have called "choppers", although it is perhaps doubtful if the users thought of them as a special class of tool. Fishing sinkers also grade into hammers in that here (as at other sites) a number of stones have been both used as hammers and prepared as sinkers. Figure 13.10 shows three of the Tai Rua sinkers.

Because of their durability, their size and their ease of recognition by the layman, stone adzes (i.e. adze heads) form the largest class of artefacts in most museum collections. More often than not, however, they comprise such a small proportion of archaeological assemblages that they cannot legitimately be related to the general typological studies that have been made of them. Such is the case at Tai Rua. No complete adzes with "diagnostic" features were recovered. There were broken portions of five unfinished adzes, and 93 flakes from the manufacture, reshaping or breakage of adzes. My interpretation of these pieces is that the people of Tai Rua had both adze "blanks" and completed adzes and carried out a very small amount of shaping and reshaping, perhaps more in the nature of maintenance than of manufacture, by flaking and grinding. The materials from which these adzes were made were diverse, and included various grades of indurated mudstones, tuffs and fine-grained arenites, andesite and greenstone. Some pieces showed an adze shape having a quadrangular section, and one piece was a hammer-dressed, rounded quadrangular poll. Two pieces were simply primary flakes with the distal end ground to a cutting edge.

In a previous reference to pieces of moulded and baked clay from Tai Rua, I suggested that they may have come from bird carcasses that had been covered in damp clay before cooking (Trotter 1965b:166). A similar method of cooking rats by Canterbury Maoris in early European times has been recorded (Hay 1915:14-15), but rat remains were very few at Tai Rua. However, one of the pieces in the present assemblage is in the shape of a flattened ball of "clay" (actually a silty mud) that has been baked (Figure 13.10:1) and most of the rest could easily be fragments of similar balls. Part of a similar ball was recently recovered from a Moa-hunter site in Kaikoura, and shapeless fragments of baked clay from other sites are not uncommon. Marks on the nearly complete ball and on other pieces appear to be impressions of grass-like leaves made while the clay was still damp.

Probable determinations of the main Tai Rua stone materials, which have been made by hand specimen comparisons only, together with closest source locations, are as follows:

Chalcedony: Moeraki-Katiki coast area, 20 kilometres south.
Orthoquartzite: Central Otago, 50 or more kilometres south west.
Porcellanite (baked mudstone): Moeraki-Katiki coast.
Obsidian: Seven flakes with green translucency and six with grey doubtless come from North Island offshore and mainland sources, while one black flake is possibly a Canterbury material.
Greywacke: local beach stones.
Argillite (various fine-grained indurated mudstones): at least some from the Nelson-Marlborough area.
Andesite: probably Otago.
Greenstone (nephrite): probably west coast of the South Island, about 225 kilometres north of Tai Rua.
Sandstone (fine-grained with abrasive quality): some appears to be from Shag Point on the coast 25 kilometres south.
Abrasive schist: inland Otago.
Fossil <u>Dentalium</u>: from limestone deposits, the nearest of which would be two or three kilometres north.

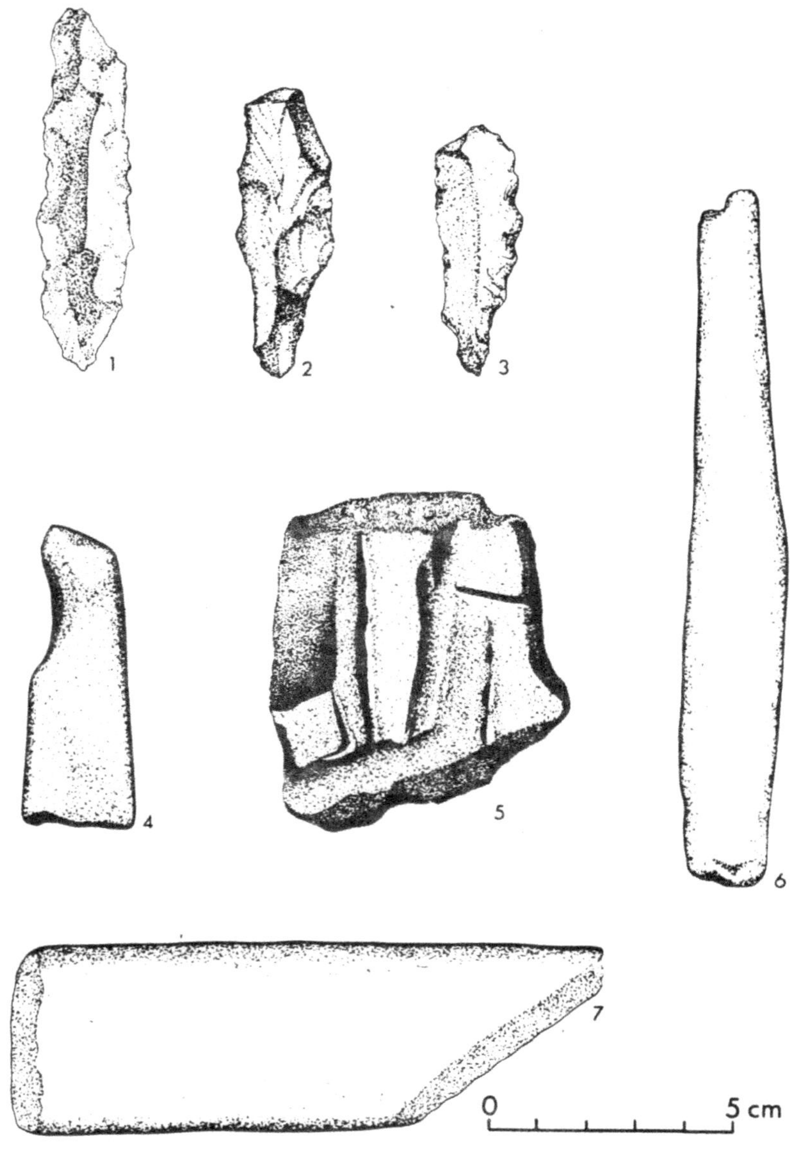

Fig. 13.9 1-3, drill points of orthoquartzite—these or similar artifacts were probably used also as chisels. 4-7, attrition tools. 6 is of schist, 4, 5, 7 are of fine grained sandstone.

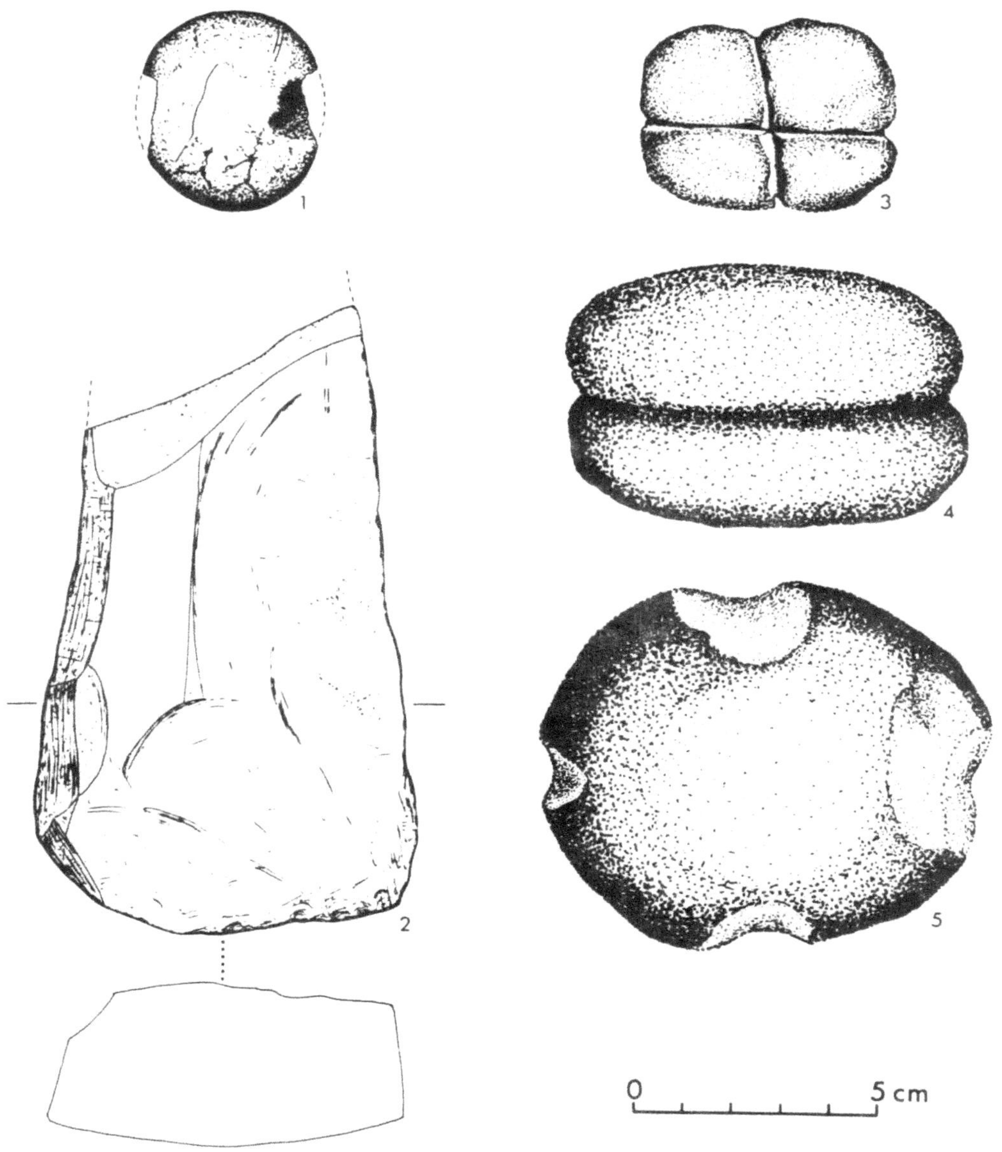

Fig. 13.10 1, flattened (and broken) clay ball; 2, part of an unfinished quadrangular sectioned adze of fine-grained litharenite; 3-5, fishing sinkers.

Faunal Remains

No detailed data on faunal remains are available due to the lack of identification facilities and available time. An extensive shell midden area occurred in Area C and there was at least one small shell midden in Area E. Scattered shell middens also occurred in Area D but were virtually absent in A and B. The species composition varied throughout the site but the average amounts over 350 square feet of Area C were as listed below (Table 13.2). Percentages are of individual whole or broken bones and shells.

Table 13.2

Catseye shells (*Lunella smaragda*)	56%
Fish bones	29%
Moa bone pieces (mostly *Euryapteryx gravis*)	each < 5%
Paua shells (*Haliotis* spp.)	
Dog bones (*Canis familiaris*)	
Limpet shells (*Cellana* spp.)	
Mussel valves (*Perna canaliculus*)	each < 1%
Oyster valves (*Ostrea* sp.)	
Turban shells (*Cookia sulcata*)	
Bird bones (listed separately)	
Cockle valves (*Chione stutchburyi*)	
Mactra valves (*Resania lanceolata*)	each < 0.2%
Seal bones	
Mussel valves (*Mytilus edulis*	
Tuatua valves (*Paphies subtriangulata*)	each < 0.1%
Shield shells (*Scutus breviculus*)	
Sea-urchin pieces (*Evechinus chloroticus*)	

Besides the molluscs listed here, fresh-water mussel shells, *Hyridella menziesi*, were found elsewhere on the site; all these species are considered to have been used for food. The above list does not indicate the relative importance of food species throughout the site as there was an area almost exclusively of moa bones in part of Area E. Some shells had been carefully placed on the midden heaps—for example seven paua were found together with the backs of some shells in the openings of others, and three *Mactra* valves were found together nearby. Large water-worn stones had been placed in the midden deposit in several places. As mentioned previously some *Dentalium* sp. shells were also present amongst the midden material and had probably been collected for ornamental use as had the *Zediloma* sp. shells found with them in Area D. Beach shells, distinguishable by wear, internal deposits and the borings of carnivorous species, occurred in the sand matrix of the site, mostly in layer 5 in association with scattered beach pebbles.

Moa bones were present in the occupational deposits throughout Tai Rua, but were concentrated in parts of Area E. In places the occupational layer here was composed almost solely of moa remains—both body and leg bones in approximately natural proportions, with some tracheal rings, gizzard stones and eggshell. There were also dog bones and faeces amongst them, but very few moa (or any other) bones had any possible dog tooth marks on them—a feature that has been noted on other sites. No attempt has been made to estimate the number of individual moas represented, but there can be little doubt that they, and shellfish, provided the bulk of the food-stuffs at Tai Rua, followed closely by fish. The place of the latter is perhaps surprising in view of the numbers of completed and unfinished fish-hooks found. The ratio of moa to dog bones was roughly ten to one, and to seal bones about twenty to one.

Sections of moa vertebrae in position of articulation were not uncommon and two almost complete necks with skulls and tracheal rings in position were found in Area E. In this area too was a group of fragments of moa egg-shell which appeared to be from a single egg.

There was considerable variation in the specific gravity of moa bones from different parts of the site; bones from the damp Area E were always heavier and denser, even after drying, than those from the rest of the site where the matrix was generally dry and sandy. This was due, not to "mineralisation" of the heavy bones, but to their high organic content which was up to 23% for bone from Area E where conditions were better for preservation, compared with as little as 8% for bone from a dry sand matrix.

Vertebrate species represented by bones at Tai Rua are listed in Table 13.3. Where given, numbers refer to the numbers of bones identified by Ron Scarlett, who examined all the small bird bones. Half as many again were considered not identifiable. Bones of the moas Euryapteryx and Pachyornis were present at a ratio of about 11 to 4, but total numbers are not available.

The composition of the bird list is interesting in that apart from the moas there is only one bone of a forest-dwelling species (tui). This may reflect the abundance of moas rather than a scarcity of small forest birds—moas would presumably provide a greater return for the energy, time and trouble required in catching them than would small birds in the same habitat. It might further be hypothesised that the small, predominantly coastal, species present were included in the diet to provide taste variety. In short I suggest that the composition of Table 13.3 may represent the choice of the inhabitants of Tai Rua rather than the availability of species.

The habitat of the still extant species indicates that most of the small bird hunting was carried out in coastal or estuarine localities, possibly the Tai Rua headland and adjacent All Day Bay lagoon to the north. The number of bones identified, 99, represents approximately two thirds of the total number found; small birds thus provided only a small proportion of the food eaten.

Very few rat bones were found, nor was there much rat gnawing evident on any midden bones. Dog and seal bones were reasonably plentiful and these animals could have provided a fair proportion of the meat eaten. There were some broken pieces of whale bone in the occupational deposit but there is no suggestion that whale flesh was eaten.

Associated with some of the small fireplace depressions in Area D, and also in Area C, was a material that I have called "consolidated ash" (Trotter 1966). Analyses of this material from Tai Rua and other North Otago sites have shown it to be largely calcium carbonate, presumably derived from burnt mollusc shells. The purity of the "ash" and the degree of burning that must have been required to produce it suggest that it was done deliberately.

Table 13.3 Bird Species

		Bone numbers
Extinct		
Moa	Pachyornis elephantopus	n.a.
Moa	Euryapteryx gravis	n.a.
Duck	Euryanas finschi	1
Hawk	Accipeter (Circus) eylesi	1
Coot	Nesophalaris chathamensis	1
Crow	Phalaeocorax moriorum	2
Living		
Blue penguin	Eudyptula minor	22
Crested penguin	Eudyptes pachyrhynchus	11
Mollymawk	Diomedea cauta	27*
Fluttering shearwater	Puffinus gavia	1
Stewart Is. shag	Leucocarbo carunculatus	8
Spotted shag	Stictocarbo punctatus	13
Paradise duck	Tadorna variegata	2
Grey duck	Anas superciliosa	5
N.Z. scaup	Aythya novaeseelandiae	2
N.Z. quail	Coturnix novaezealandiae	1
Banded rail	Hypotaemidia (Rallus) philippensis	1
Tui	Prosthemadera novaeseelandiae	1
		99

*Seven of these were artefacts

Mammals	
Seal	?Arctocephalus forsteri
Dog	Canis familaris
Rat	Rattus exulans
Fish	
Barracouta	Thyrsites atun
Parrot-fish	Pseudolabrus sp.
(Others not identified)	

Dating

Initially (in the late 1950s) the occupation at Tai Rua was dated at about 500 years before present. This estimation was made on the basis of comparison of artefacts types with dated sites at Wairau Bar and in South Otago, and on the species of extinct birds represented there. This estimate coincided with radiocarbon dates that were obtained some time later.

Carbon isotope analyses were made of nine samples of bone collagen, bone carbonate, shell and charcoal by the Institute of Nuclear Sciences, partly to obtain radiocarbon dates for the occupation, and partly to compare the results from different materials (see Trotter 1968b, Rafter et al. 1972). The results are briefly listed below; ages in radiocarbon years before present have been calculated according to the "old" half life with reference to the appropriate New Zealand standards for bone and shell (see Rafter et al. 1972 for details of these standards) and to the 0.95 NBS oxalic acid standard for charcoal. They have been corrected for "industrial effect" but not for conversion to calendar dates.

NZ 558	Moa bone carbonate	Recent
NZ 559	Moa bone collagen	503 ± 32 BP
NZ 578	Moa bone collagen	503 ± 32 BP
NZ 749	Marine shell (Haliotis)	485 ± 32 BP
NZ 750	Charcoal	831 ± 33 BP
NZ 751	Moa bone carbonate	Recent
NZ 752	Moa bone collagen	543 ± 32 BP
NZ 765	Moa bone carbonate	Recent
NZ 766	Moa bone collagen	393 ± 37 BP

All samples came from the main occupational layer (i.e. number 5) although from different parts of the site. NZ 559 was obtained from a dry sandy matrix and the bone had a low specific gravity, whereas NZ 578 was heavy bone from a dense "greasy" matrix; both gave the same radiocarbon result.

Carbonate samples NZ 558, 751 and 765 show the usual atmospheric contamination due to nuclear bomb tests, while NZ 766 was analysed to determine the result of long storage and handling.

At the time the samples were processed, 1965-66, the difference between the charcoal and the animal remains proved to be greater than expected. Subsequent dating from many South Island sites however, has shown that this is a normal situation, and that charcoal dates can be expected to be several centuries older than those from bone collagen or shell (Trotter 1968b; Rafter et al. 1972; McCulloch & Trotter 1975). Because of this charcoal error, the shell and bone collagen dates must be preferred for reliability.

From these results the date of the prehistoric occupation of Tai Rua can be placed at about 500 radiocarbon years before present. Although northern hemisphere research indicates that this age is not equivalent to 500 calendar years, the absence of specific data for New Zealand makes any conversion unwise.

Moa bone from Tai Rua was also used in experimental relative dating methods. Samples contained 0.19%, 0.20%, 0.20% and 0.24% of fluorine which is much as expected, but the fluorine content of archaeological bones has been found to be too variable under average New Zealand conditions to be useful even for relative dating (Trotter & Malthus 1967). Other chemical analyses made through the courtesy of Professor F. B. Cousins of the Otago Dental School included measurements of total organic content, nitrogen, phosphorus and calcium for bones from Tai Rua and a number of other sites. Experimental work on thermoluminescent dating was also carried out at Tai Rua (and elsewhere) with at least promising results (Driver 1973).

Conclusions

I do not feel that many positive conclusions about the people of Tai Rua can be drawn from the material recovered and analysed so far. Information was obtained on only a very few of the occupants' probable activities or on other aspects of their lives. More information, particularly regarding their diet and their shelter, lies there in the ground and could be obtained by present-day techniques, but hopefully, as mentioned earlier, the techniques of future decades will yield still more. For this reason I have resisted the strong temptation faced while writing this report, to go back and extend the excavations to tidy up a number of loose ends, to look at the distribution of post-holes, the hearths, middens and other structures, and to obtain data on food use and, if possible, the size of the community and the length of their occupation.

From the information presently available we can see that a small group of people lived for a time at Tai Rua some 500 radiocarbon years before present. Various activities such as butchering, cooking, refuse dumping and lighting fires for warmth tended to be carried out in specific parts of the site. Protein food was largely the flesh of moas, shellfish and fish (in that order), with dogs, seals and small birds providing a smaller proportion of their diet. Various tools, particularly cutting implements, were made from rock materials from a variety of locations, principally within 50 kilometres of Tai Rua, but with small amounts from as far away as the West Coast, northern Marlborough and the northern half of the North Island. Most of the stone tools (flakes of siliceous rock or greywacke) seem to have been intended for cutting soft material e.g. flesh; some were designed for drilling or chiselling bone, and a few were adzes for shaping wood or possibly digging in the ground. Other artefacts fashioned from rock included hammers and sinkers. Fish-hooks comprised the most prominent class of artefacts that were made of bone. There is some evidence that the emphasis was on manufacturing rather than using fish-hooks at Tai Rua, and a variety of different designs were made.

Although the artefact assemblage from Tai Rua differs greatly from that of Wairau Bar, which is commonly thought of as a "type site" for the Moa-hunter period of Polynesian culture in New Zealand, this may well have been due to factors other than solely an evolutionary change during the period of 150-200 radiocarbon years between their occupations (see Trotter 1975b). Tai Rua may have been a less permanent settlement, for example, with an economy reflected more by fish-hook production than by the types of activities that were carried out at Wairau.

While there is a greater variety of fish-hooks than is usual from most Moa-hunter sites, the general types of these and other artefacts are not inconsistent with artefact types from other early Polynesian sites throughout the country.

Acknowledgements

I am grateful to John Brown and to the Waitaki County Council for permission to excavate at Tai Rua, and to members of the North Otago Scientific and Historical Society for the work they did there. My thanks go also to Peter Gathercole for help with data from his investigations of 1960-62, to Roger Duff for advice and encouragement, to Athol Rafter for radiocarbon dates, to

Beverley McCulloch for drawing the bone artefacts, and to Ron Scarlett for bird bone identifications.

Bibliography

Driver, P. M. 1973. Thermoluminescent Dating in New Zealand: A Preliminary Study. Physics Honours Project, University of Canterbury. Unpublished ms.

Duff, R. S. 1950. The Moa-hunter Period of Maori Culture. Canterbury Museum Bulletin 1. Government Printer, Wellington.

Gathercole, P. 1961. Excavations at Tai Rua, Otago, 1961. N.Z. Arch. Assn. Newsletter 4:32.33

Hay, James 1915. Reminiscences of Earliest Canterbury. Christchurch Press Company, Christchurch.

Hjarnø, J. 1967. Maori Fish-hooks in Southern New Zealand. Rec. Otago Mus. Anthropology 3.

McCulloch, B. A. and Trotter, M. M. 1975. The First Twenty Years: Radiocarbon Dates for South Island Moa-hunter Sites 1955-74. N.Z. Arch. Assn. Newsletter 18:2-17.

McDonald, J. C. 1962. White Stone Country. North Otago Centennial Committee, Oamaru.

Otago Anthropological Society 1960. Fieldwork in Otago, 1959-1960. N.Z. Arch. Assn. Newsletter 3:13-18.

Rafter, T. A., Jansen, H. S., Lockerbie, L. and Trotter, M. M. 1972. New Zealand Radiocarbon Reference Standards. Proceedings of the Eighth International Radiocarbon Dating Conference H29-80. Royal Society of New Zealand, Wellington.

Semenov, S. A. 1964. Prehistoric Technology. Cory, Adams & Mackay, London.

Skinner, H. D. 1942. A Classification of the Fish-hooks of Murihiku. J. Polynes. Soc. 51:208-221; 256-286.

Stevenson, G. B. 1947. Maori and Pakeha in North Otago. A. H. & A. W. Reed, Wellington.

Trotter, Michael M. 1959. Archaeological Investigations in North Otago. N.Z. Arch. Assn. Newsletter 2:10-13.

Trotter, Michael M. 1965a. The Barbed Fish-hook: Its Place in the Murihiku Cultural Sequence. J. Polynes. Soc. 74:347-355.

Trotter, Michael M. 1965b. Avian Remains from North Otago Archaeological Sites. Notornis 12:176-178.

Trotter, Michael M. 1966. Consolidated Ash from North Otago Archaeological Sites. N.Z. Arch. Assn. Newsletter 9:48-52.

Trotter, Michael M. 1968a. North Otago Archaeological Sites (Part 1). N.Z. Arch. Assn. Newsletter 11:94-102.

Trotter, Michael M. 1968b. On the Reliability of Charcoal for Radiocarbon Dating of New Zealand Archaeological Sites. N.Z. Arch. Assn. Newsletter 11:86-88.

Trotter, Michael M. 1970. North Otago Archaeological Sites (Part 2). N.Z. Arch. Assn. Newsletter 13:135-142.

Trotter, Michael M. 1972. A Moa-hunter Site Near the Mouth of the Rakaia River, South Island. Rec. Cant. Mus. 9:129-150.

Trotter, Michael M. 1975a. Archaeological Investigations at Redcliffs, Canterbury, New Zealand. Rec. Cant. Mus. 9:189-220.

Trotter, Michael M. 1975b. Further Excavations at Wairau Bar, New Zealand. Asian Perspectives 18:75-80.

Trotter, M. M. and Malthus, R. S. 1967. Fluorine Analysis in New Zealand Archaeology. N.Z. Arch. Assn. Newsletter 10:151-157.

THE NEW ZEALAND OCTOPUS LURE: FACT OR FICTION

Helen M. Leach

Just as the members of the order Octopoda occur in all major oceans and seas, a number of tricks and devices for catching them are also world wide in their distribution. Probing their lairs with a pointed stick, or simply by hand, and the dispatching of the prey by turning the body inside out have been recorded from the Mediterranean to the Pacific (Hornell, 1950:130). In Italy and Japan strings of earthenware pots are lowered to encourage small octopuses to make their homes in them, while on Easter Island stone shelters are constructed to improve the ease of finding and capturing them (Métraux, 1940:191). Polynesians caught the octopus by hand or with spears and/or gaffs in almost every island group for which ethnographic records exist. A more complicated device (Fig. 14.1), erroneously referred to as a " cuttlefish bait" (Beasley, 1921) or "squid lure" (Buck, 1957:359), was collected from several Polynesian islands by early European explorers. Collating museum specimens and missionary accounts, Beasley (1921) produced a detailed paper which not only established the octopus lure as a consistent artefact type but drew an important conclusion from its distribution:

> "Ten groups of islands are shown to have more or less knowledge of this highly specialized form of fishing, and with one exception all are either pure Polynesian or contain well-known Polynesian influences... Some consideration of the distribution of these baits may be of help in the study of the early Polynesian migration." (Beasley, 1921:100)[1]

Until Buck compiled Hawaiian accounts of octopus fishing, no writer had offered any explanation of why this more complicated device supplemented the use of the spear. The answer was quite straightforward: spearing was only practicable in shallow water (1-2 fathoms), while the lure could extend octopus fishing to a depth of 120 fathoms (Buck, 1957:359).

Ethnographic descriptions of octopus lure fishing in West Polynesia frequently refer to the lure as representing a rat, and relate a fable to account for the octopus's fury at seeing it. The story related on Ngau (Fiji Is.), Tonga, Niue, and Lifu (Loyalty Is.) involves the rat obtaining a passage across the ocean on the octopus's back but failing to show his gratitude. In one version the rat mocks the octopus's bald head, in another it is seasick over it, and in another it commits the ultimate insult, defaecating on the head of the octopus, and once safely ashore telling the octopus to put a hand on its head (Beasley, 1921:101-2). The Samoan version is by contrast quite simple, referring to a fight between the two creatures, in which the rat was victorious.

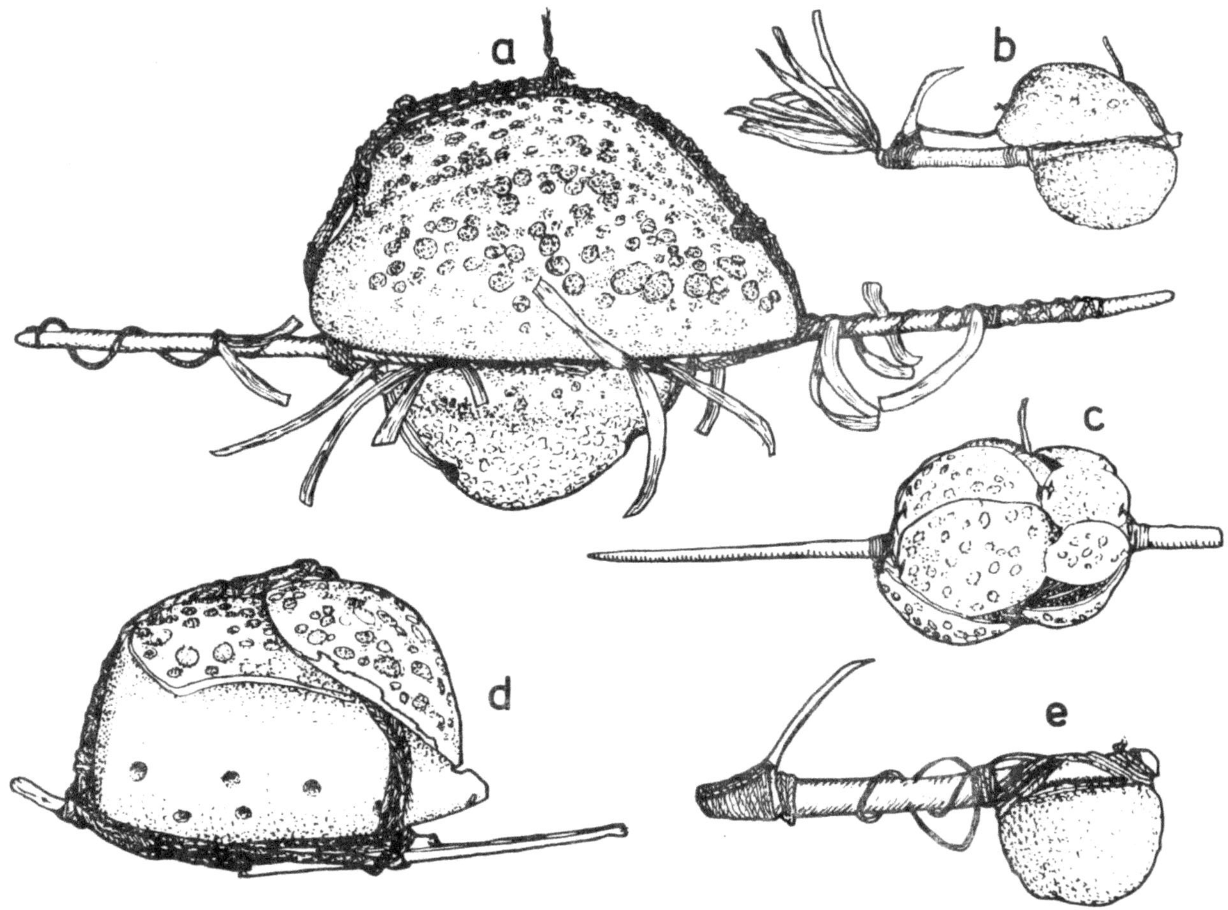

Fig. 14.1 Ethnographic Specimens of Octopus Lures. a. Niue. Otago Museum D24.295. Length 25 cm. b. Hawaii. After Beasley, 1921. c. Tahiti. Otago Museum D31.126. Length 33 cm. d. Tonga. Otago Museum D24.721. Length 13 cm. e. Hawaii. Kilo type. After Beasley, 1921.

From a behavioural point of view it is inconceivable that an octopus should recognize a rat, since their habitats are necessarily quite separate. What then is the attraction of the lure? An answer to this question can only follow the isolation of the key elements of the Polynesian octopus lure. Cowry shells, either whole or in sections, or as combinations of whole shells and pieces, occur on octopus lures throughout Polynesia.

The spotted or tiger cowry (Cypraea tigris) seems to have been preferred (Suggs, 1961, 90-91). Cowry shell was not essential, however, to the success of the lure for one of two Hawaiian methods, known as kilo, utilised the other usual components (shank, stone sinker and hook) without any shell. Buck noted that the fisherman had to be able to see (kilo) the octopus at a depth of up to 10 fathoms in order to jerk the line when the octopus drew the stone beneath its mantle to explore it with its mouth (Buck, 1957:357-8). Even a sinker can be omitted from the lure rig if the weight of the shells is sufficient, as in Niue (Beasley, 1921:105-6) or Tahiti (ibid.:108-9). Alternatively, the cowry shell could be filled with pebbles as suggested by an archaeological specimen from the Marquesas (Suggs, 1961:90). The shank occurs on all known museum specimens but varies from a flexible or stout rootlet to a straight piece of rounded and polished wood, or a long bird wing bone wrapped in braided sennit (Beasley, 1921). The hook was positively recorded in Hawaii only (see below). Strips of coconut or banana leaves attached beneath the cowry shell, along the 'tail' portion of the shank and/or on the end of the rod were common in West Polynesia, and were developed into a distinctive 'hackle' in Hawaii (Buck, 1957:357; Emory, Bonk and Sinoto, 1959:11, Figure 5). It may be argued then that the essential features of the lure are the shank, a round mass towards one end of the shank (stone, cowry shell or both), and the cord. The non-essential elements may nevertheless have played an important role in increasing the lure's attractiveness, as in the case of the tiger cowry shell and the leaf strips. The hook presumably increased the device's holding power, especially in deep water octopus fishing.

Several studies of octopus behaviour in aquaria have now been made, and these are of some relevance to the archaeological problem posed by octopus lures. A leading researcher into cephalopod brain structure, M. J. Wells commented that "Cephalopods have the largest brains of all invertebrates and their behaviour is comparable in many respects with that of the higher vertebrates" (Wells, 1962:2). They have been found to have good image-forming eyes, and they hunt their prey by sight. The octopus is also capable of detecting sound waves but there is no evidence that it reacts to sound. As well, the octopus seems able to taste things by touch, using chemo-tactile sensors on its arms (ibid.:1, 5). It pays most attention to moving objects, seldom attacks a stationary object, and is extremely cautious of unfamiliar objects. Anything that is both moving and larger than the octopus provokes a fear response, although octopuses in tanks soon become accustomed to the appearance of an experimenter.

Home-making is a common activity with moveable objects, especially stones, being collected up around the entrance of the burrow. Thus it is possible that the stone portion of the kilo lure of Hawaii is seen by the octopus not as potential food but as suitable material for concealing its home entrance.

Crabs form a major portion of the diet of the octopus species and as they elicit a rapid attack response they are most commonly used in learning experiments (ibid.:52; Young, 1956). An adult octopus will consume 20 or 30 small crabs in a day (Walls, 1962:15). Their attitude to the hermit crab is particularly significant to this study for they will learn to take specimens with anemones attached to the shell, despite the stings they receive. It has been found that "shells and anemones were later discarded intact together with the hard parts of the crabs." Of course the most poisonous types of anemone are left well alone.

It is possible that the cowry lure is seized by the octopus simply because the octopus eats cowries. Several factors make this seem unlikely. When the cowry has its mantle fully extended it covers virtually the whole shell and the spots are hardly visible under the covering of flesh and bluntish papillae (Roughley, 1936:115). When the animal is withdrawn into the shell, the exceptional shell thickness should protect it from crushing. In addition, the octopus is not known as an important predator of univalves.

The octopus may, however, recognize the cowry lure as a hermit crab inhabiting a shell and able to be prized out. Its eagerness to take the lure then becomes explicable. Because the hermit crab cannot withdraw completely into the shell, various appendages such as claws protrude from it. In the octopus lure these are simulated by the shank and leaf strips, and the scuttling movement of the crab is simulated by jerking the lure along the bottom (Beasley, 1921:108; Suggs, 1961:92; Buck, 1957:362; Fornander, 1919:180). In contrast the kilo lure was not moved until the octopus had enveloped it; this reinforces the view that the octopus did not regard it as food but as 'building material'.

Regional variation in tropical Polynesian or related octopus lures can be summarised as follows:

Recent Examples

Fiji (Beasley, 1921:106-7; Otago Museum specimen 010.334).

(a) 2 cowry plates, both with single perforations, attached to conical stone sinker, thin shank.

(b) 2 cowry shells of equal size with bases removed enclosing egg-shaped lump of coral, 2 pairs smaller cowry shells filling in front and rear space, short rootlet 'tail' (Ngau Is.).

Tonga (Beasley, 1921:103-4; Otago Museum D23.990, D24.721, D24.722, D57.361)

2 cowry plates, 1 or both with single perforations, attached to conical stone sinker occasionally with flattened area to accommodate edge of rear cowry plate, thin shank, leaf strips known - 1 sinker drilled with shallow holes imitating cowry spots?

Samoa (Beasley, 1921:104-5; Hornell, 1950:131 & fig; Otago Museum D41.324).

(a) 2 cowry plates, both with single perforations, attached to conical stone sinker, thin shank, leaf strips common right along shank.
(b) Wickerwork bait made in shape of rat containing small stones.

Lifu, Loyalty Island (Beasley, 1921:103)

1 cowry plate with single perforation, attached to oval lump of coral rubbed down, thin shank, leaf strips beneath body.

Niue (Beasley, 1921:105-6; Otago Museum D24.295)

2 whole cowry shells lashed base to base, 2 perforations on each shell at front and rear, upper shell larger than lower, no sinker, very long thin shank, leaf strips common.

Fakaofu, Union Island (Beasley, 1921:106-7).

Cowry body, shank with leaf strips.

Society Island (Beasley, 1921:108-9; Handy, 1932:101; Rappaport, Rappaport & Green 1967:195; Otago Museum D31.126).

(a) Egg-shaped collection of cowry backs, each with double perforations, no internal sinker, substantial pointed shank 30-35 cm long, recorded 1820, 1838.
(b) Large complete cowry with 8 attached perforated cowry plates, no internal sinker, slender bone shank c. 25 cm long with 'tail' end wrapped in sennit with 2-3 fibre tails, front of shank covered with a small Turbo shell, recorded 1769, possible bone hook suspended beneath.

Marquesas Islands (Linton, 1923:402; Suggs, 1961:90).

(a) White stone sinker and lure above 3 hooks lashed back to back, no native knowledge of 'coffee-bean' shaped sinker (Linton).
(b) Coffee-bean sinker, cowry shell lure with 1 or 2 perforations, cowry plate 'flapper', shank, hook (Suggs).

Hawaii Islands (Fornander, 1919:180; Beasley, 1921:111-4; Buck, 1957: 356-63; Emory, Bonk & Sinoto, 1959:7).

(a) Kilo method, rough sinker with flattish surface, shank 13 cm long, bone hook with backwards extension, tufts of leaves camouflaging hook.
(b) Cowry lure utilising complete shell with 2 perforations front and rear, 'coffee-bean' sinker, shank 15-23 cm long, bone hook point with basal perforation or backward extension, toggle adjacent to rear perforation of shell, hackle.
(c) 2 cowry shells, one mounted over 'breadloaf' sinker, shank, 1-2 hooks sometimes iron, toggle, hackle.

Because of the separation and decay of the various components in the ground, archaeological finds of the Polynesian octopus lure (Figure 14.2) are

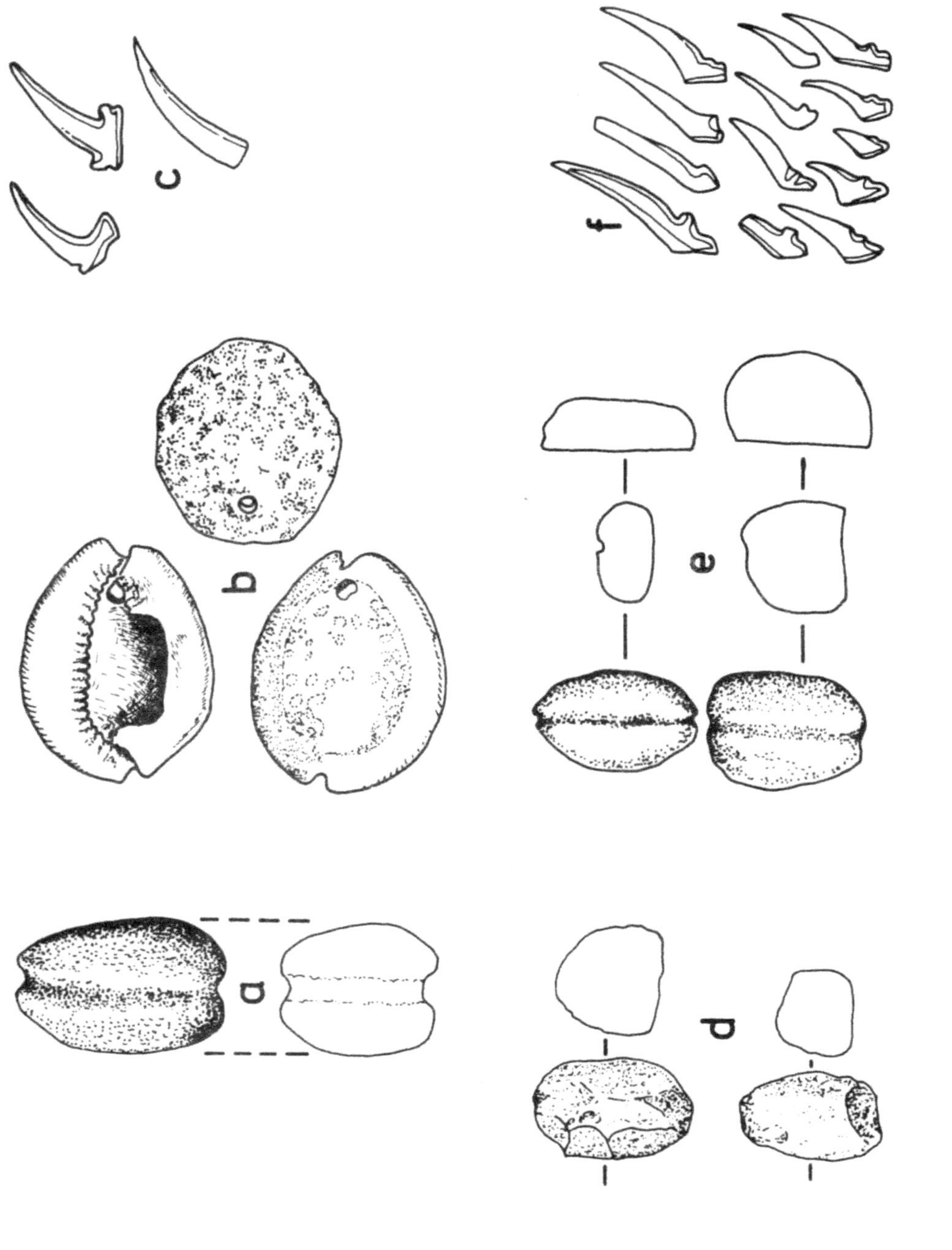

Fig. 14.2 Archaeological Evidence of Possible Octopus Lures.

- a. Sinker. Marquesas Is. After Suggs, 1961: Figure 28.
- b. Perforated cowry shells. Marquesas Is. After Suggs, 1961: Figure 26.
- c. Possible hooks. Marquesas Is. After Suggs, 1961: Figure 26.
- d. Sinkers. Samoa, Su-Va-1. After Green & Davidson, 1969: Figure 58.
- e. Sinkers. Hawaii. After Emory & Sinoto, 1961: Figure 49.
- f. Hooks. Hawaii. After Emory & Sinoto, 1961: Figure 42.

subject to problems of identification and interpretation. When is a stone sinker an octopus lure sinker, when is it a line or net sinker? How can one distinguish an octopus lure hook from a bonito hook? Was the cowry shell perforated for functions other than attachment to an octopus lure?

Taking advantage of the large Bishop Museum collection (for example, 35 cowry lures - Buck, 1957:359), Emory, Bonk and Sinoto (1959) believed they were able to identify hook points, shanks, cowry lures, sinkers and toggles, found in 24 excavated Hawaiian middens and cave sites. In all they listed 36 points, 2 shanks, 37 lures, 42 sinkers, and 37 toggles from archaeological deposits. Although the overall numbers of each element are fairly consistent (with the exception of shanks), only 7 sites contained the basic assemblage of hook points, cowry lures and sinkers. Another 7 sites produced only hooks, raising the problem of the fate of the cowry lures and sinkers. This anomaly casts doubts on the identification of some of the points as octopus lure points. The statement that the average length of octopus lure points is 43.1 mm, and of bonito points 29.1 mm (Emory, Bonk and Sinoto 1959:18) does not specify whether the averages include ethnographic specimens or have been calculated from archaeological examples alone.

Both size and angle (not stated) were used to distinguish octopus lure points from bonito points. The criteria for identifying octopus lure sinkers seem to have been their adherence to 1 of 4 coffee-bean shapes or 1 of 2 bread-loaf shapes. No mention is made of the rougher sinkers which Buck recorded in use on kilo lures. Cowry shells were identified as octopus lure cowries from the position of their perforations. These criteria are still used to identify components of octopus lures. Kirch, for example, found the base of a large point with backward extension in the early Halawa Dune Site (600-1200 A.D.) describing it as "an unfinished octopus-hook point" (Kirch and Kelly, 1975:34, Figure 17). A broken, shaped, canine molar root from the same site was thought to be an octopus lure toggle. No coffee bean sinkers were found, only 2 grooved 'line' sinkers. If the identifications of points and toggle are correct they substantially alter the conclusion of Emory, Bonk and Sinoto that in the Hawaiian sequence

> "elaborated toggles of the U and arched shapes as well as the bone points, appear to be of the middle and later periods" (Emory, Bonk and Sinoto, 1959:40-1).

Identification of octopus lure components in Marquesan archaeological sites appears to have been influenced by Hawaiian experiences. In 1923 Linton described a coffee-bean sinker in a Hiva Oa collection as a possible import via missionaries from Hawaii. His claim that Marquesans had no knowledge of this lure was contradicted by Suggs:

> "Squid lure sinkers are found throughout the Marquesan sequence from the settlement period to the present" (Suggs, 1961:90).

Suggs' 'squid lure sinkers' were described as plano-convex in section, ovoid to ellipsoid in plan, and bearing a groove along the convex surface sometimes continuing on to the flat side. In contrast the Hawaiian sinkers studied by Emory, Bonk and Sinoto (1959:27, Figure 11) have the groove most commonly on the flat surface sometimes extending on to the curved back.

It is difficult to judge from his report how Suggs believed the Marquesan lure was made up. He describes cowry shells with single or double perforations and base removed from 4 sites, plano-convex sinkers from 3 sites, and cowry plates or 'flappers' from 6 sites. Although he twice refers to the "hook shank" (ibid.:90, 92) there is no attempt to equate any hook type known archaeologically with the octopus lure hook, nor to explain the absence of octopus lure hooks in the sites. Were Marquesan cowry shells mounted on the back of the sinker as in West Polynesia or opposite it as in Hawaii? Since the Hawaiian cowry can be left whole (except for the perforations) without interfering with the lashings or shank, it seems likely that Marquesans went to the extra effort of breaking out the base of the whole shell and cutting shell plates because these were to be mounted directly on the sinker. The tendency for Marquesan sinkers to be grooved on their convex surface rather than on the flat side supports this interpretation, for without a groove the cord securing the sinker would interfere with the close fitting of the cowry shell pieces.

Sinoto's Marquesan research led to the discovery of 7 specimens of an earlier type of presumed octopus lure sinker:

> "Those of the coffee-bean type were found throughout the Nukuhiva sequence, but none in the excavations in Hane. Instead, Hane sinkers appear to be of a type ancestral to present day West Polynesian octopus sinkers of the conical or top shape, and also to the later East Polynesian coffee-bean sinkers. Thus Hane sinkers are conical in shape with one side and end flattened and with or without a longitudinally encircling groove." (Sinoto, 1966:299).

The implication of this find is that the coffee-bean type developed in the Marquesas group and was subsequently introduced to the Hawaiian Islands.

As has been shown ethnographic specimens of the octopus lure from Tahiti do not possess a stone sinker. Suggs claimed that there was a tendency in the Marquesas for the stone sinker to be dropped in favour of filling the cowry shell with pebbles (Suggs, 1961:92). The same tendency may have been present in the Society Islands where several coffee-bean sinkers have been recovered from surface collections. In 1964 Emory and Sinoto attributed these to "influence from Hawaii in post-European times" or "sporadic contact with Marquesans" (Emory and Sinoto, 1964:158). Of the 4 known from Tahiti, 3 had a groove over the top only, in the Marquesan rather than Hawaiian style. Excavations on Mo'orea produced a few items assigned to octopus lures. One was a perforated cowry shell plate, another was a large (65 mm) unfinished hook point which may have articulated with its shank at a wide angle (Rappaport, Rappaport and Green, 1967:195). The latter identification was unexpected since the existence of a octopus lure hook was not known outside Hawaii. Of course Suggs had referred to octopus hook 'shanks' in the Marquesas, and on the grounds of hook size and openness of angle, several of his illustrated lure points might be interpreted as octopus lure points, for example his biflanged point (45 mm long), L-shaped point (51 mm long), or inset point (63 mm long) (Suggs, 1961:Fig. 26).

In West Polynesia, archaeological finds of octopus lure components have been confined to Samoa. They consist of an appropriately shaped but imperforate piece of cowry shell from a coastal midden (Davidson, 1969:244), and 2 stone sinkers from Vailele 1 which prompted the following comments from Green:

> "The early [octopus lure] sinkers from Samoa are of a type which I propose to call cowrie-shaped sinkers, because their form approximates that of the complete cowrie shell that is usually employed in whole or in part in making the lure portion of these objects. Stone sinkers of this type have a flattened base with the body at one end being broader and rising more steeply than at the other end... In addition the more steeply rising end is often partially flattened. Some may even have traces of a groove on the upper surface along the centre line of the sinker. Octopus lure sinkers of this type are known from these early Samoan examples and from those of the Settlement and Developmental periods in the Marquesas." (Green:1969:135).

> "The identity in form with the early octopus lure of the Marquesas but not the later octopus lure sinkers of Samoa, Tonga, Hawaii, or the Marquesas, and the complete lack of sinkers of this type from any Tongan or other Lapita contexts, appears to me to have some importance. It may well be that the attachment of the large stone sinker to this lure is a Polynesian innovation which had its beginnings in Samoa" (Green, 1974:269).

Combining the archaeological and ethnographic evidence, several other interesting suggestions may be added to Green's:

1. That the coffee-bean sinker developed out of the cowry-shaped sinker in the Marquesan group and subsequently spread to Hawaii and possibly the Society Islands.
2. That the long, wide-angled hook was not, as originally thought, a Hawaiian invention but occurred in the Marquesas and Society Islands. It may possibly have evolved with the coffee-bean sinker and been distributed with it.
3. That the practice of lashing the sinker on opposite sides of the shank from the cowry lure was a Hawaiian invention.
4. That the notion of the lure representing a rat arose in West Polynesia after the settlement of East Polynesia.

Some of the major difficulties in the identification of octopus lure components in sites have already been discussed. It is apparent, too, that the only archaeological examples so far reported have been noticed by archaeologists with Hawaiian experience. Even in Hawaii with its continuity of material culture from prehistoric to historic eras the various toggle and hook types have not always been identified with certainty. It also seems likely that many rougher examples of sinkers have been mislabelled as line or net sinkers.

New Zealand archaeologists might be excused for failing to recover evidence of octopus lures in this country for several reasons. Firstly, the genus Cypraea to which most of the cowries belong, does not live in New Zealand's cool coastal waters. Secondly, neither coffee-bean nor bread-loaf shaped sinkers have been recovered in 'significant' numbers from New Zealand sites; odd examples which approximate these shapes are presumed to do so by accident. Thirdly, the Polynesian octopus lure was never seen in use in New Zealand by European recorders.

At this point it would be easy to conclude that the Polynesian octopus lure was never made in New Zealand. However, there are several important examples of artefacts, such as the reel, chevroned pendant, various adzes and the large harpoon, which were manufactured in New Zealand in the first few centuries of Polynesian occupation and subsequently dropped out of use. A type of octopus lure might be added to this list since it was certainly present in the Marquesas and probably in the Society Islands at the time New Zealand was settled. We can only guess at the details of the octopus lure which the first New Zealanders had known in their East Polynesian homeland and which might serve as a model for a locally produced lure. It probably used 1 or 2 perforated cowry pieces lashed on to a rough partially-grooved sinker, in shape somewhere between a cowry, cone, and a coffee-bean, and it may have possessed a bone hook, at the end of its wooden shank.

On their arrival in New Zealand, the Polynesians demonstrated conservatism in maintaining their mixed horticulture and marine exploitation while at the same time modifying the design of fishing equipment to suit new raw materials. It might be expected therefore that the octopus lure which allowed the extension of octopus fishing beyond the immediate coastal reef would continue in production perhaps for a few centuries. It is clear from the evidence of trolling lures that other forms of line fishing from canoes persisted.

The large local octopus which is so widespread that Polynesians must have encountered it almost at first landfall, is known as Octopus maorum. It is believed to be derived from the wide-ranging Pacific and Indo-Pacific species Octopus macropus (Dell, 1952:145). In size, O. maorum is comparable to O. macropus. Dell gives length measurements between 207 and 1482 mm (ibid.) for O. maorum. Like its cosmopolitan cousin, the large New Zealand octopus and the small Robsonella australis feed on crustaceans. Indeed, Dell comments that "There is noevidence to show that these forms ever feed upon species of Mollusca in this country" (ibid.:149). Morton and Miller (1968:549-50), however, list Octopus maorum as an enemy of the scallop, Pecten novaezelandiae, as well as a consumer of crabs. If they possess the same preference for hermit crabs as the other octopus species display, the most suitable shell to replace the cowry in the octopus lure would surely be one commonly inhabited by the New Zealand family of hermit crabs (Paguridae). The common Pagurus novaezelandiae " uses chiefly small cerithiid shells when young, then transfers to topshells and cats eyes, and later Cominella or even larger whelk shells" (ibid.:94). Among the latter could be Penion, Neothais, Haustrum or even Struthiolaria. Since these are quite common components of Archaic middens, a perforated or roughly shaped example might easily be missed. Similarly, a grooved sinker of cowry or coffee-bean shape might be classified as a line or net sinker simply because an octopus lure sinker has up till now not been expected in a New Zealand site. Until appropriately modified gastropod shells are recovered in association with sinkers, evidence of hooks may provide the strongest support for the hypothesis. The Hawaiian examples show a tendency to greater length and more open angle between hook and shank than the points of bonito lures. The purpose of the hook appears to be to prevent the octopus releasing its hold of the lure and slipping off the rig on the long haul to the surface. Being an animal without an internal skeleton it cannot exert any leverage or force once its grip on the sea floor has been lost. With its mantle impaled on the bone point there would be little chance of the point breaking as it would with a thrashing fish.

An exceptionally long and slender type of bone point has been recovered from several sites in Otago and Southland (Figure 14.3). Unfortunately none of these is accurately dated or adequately recorded. In broad terms, however, the associated material is Archaic. In Hjarno's classification (1967:26) this hook is Type C 1C. He noted its presence in Shag River, and the lower levels of Little Papanui and Papatowai. The most interesting feature of this point, apart from its length and slender appearance, is the area of articulation with the shank, which is bevelled to a distinct edge, not flattened as in the minnow lure point or large composite bait hook. Presumably this enabled it to fit into a groove or slot in the lure shank. If the preferred method of articulation of a hook with a bone or stone minnow lure or bone shank was a flat platform, it seems likely that the bevelled hook was intended to be coupled with a wooden shank which can be grooved without difficulty. As has already been shown, most octopus lures utilize a wooden shank.

A number of sinkers are illustrated from the sites of Little Papanui and Shag River (Figure 14.4). If found elsewhere in Polynesia they might be described as coffee-bean, incipient coffee-bean, or conical octopus lure sinkers. Traditionally, any conical object found in New Zealand is called a top so the Shag Shag River examples which are man-modified sandstone concretions were undoubtedly classified as tops by both Skinner and Teviotdale. Other conical stone objects in the Otago Museum have also acquired this description for the purposes of display; for example D21.416 is displayed as a Maori top while the Register describes it as "part of squid hook? Tonga". A "top" possibly from the Bay of Plenty (D.19.164) is also referred to as a "pounder" from "?Polynesia". For sure, wooden whip tops were seen in use by early recorders in New Zealand, but it appears that no-one ever saw a stone top being spun. Those described by Best (1925:88) were identified from Museum col lections. A comparison of a dismantled Tongan lure rendered in agate with the largest Shag River specimen shows several points of similarity: asymmetry (a handicap for a top), flattening of one side, flattened end (Figure 14.5). Unfortunately Teviotdale did not recover shell from his excavations so it will never be known if roughly perforated whelk shells or shell fragments were associated with these possible sinkers. Similarly the assymetrical grooved sinkers of Little Papanui may have been found with pieces of worked shell.

In future, New Zealand archaeologists should not assume that all conical objects of stone are tops, that grooved stones are solely line or net sinkers, or that all hooks are meant for fish. Archaeological evidence from tropical Polynesia coupled with information on octopus distribution and diet suggest that early New Zealanders may have tried to manufacture a local form of octopus lure for offshore fishing. It if existed at all, it will be positively identified not from its individual components (which may have been discovered already in a hundred Archaic sites) but from the repeated association of these elements.

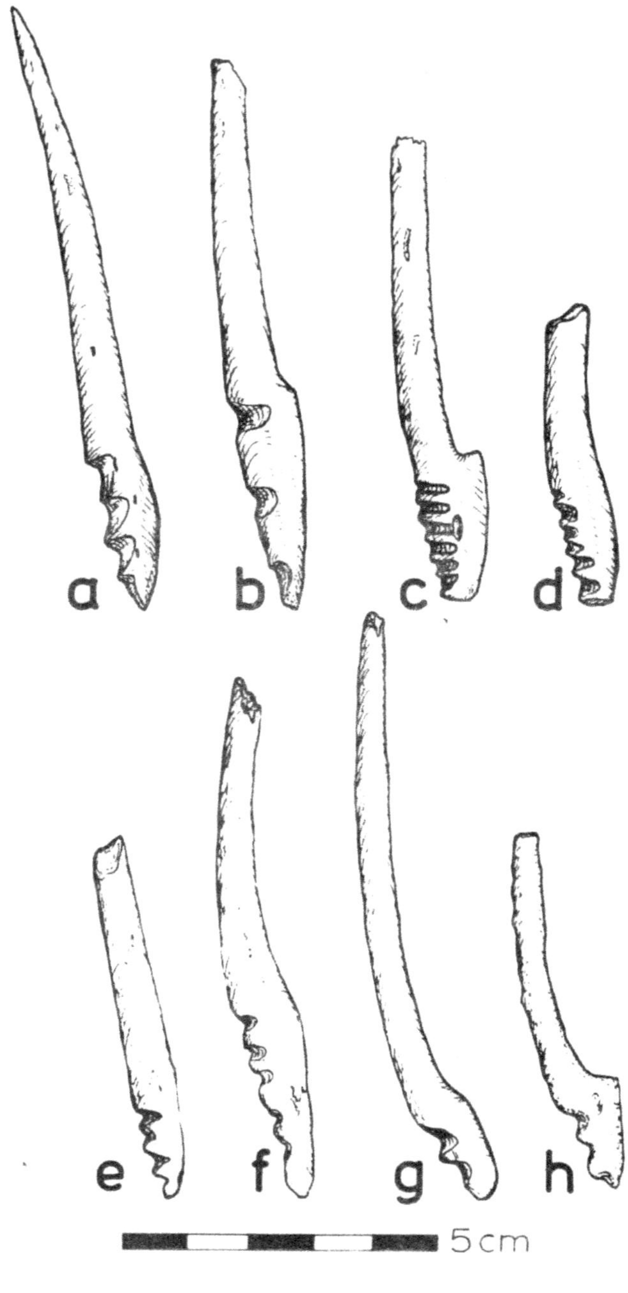

Fig. 14.3 Possible Octopus Lure Hook Points from Otago, New Zealand.

a. Little Papanui. Otago Museum D29.226.
b. Little Papanui. Otago Museum D29.228.
c. Little Papanui. Otago Museum D29.229.
d. Little Papanui. Otago Museum D77.2.
e. Shag River Mouth. Otago Museum D77.3.
f. Shag River Mouth. Otago Museum D27.1167.
g. Shag River Mouth. Otago Museum D27.1164.
h. Shag River Mouth. Otago Museum D77.4.

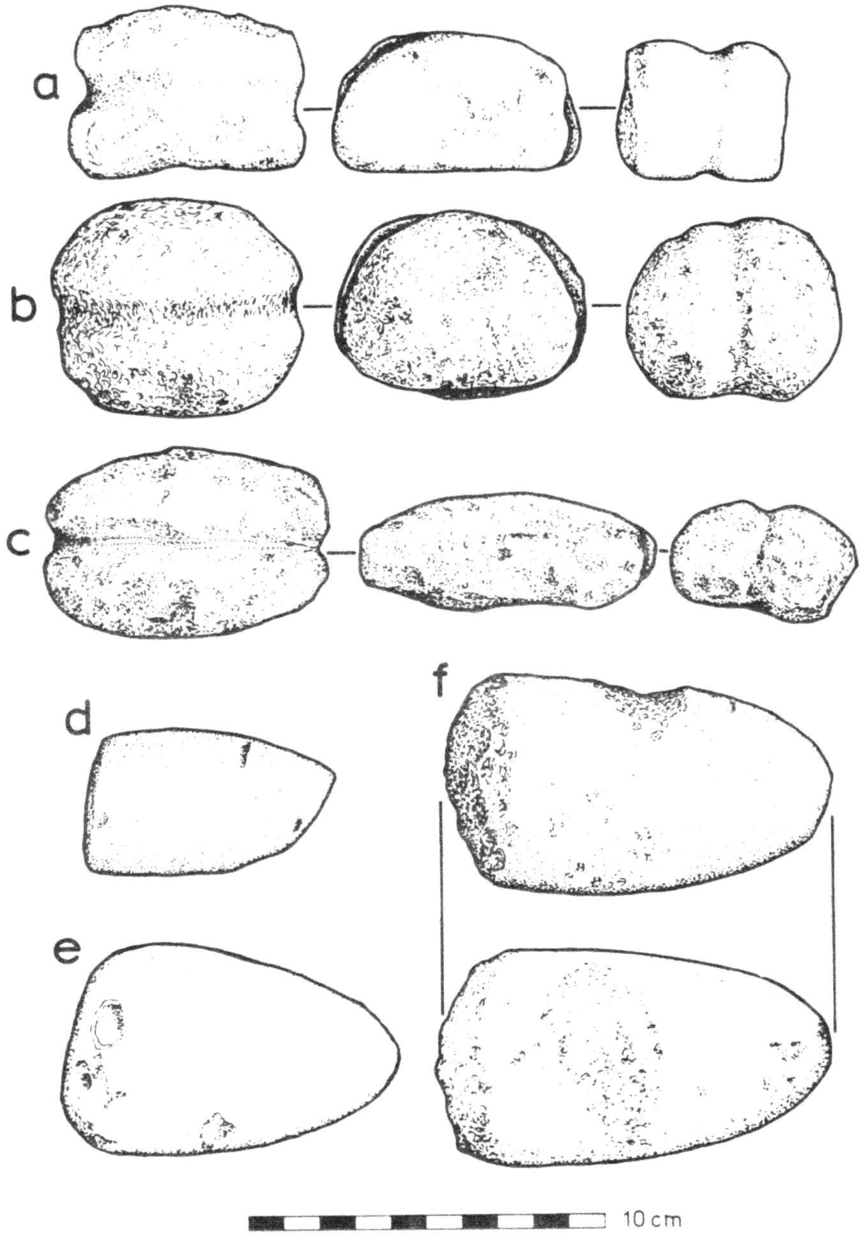

Fig. 14.4 Possible Octopus Lure Sinkers from Otago, New Zealand.

a. Little Papanui. Otago Museum D32.1887 (?)
b. Shag River Mouth. Otago Museum D30.1070.
c. Shag River Mouth. Otago Museum D30.1066.
d. ?Shag River Mouth. Otago Museum D77.5.
e. Shag River Mouth. Otago Museum D27.1334.
f. Shag River Mouth. Otago Museum D29.4382.

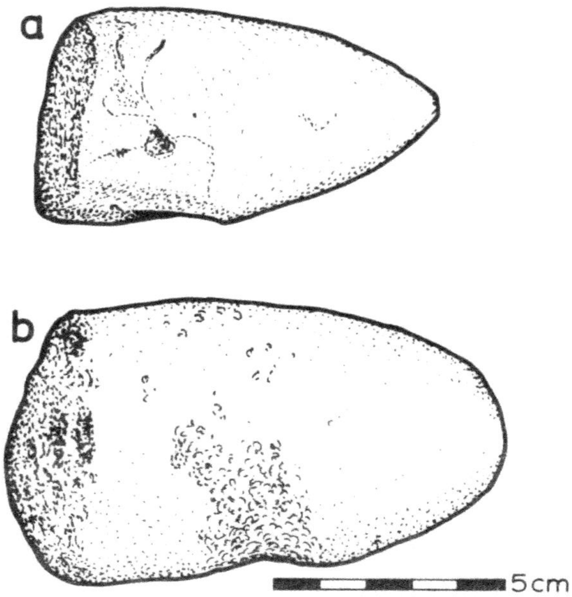

Fig. 14.5 A Comparison of a Tongan Octopus Lure Sinker and a Possible Sinker from New Zealand.

a. Tonga. Otago Museum D57.361.
b. Shag River Mouth. Otago Museum D29.4382.

NOTE

1. Beasley in fact lists only nine: Lifu (Loyalty Is.), Tonga, Samoa, Niue, Ngau (Fiji Is.), Fakaofu (Union Is.), Funafuti (Ellice Is.), Tahiti, Hawaii.

BIBLIOGRAPHY

Beasley, H. G., 1921. Some Polynesian Cuttlefish Baits. Journal of the Royal Anthropological Institute 51:100-14.

Best, E., 1925. Games and Pastimes of the Maori. Dominion Museum Bulletin 8.

Buck, P., 1957. Arts and Crafts of Hawaii. Bernice P. Bishop Museum Special Publication 45.

Davidson, J. M., 1969. Excavation of a Coastal Midden Deposit, Su-Lo 1. Report 15 IN: Archaeology in Western Samoa, Vol. 1, (eds.) R. C. Green & J. M. Davidson. Auckland Inst. Mus. Bull. 6.

Dell, R. K., 1952. The Recent Cephalopoda of New Zealand. Dom. Mus. Bull. 16.

Emory, K. P., Bonk, W. J., Sinoto, Y. H., 1959. Hawaiian Archaeology: Fishhooks. Bernice P. Bishop Museum Special Publication 47.

Emory, K. P. and Sinoto, Y. H., 1961. Hawaiian Archaeology: Oahu Excavations. Bernice P. Bishop Museum Special Publication 49.

Emory, K. P. and Sinoto, Y. H., 1964. Eastern Polynesian Burials at Maupiti. J. Polynes. Soc. 73:143-60.

Fornander, A., 1919. Hawaiian Antiquities and Folk-lore. Memoirs of the Bernice P. Bishop Museum 6(1).

Green, R. C., 1969. Excavations at Va. 1, 1963-64. Report 7b IN: Archaeology in Western Samoa, Vol. 1, (eds.) R. C. Green, and J. M. Davidson. Auckland Inst. Mus. Bull. 6.

Green, R. C., 1974. A Review of Portable Artifacts from Western Samoa. Report 40 IN: Archaeology in Western Samoa, Vol. 2, (eds.) R. C. Green, and J. M. Davidson. Auckland Inst. Mus. Bull. 7.

Handy, E. S. C., 1932. Houses, Boats, and Fishing in the Society Islands. Bernice P. Bishop Museum Bulletin 90.

Hjarnø, J., 1967. Maori Fish-hooks in Southern New Zealand. Rec. Otago Mus. Anthropology 3.

Hornell, J., Fishing in Many Waters. Cambridge University Press.

Kirch, P. V. and Kelly, M. (eds.),1975. Prehistory and Ecology in a Windward Hawaiian Valley: Halawa Valley, Molokai. Pacific Anthropological Records 24.

Linton, R., 1923. The Material Culture of the Marquesas Islands. Memoirs of the Bernice P. Bishop Museum 8 (5).

Métraux, A., 1940. Ethnology of Easter Island. Bernice P. Bishop Museum Bulletin 160.

Morton, J. and Miller, M., 1968. The New Zealand Sea Shore. Collins. London.

Rappaport, R. A., Rappaport, A. and Green, R. C., 1967. Descriptive and Comparative Analysis of Portable Artifacts Recovered from Coastal Excavations IN: Archaeology on the Island of Mo'orea, French Polynesia (eds.) R. C. Green et al. Anthropological Papers of the American Museum of Natural History 51 (2).

Roughley, T. C., 1936. Wonders of the Great Barrier Reef. Angus and Robertson. Sydney.

Sinoto, Y. H., 1966. A Tentative Prehistoric Cultural Sequence in the Northern Marquesas Islands, French Polynesia. J. Polynes. Soc. 75: 287-303.

Suggs, R. C., 1961. The Archaeology of Nuku Hiva, Marquesas Islands, French Polynesia. Anthropological Papers of the American Museum of Natural History 49 (1).

Wells, M. J., 1962. Brain and Behaviour in Cephalopods. Heinemann.

Young, J. Z., 1956. Visual responses by Octopus to crabs and other figures before and after training. Journal of Experimental Biology 33:709-29.

AXE HEADS AND ZOOMORPHS IN THE SOLOMON ISLANDS

Jim Specht and Sheila Specht

Introduction

In 1967 Ron Scarlett visited Papua New Guinea to assist us with archaeological surveys and excavations on Buka and Sohano Islands in the Province of the Northern Solomons. During that season, but after his departure, we recorded a stone axe blade of unusual form which resembled one previously reported from Toiminapu on Bougainville Island (Casey, 1939). These blades, together with four others of similar style also found in the Solomon Islands' chain, form a distinctive and singular group among Oceanic percussion tools. In his original report on the Toiminapu specimen Casey compared certain decorative elements of that blade with bird forms, and in particular with bird-shaped stone objects from the islands to the west of Bougainville. This interpretation appears to have been accepted without question in subsequent literature (e.g. Riesenfeld, 1950; 1955; Pretty, 1965; Golson, 1972a: 968) especially following the report of a stone bird figure on Bougainville Island (Neich, 1971). It seems fitting to provide the first published account of the additional specimens in a volume dedicated to a man among whose many interests birds, and the correct interpretation of their remains, have played an important role.

This paper is a descriptive review of these stone blades, with a brief consideration of the identity of the zoomorphic elements. None of the six specimens was found in an archaeological excavation; hence, little can be said of their age or cultural associations. This must cast doubt on the significance, and perhaps relevance, of previous comparisons with material originating from beyond the immediate area of the northern Solomons. However, evidence from archaeological sources is cited to give an estimate of their possible age, while their rarity and distribution pattern provide grounds for limited speculation on their possible cultural significance. For convenience, each blade is identified by its attributed locality of discovery, except that in the Auckland Institute and Museum, which is provenanced only as originating from the 'Solomon Islands'; this specimen will be called the 'Auckland blade'.

1. The Toiminapu blade

The original of this blade was held in the Rabaul Museum, the contents of which were lost or destroyed during the second World War. Fortunately, a copy had been made by E. W. P. Chinnery for the National Museum of Victoria in Melbourne and a description was published by Casey (1939: 144-5).

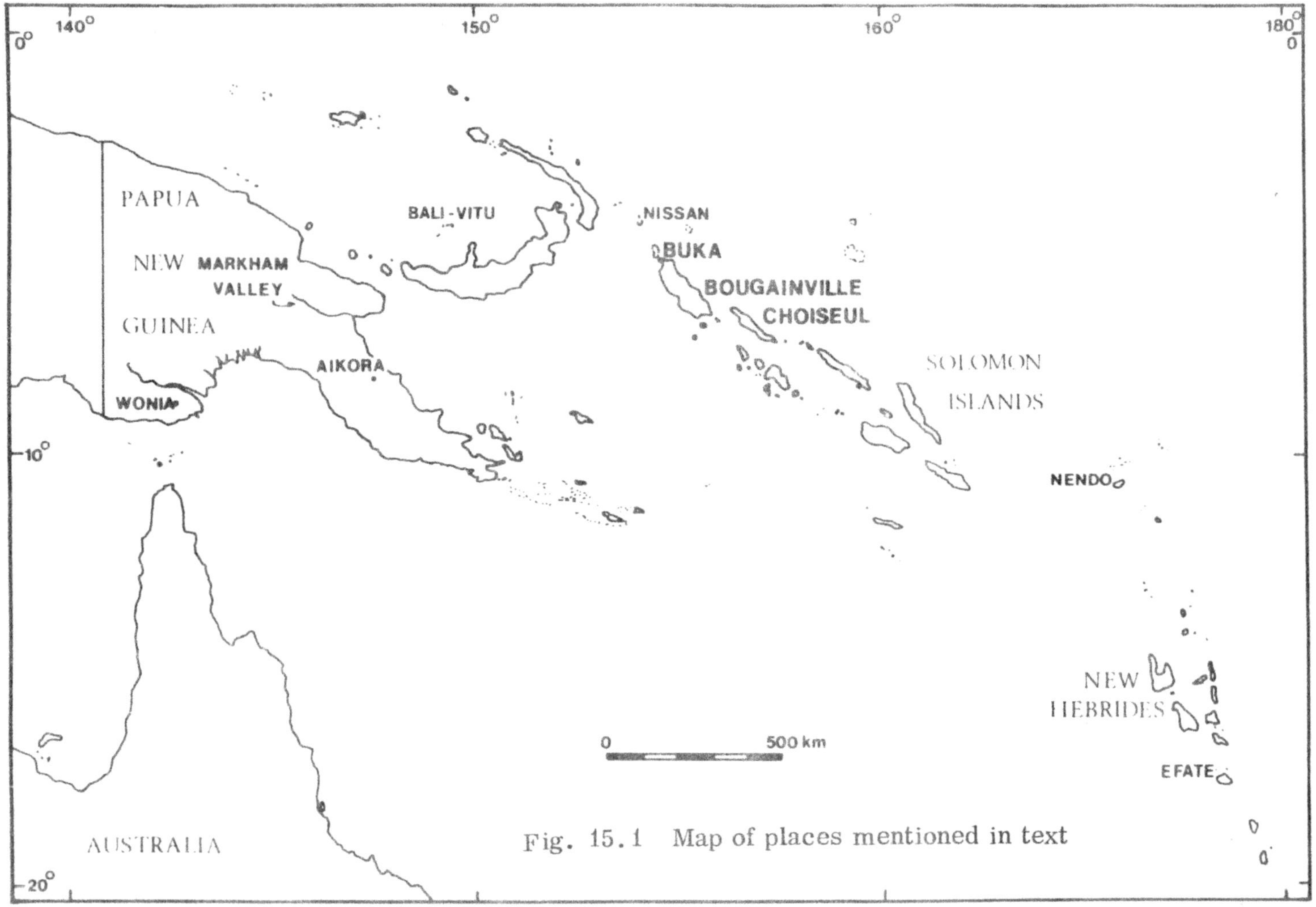

Fig. 15.1 Map of places mentioned in text

This axe blade was found in or before 1938, for the cast held by the National Museum of Victoria (reg. no. X43604) was registered in December 1938. The discovery was made by W. A. L. Clarke, of Toiminapu Estate on western Bougainville (Swadling and Phillips, no date: 7) (fig. 15.1). According to Casey (1939: 144), the findspot was within the plantation area, but is not further identified in detail. It was made from a dark grey volcanic rock, apparently hammer-dressed and ground to form.

The blade has two side lugs and a laterally-expanded poll, with decorative elements on the lugs, poll and the area between the lugs on both main surfaces (fig. 15.2). The cutting-edge is medial in section and strongly rounded in plan; the cross-section just below the lugs is lenticular. The poll has two grooves on each face, with the areas bordering the grooves serrated to form bosses. There are four relief bosses on each face between the lugs. Below one lug there are two very faint grooves.

The lugs have small, upward extensions which terminate in blunt points with a groove lightly worked to form what looks like a mouth. Each lug carries two bosses, one on each side and set in line with those across the main surfaces. These bosses appear to have been encircled by shallow grooves, perhaps made by grinding with a hollow cylinder. The overall effect is that of an animal head.

2. <u>The Malasang blade</u>

This blade, now in the Papua New Guinea National Museum and Art Gallery, Port Moresby, was sold to us by a man from Malasang village, on the east coast of Buka Island in 1967 (fig. 15.1).

The findspot (site DCO) lies within the Kessa area of Malasang, about 5 metres east of the coastal road. This area is on top of a limestone plateau formed by an elevated coral reef formation capped by a stiff red clay. The blade was said to have been found on the ground surface among weeds and grass growing round the base of coconut palms. Examination of the area revealed a few plain sherds and decorated sherds in the Mararing style, dated approximately A.D. 1200-1800 (Specht, 1969) though it is impossible to determine whether they were associated with the blade. A road ditch, 3 metres from the alleged findspot, revealed no occupation deposit, and it is therefore doubtful whether the sherds and blade were ever buried below ground surface here.

The blade is made from a light grey rock similar to that used in recent times to manufacture stone pestles; a tuffaceous sandstone found in central and western Buka (Specht, 1974a). It was fashioned by hammer-dressing and grinding. The tool has suffered considerable damage, and its complete form cannot be reconstructed (fig. 15.3).

The blade has two side lugs and a laterally-expanded poll. The cutting-edge is flared and, though damaged, clearly strongly rounded. The edge itself appears to have been medial, and the cross-section just below the lugs is lenticular. In profile the poll is formed by three ridges, each serrated

Fig. 15.2 Toiminapu blade (length 140 mm)

Fig. 15.3 Malasang blade (length 186 mm)

to form bosses. Between the lugs on each face is a row of four prominent bosses, and between these and the poll are two pairs of bosses placed vertically.

The lugs have a flattened oval cross-section. Both are badly damaged, and their original forms are uncertain. One terminates in five bosses, and has a vertical extension seemingly curving inwards. At the base of this extension are two further bosses.

3. The Hutjena blade

This blade is held by the Hutjena High School on the east coast of Buka Island (fig. 15.1). It was found in 1968, subsequent to our visit to Buka, and we have been unable to inspect it in person. Photographs were supplied by the then Headmaster of the school, Mr. N. Murray. Although found on Buka, the exact provenance of this specimen is unknown. It is apparently of a light grey rock similar to the Malasang blade, and made by the same technique.

This blade is extensively damaged, but sufficient remains to indicate that it must have been similar to the Malasang blade (fig. 15.4). The cutting-edge seems to have been medial in section, and markedly rounded and flaring in plan. The cross-section below the lugs appears to be lenticular. The poll, heavily damaged, is laterally expanded, though it is not clear whether it carri grooves and bosses. Of one lug only the base remains, but the other is better preserved. There are four bosses across the surface between the lugs.

The surviving lug appears to have a more rounded cross-section than those on the Malasang specimen, but with a similar upward vertical extension. There is a boss on each surface at the base of the extension, which appears to terminate in a grooved point. There are no bosses on the surfaces between the lugs and the poll.

4. The Choiseul blade

This hafted blade, from Nuatambu, Choiseul, was brought to our attention by Mr. D. Newton, Metropolitan Museum of Art in New York, and photos were supplied by the Solomon Islands Information Services, through the Solomon Islands Museum; for these details of the blade we thank Ms. A. Craven, Curator (Training) and Mr. D. Miller of the Solomon Islands Museum in Honiara.

This tool was bought by the Solomon Islands Museum in 1971 (reg. no. 71.45.1 and 2) from a man named Karaso, who was then working for the mining company Conzinc—Riotinto of Australia. It was originally owned by Golomolo of Choiseul, and is said to have been made by his grandfather. It is possible this means his grandfather made the haft, which is straight with a U-shaped fork to receive the blade. The blade is bound in with cane over the upper side only, and sits in a bed of gum. It was made by hammer-dressing and grinding.

Fig. 15.4   Hutjena blade (length extant 170 mm)

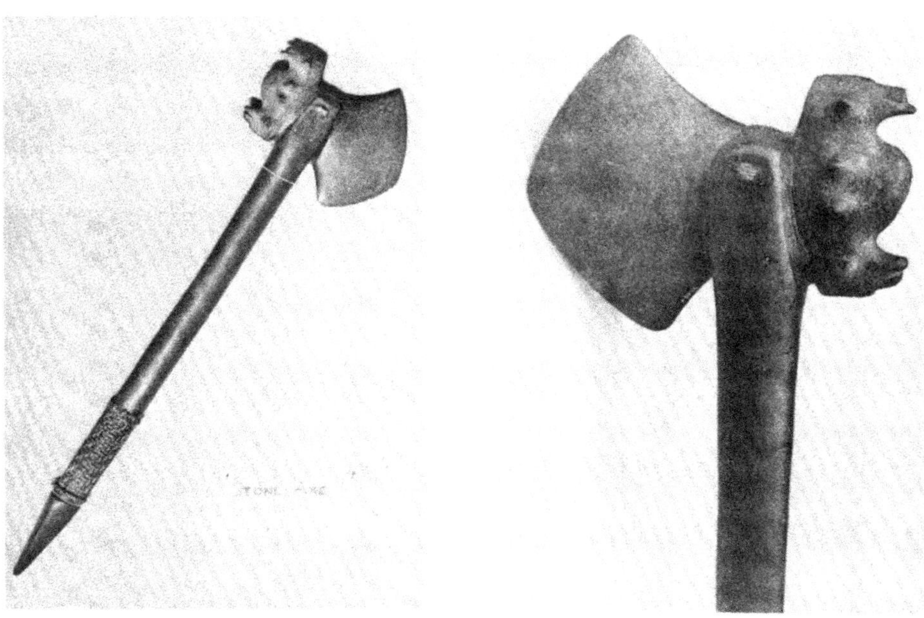

Fig. 15.5   Choiseul blade (length extant 190 mm)

With the exception of what appears to be minor damage on one lug, this specimen is complete (fig, 15.5). It appears to have a medial cutting-edge, strongly rounded to a blunt point in plan. The cross-section appears to be lenticular. The poll is low, rounded and asymmetrical, without bosses or grooves. On each face between the lugs are two bosses, with a further two bosses on the lugs themselves. The upward extensions of the lugs are short and grooved, and terminate in a stepped fashion.

Information supplied to the museum at the time of purchase states that this kind of hafted tool (maza) was used in war; the lugs represent dog-hands which were said to smell out the enemy.

5. The Auckland blade

This specimen is held by the Auckland Institute and Museum, and I am grateful to Ms. J. Davidson, latterly Archaeologist at the museum, for photographs and information. The blade was donated in 1961 by Mr. G. V. Binet (reg. no. 36295), whose father, Rev. Binet, was a Methodist missionary in the Solomon Islands. Its place or origin is listed only as 'Solomon Islands'.

This specimen, too, is badly damaged, but clearly belongs to the same style as those described above (fig. 15.6). The cutting-edge appears to have been flaring, and probably medial and the cross-section is lenticular. The poll has lateral extensions, of which only part of one survives. There are two bosses on each face of the poll. The line of the base of the lugs is carried across both surfaces, forming a low but visible ridge. Two bosses are placed on each face just above the ridge, and between them another, more faintly defined, ridge runs vertically to the poll. The lugs have upward extensions curving inwards, and carry a boss on each face.

6. The Simbo blade

This was reported to us by Mr. D. Miller, Archaeologist at the Solomon Islands Museum, in 1978. We are grateful to Mr. Miller for providing information, including a sketch of the blade, and for permission to discuss it here.

The blade was found on an old fortress site (SI-SN-Z-39) at the southern end of Nusasimbo, the smaller of the two islands which constitute Simbo. It is made from a coarse grey rock. There is damage on each margin of the cutting-edge and on one lug. The cutting-edge is medial and strongly flaring in plan. The poll is expanded, with rounded form, and has four bosses on each face. Between the lugs there is a pair of bosses, and a single boss on each face of the lugs. The lugs themselves extend upwards, apparently terminating in a blunt point.

With only two complete specimens, and one of those hafted, it is impossible to list the full range of dimensions of these blades. The following table, however, shows that the first reported example, the Toiminapu blade, is in fact the smallest; measurements for the Simbo blade are taken from a sketch supplied by D. Miller, and are approximate only. (Dimensions in parentheses indicate minimum values; all dimensions in millimetres).

Fig. 15.6 Auckland blade (length 190 mm)

	Toiminapu	Malasang	Hutjena	Choiseul	Auckland	Simbo
length	140	186	(170)	(190)	190	200
cutting-edge width	93	150	(125)	(120)	147	(120)
width across lugs	93	(106)	(85)	(95)	105	120

Discussion

Although the six blades form a distinctive set in terms of decoration, lug form, and poll embellishment, they are closely related to a major series of stone blades from the Solomon Islands which is characterised by butt differentiation consisting of lugs, notches and/or shallow waisting, either singly or in combination (Rolston, 1945; Specht, 1969; Gerard, 1973; Swadling and Phillips, n.d.; Neich, 1971). One group within this series has double lugs, one on each margin, with or without shallow marginal notches or waisting. Such blades usually have a pointed to bluntly pointed poll with a cutting-edge which is medial in profile and markedly flaring and rounded in plan. Some blades in this group have an expanded, or thickened, but undecorated poll (Gerard, 1973: fig.1 and plate 1A). The general morphological similarity between this group and the six decorated blades suggests that the latter are a sub-group distinguished by their decorative elements and the elaboration of their lugs and polls (cf. Swadling and Phillips, no date: 14-5). Examination of the extensive collections of stone blades from the Solomon Islands made by Voyce, Waterhouse and others, and now housed in the Auckland Institute and Museum, the Otago Museum, and The Australian Museum, indicates that the distribution of the undecorated double-lugged blades matches that of the four decorated blades for which provenances are recorded: the northern Solomon Islands especially Buka and Bougainville Islands.

In her study of hafted stone percussion implements in Melanesia, Crosby (1973, vol. I : 83) was unable to locate any hafted specimens of double-lugged blades from the northern Solomons in the museums covered by her study. It would appear, therefore, that the hafted Choiseul blade may be unique. It is hafted as an axe. Even though the hafting could be a late event in the blade's history, two morphological considerations favour the extension of this kind of hafting to all double-lugged blades:

(1) the cutting-edges are medial in profile;
(2) the cutting-edges are flared and rounded in plan.

Hafted examples of true adzes in Melanesia usually lack these characteristics. Blades with flaring cutting-edges are generally hafted as axes; e.g., the so-called 'Mt. Hagen axes' of the central highlands of New Guinea.

If we accept that the double-lugged blades from the northern Solomon Islands were hafted as axes, what kind of haft was used? The Choiseul blade has a solid wooden handle with a deep U-shaped notch at one end into which the blade is set and secured with cane binding and gum (fig. 15.5), similar to the hafting method for a steinhammer from Nissan Island to the north of Buka (Krause, 1906: 123, fig. 87). Parkinson, however, mentions only the use of rattan cane bound around the axe blades of the Buka—north Bougainville areas (Parkinson, 1907: 499 and 501, fig. 81; cf. 1898 : 25). Parkinson also draws comparison between this method and that used on the Bali-Vitu Islands, north of New Britain, and on New Britain itself from the Willaumez Peninsula to South Cape on the south coast (Parkinson, 1907: 239). In the New Britain area this method of hafting seems restricted to axe blades which have marginal notches or low, indistinct lugs. At this stage it is impossible to suggest which method of hafting was employed for all of the double-lugged blades of the northern Solomons, though the solid wooden haft of Nissan and the Choiseul blade are so similar that this technique may once have been widespread in the region.

In his original description of the Toiminapu blade, Casey (1939: 144-5) interpreted the lugs as representing bird heads. Subsequent authors appear to have taken his statement at face value (e.g. Riesenfeld, 1955; Golson, 1968: 7; Golson, 1972a : 968, 1972b : 588), and on occasion have cited it in comparisons with undoubted aviforms (Pretty, 1964; Neich, 1971). We suggest, however, that Casey's identification is suspect. Given that the working of fine detail in stone may be extremely difficult, particularly when dealing with small lugs on these blades, we cannot isolate one attribute of the Toiminapu lugs which can be assigned without qualification to a bird and to no other animal form. A similar comment can be made about the Choiseul blade, the lugs of which have been identified by people from Choiseul as representing dogs; as Ms. Craven of the Solomon Islands Museum has pointed out to us, there is nothing specifically dog-like about these lugs.

Stone zoomorphs identified as birds have been reported from many parts of the northern Melanesian area, especially from New Guinea (e.g. Höltker, 1951; Schmitz, 1956, Schmitz, 1966; Mc Carthy, 1949; Bulmer and Bulmer, 1962; Neich, 1971). Some of these identifications cannot be challenged, for many have wings which betray their avian origins. Newton (1967: fig. 97) suggests that some of these figures may have been used as stoppers for sacred flutes, such as are used in various parts of the East Sepik Province of Papua New Guinea. Most authors, however, identify the majority of these figures as fragments of stone peselt handles, bearing in mind their close similarity with two complete bird-handled pestles from the Aikora River (Barton, 1908) and Wonia (Pretty, 1965) in Papua New Guinea. Casey (1939 : 143-5) was well aware of the stone figures, and compared them with the lugs of the Toiminapu blade.

Among the undoubted bird-forms, the eyes may be depicted in a variety of ways; as circular or lozenge-shaped raised bosses, as circular hollows, or as circular grooves ground-out by a hollow cylinder. The latter technique is that used for the Toiminapu blade and the Bougainville figurine. In many cases the eyes are not delineated at all. McCarthy (1949 : 159 and plate X, fig. 14) has identified one stone animal head with bosses as eyes to represent an echidna or a possum. If this is so, then eye form cannot be used to identify bird figures. Neither can mouth form be so used. On the Toiminapu and Choiseul blades and the Bougainville figurine, the mouth is indicated by a shallow groove; but on many other figures, including definite bird figures, the mouth is not indicated. The identification of most figures as birds is based on overall body form, including wings where present. On the blades described here, only the head portion of the animal is present, so that the identification is based on a part of the body which is in itself undiagnostic of birds. The only attribute apparently shared by all of the decorated blades and the definite bird figures is the manufacturing technique: hammer dressing and grinding. This technique is found on a wide range of stone objects throughout Melanesia, and currently does not appear to be characteristic of a particular group of objects or time period. It is a technique suited to certain kinds of rocks and as such cannot be used to define any kind of cultural grouping, though some authors have sought to do so (Pretty, 1964).

If, however, we allow the possibility that the lugs do indeed represent bird heads, which bird is thus depicted? A brief review of bird forms used in recent ethnographic art from the Solomon Islands suggests that the birds mainly represented are the hornbill and frigate bird. None of the attributes of the lugs specifically resembles these birds. It may be objected that to attempt such a fine identification is unreasonable, but attempts elsewhere in the New Guinea region appear to have been successful (e.g. Pretty, 1964; White et al., 1970).[2] There are many other bird-like forms incorporated into stone objects, particularly from the New Guinea mainland, for which precise identifications have not been possible.

The only indications regarding possible age relate to the Malasang and Choiseul specimens. In each case the evidence implies a late prehistoric date. However, it is difficult to accept at face value the statement that the Choiseul blade was made by the owner's grandfather. Although we do not know when his grandfather died, he is unlikely to have been born much earlier than 1830-1850, a time-depth well within the period of contact with Europeans. Had such blades been used in the mid-19th century, it is surprising that they were not mentioned by European observers. It is possible that the wooden shaft was added by the owner's grandfather to an older stone blade. Although the Malasang blade was found on the ground surface with sherds of the Mararing Style of pottery (AD1200-1800), there is no indication that they were deposited at the same time.

More useful comparisons can be made with dated archaeological materials excavated further south in the Solomon Islands themselves and in the neighbouring New Hebrides group. Most of this information became available after the comparisons between the Toiminapu blade and the bird figures of the New Guinea mainland were published. In the New Hebrides, Garanger (1971; 1972) has defined an early prehistoric pottery style which he has termed the 'Manga'asi style', after the Manga'asi site on Efate Island. At this site he found 93 pottery handles in Horizons II and III between 100 and 150 cm below ground surface (1971 : 56). These handles terminate in zoomorphic heads which taper to what appear to be mouths (Garanger, 1971 : 56, fig. 2; 1972 : fig. 134). Radiocarbon-14 dates for Horizon III indicate a first millenium B.C. date (samples GX-0963 and GX-0964, uncalibrated). There are no dates for Horizon II, but a date for Horizon I suggests that Horizon II accumulated before about 650 years ago. Garanger has compared the pottery zoomorphs with figures modelled in clay on the rims of Atzera pottery from the Markham Valley of Papua New Guinea, some of which are identified as heads of flying foxes (Holzknecht, 1957.

He also noted (1971: 65) that 'some motifs are reminiscent of the "birds' eyes" of the stone pestles and mortars of the Highlands' of New Guinea. These Manga'asi zoomorphs display no attributes which can be associated exclusively with birds, though some terminate in curved, beak-like forms (Garanger, 1971 : 56, fig. 2, centre bottom; 1972 : fig. 134, nos. 9, 17). One handle, however, is virtually identical to the lugs of the Choiseul blade (Garanger, 1972 : fig. 134, no. 19).

A three-dimensional zoomorphic head in pottery has been found at the BS-SZ-8 Lapita pottery site on Nendö Island (Santa Cruz) in the south-east Solomon Islands, dated to the late second millenium B.C. (Green, 1974: fig. 3; 1976: 264). The eyes and mouth appear to be incised, and the mouth has a bluntly pointed end. This zoomorph resembles the handles from Manga'asi, though there are major differences between the two associated potteries. In general form, but not treatment of the eyes, the Nendö specimen can be compared with the lugs of the Toiminapu and Choiseul blades.

The lugs on the Toiminapu and Choiseul blades can thus be compared with zoomorphs on pottery handles from two dated archaeological sites. If these comparisons are accepted, a revised age estimate for the blades can be offered. An upper limit of about 2500 to 3500 years ago is indicated by radiocarbon dates for the Nendö site and for Horizon III at Manga'asi; in the absence of dates for Horizon II at Manga'asi, a lower limit cannot be set, but can reasonably be expected to be greater than 650 years ago.

The zoomorphic handle from the Nendö site is the only one yet known from a Lapita pottery context. On the other hand, such zoomorphs appear quite common at Manga'asi itself. There is a similarity between the relief bosses of of the Manga'asi pottery and the bosses of the five decorated stone blades. Prehistoric pottery related to or comparable with the Manga'asi pottery has been reported from southern Bougainville (Golson, 1972b: 564-6, 574) and from Buka Island (Specht, 1969), a distribution which also overlaps with the known range of the decorated blades. It is tempting to see in these stylistic similarities and overlapping distribution an historical link between the decorated blades and the Manga'asi and related potteries. However, double-lugged stone blades seem absent from Manga'asi sites, and from surface collections in the New Hebrides in general.

At present only six decorated stone blades with double lugs are known from island Melanesia, and these all from the Solomon Islands. They form but a minute proportion of the thousands of stone blades from that island group held in museums. This extreme rarity cannot be attributed solely to inadequate sample size, since the Voyce, Waterhouse and O'Reilly collections alone total over 4,000 blades. The close similarity between the decorated blades from Buka to Simbo argues against idiosyncratic production by a single stone-worker or group of stone-workers, unless the blades were distributed from a central source through an exchange network. The extensive movements of goods, and people, within the northern Solomon Islands is well documented (see summaries in Specht, 1974; Blackwood, 1935; and Terrell and Irwin, 1972 : 340), but acceptance of this explanation for the distribution of the decorated blades would present a problem. These blades are probably a sub-group of a much larger series of butt-differentiated blades known from the northern Solomon Islands, of which the decorated blades form only a small percentage. It is most economical to explain the distribution of this larger series in terms of many different production centres within what might be regarded as a prehistoric 'areal culture' (cf. Schwartz, 1963). In the absence of petrological identifications of probable places of origin for the decorated blades, they may have been produced at one or several points within this postulated prehistoric areal culture. Their rarity further suggests that their production may have been rigidly controlled. What their function was, we may never know, but it is tempting to see them used in a context in which animals, possibly birds, played some role.

NOTES

1. This field work was carried out by J. Specht, with the assistance of S. Specht, as part of a doctoral programme at the Australian National University, Canberra, and was wholly funded by that institution.

2. The zoomorphic stone head described by Pretty in his 1964 paper is poorly documented as to its place of origin. It was part of a collection of artefacts from New Guinea and the Solomon Islands received by the Queen Victoria Museum and Art Gallery, Launceston, Tasmania in 1963. Pretty (1964 : 184) argues for a possible New Guinea origin, a view with which we would agree.

3. Riesenfeld (1950; 1955) has introduced 'Bronze age' inspirations or prototypes into the discussion of the Toiminapu blade. Golson (1972b: 582) rightly points out that such inspiration 'can no longer be assumed but must be argued within the limits of inference that a growing body of archaeological data is beginning to establish'. We see no value in pursuing putative 'bronze age' influences in Pacific Islands art, but note the imaginative attempt by Mead to identify 'a sub-stratum that is traceable to very remote origins in Asia' (Mead, 1972: 740).

ACKNOWLEDGEMENTS

We wish to thank the people of Buka Island for their assistance and hospitality during our fieldwork in 1967 and Mr. N. Murray, then Headmaster of Hutjena High School, for sending negatives of the Hutjena blade. Photographs of the Auckland blade were supplied by the Auckland Institute and Museum through the assistance of Ms. J. Davidson, and of the Choiseul blade by the Solomon Islands Information Service through Ms. A. Craven, Curator (Training) at the Solomon Islands Museum. The illustrations of the Toiminapu blade are taken from a cast made from a cast of the original specimen. The Malasang blade is now in the National Museum and Art Gallery, Papua New Guinea; the present whereabouts of the Hutjena blade is not known.

BIBLIOGRAPHY

Barton, F. R., 1908. "Note on stone pestles from British New Guinea". Man, 8:1-2.

Blackwood, B., 1935. Both Sides of Buka Passage. The Clarendon Press, Oxford.

Bulmer, R. N. H. and Bulmer, S., 1962. "Figurines and other stones of power among the Kyaka", J. Polynes. Soc. , 71:192-208.

Casey, D. A., 1939. "Some prehistoric artifacts from the Territory of New Guinea". Memoirs of the National Museum of Victoria, 11:143-50.

Crosby, E., 1973. A Comparative Study of Melanesian Hafted Edge—Tools and other Percussive Implements. 2 vols. Unpublished PhD thesis, Australian National University Canberra.

Garanger, J., 1971. "Incised and Applied—relief Pottery, its Chronology and Development in Southeastern Melanesia, and Extra Areal Comparisons". In: R. C. Green and M. Kelly (eds.). Studies in Oceanic Culture History, volume 2:53-66. Pacific Anthropological Records, No. 12.

Garanger, J., 1972. Archéologie des Nouvelles—Hébrides. Publications de la Société des Océanistes, No. 30, Paris.

Gerard, B., 1973. "Haches de Bougainville". Journal de la Societe des Oceanistes, 29:289-311.

Golson, J., 1968. "Archaeological Prospects for Melanesia". In: I. Yawata and Y. H. Sinoto (eds.), Prehistoric Culture in Oceania: 3-14, Bishop Museum Press, Honolulu.

Golson, J., 1972a. "Prehistory". In: P. Ryan (ed.), Encyclopaedia of Papua New Guinea, volume 2:961-80. Melbourne University Press, Melbourne.

Golson, J., 1972b. "Both Sides of the Wallace Line: New Guinea, Australia, Island Melanesia and Asian Prehistory". In: N. Barnard (ed.), Early Chinese Art and its Possible Influence in the Pacific Basin, volume 3: 533-96. Intercultural Arts Press, New York.

Green, R. C., 1976. "Lapita Sites in the Santa Cruz Group". In: R . C. Green and M. M. Creswell (eds.), Southeast Solomon Islands Cultural History: a preliminary survey; 245-65. The Royal Society of New Zealand, Bulletin 11.

Höltker, G., 1951. "Die Steinvögel in Melanesien". In: Südseestudien, Gedenkschrift Zur Erinnerung an Felix Speiser: 234-65. Museum für Völkerkunde, Basel.

Holzknecht, K., 1957. "Über Töpferei und Tontrommeln der Azera in Ost-Neuguinea". Zeitschrift für Ethnologie, 82:97-111.

Krause, F., 1906. "Zur Ethnographie der Insel Nissan". Jahrbuch des Städtischen Museum für Völkerkunde, Leipzig; 1:45-159.

McCarthy, F. D., 1949. "Some Prehistoric and recent stone implements from New Guinea". Records of The Australian Museum, 22 (2):155-63.

Mead, S. M., 1972. "Formal and Iconic Analysis of Polynesian Hafted Adzes: A Preliminary Statement of Interrelationships within Oceania". In: N. Barnard (ed.), Early Chinese Art and its Possible Influence in the Pacific Basin, volume 3:723-86. Intercultural Arts Press, New York.

Neich, R., 1971. "A Prehistoric Stone Bird from Bougainville and its Relationship to Northern Solomons Implements ". Dominion Museum Records in Ethnology, 2 (7):75-82.

Newton, D., 1967. New Guinea Art in the Collection of the Museum of Primitive Art. Museum of Primitive Art, Handbook No. 2, New York.

Parkinson, R., 1898/99. "Zur Ethnographie der nordwestlichen Salomo Inseln. Abhandlungen und Berichte, Zoologisches and Anthropologisch—Ethnographisches Museum Zu Dresden, 7 (b):1-35.

Parkinson, R., 1907. Dreissig Jahre in der Südsee. Strecker und Schroeder, Stuttgart.

Pretty, G. L., 1964. "A Stone figure of a Crested Cockatoo from Melanesia". Man, 64:193-4.

Pretty, G. L., 1965. "Two Stone Pestles from Western Papua and their Relationship to Prehistoric Pestles and Mortars from New Guinea". Records of the South Australian Museum, 14:119-30.

Riesenfeld, A., 1950. "Some probable Bronze Age influences in Melanesian culture". Far Eastern Quarterly, 9 (3):226-30.

Riesenfeld, A., 1955. "Bronze Age influence in the Pacific". Internationales Archive für Ethnographie, 47:215-55.

Rolston, R., 1945. "Notes on some Melanesian woodworking implements". J. Polynes. Soc., 54:212-7.

Schmitz, C. A., 1956. "Two 'prehistoric' stone objects from the Huon Peninsula". Mankind, 5:128.

Schmitz, C. A., 1966. "Steinerne Schalenmorser, Pistille, und Vogelfiguren aus Zentral-Neuguinea". Baessler-Archiv (N. S.) 14:1-60.

Schwartz, T., 1963. "Systems of Areal Integration: Some considerations based on the Admiralty Islands of Northern Melanesia". Anthropological Forum, 1:56-97.

Specht, J. R., 1969. Prehistoric and Modern Pottery Industries of Buka Island. T. P. N. G. Unpublished PhD. thesis, Australian National University, Canberra.

Specht, J., 1974a. "Stone Pestles on Buka Island". Papua New Guinea". Mankind, 9:324-8.

Specht, J., 1974b. "Of Menak and Men: trade and the distribution of resources on Buka Island, Papua New Guinea". Ethnology, 13:225-37.

Swadling, P. and Phillips, C., No date. "The Voyce Collection of Bougainville Axe-adzes". M. S. (Unpub.) Southeast Solomon Islands Cultural History Project. Files, Anthropology Dept., University of Auckland.

Terrell, J. E. and Irwin, G. J., 1972. "History and Tradition in the northern Solomons: an analytical study of the Torau migration to southern Bougainville in the 1860s". J. Polynes. Soc., 81:317-49.

CULTURAL PROCESSES LIMITING DIVERSITY IN NEW ZEALAND PREHISTORY

Garry Law

Introduction

We have long recognised that at about 1800 A.D. there was a distinctive culture which can be called New Zealand Maori, and that it was to a large extent, homogeneous, having internal cultural variety which is but minor when compared with the external differences. These external differences are in part a result of environment but also occur in cultural elements which are relatively environment-free. In 1800 A.D. Maori New Zealand and, say, Rarotonga, had been effectively separated for at least 600 years. The differences of social organisation, and particularly language and art styles applied in carving wood, reflect this separation. The same statement can be made of the arts and language of the Chatham Islands, although the time depth of this separation is not as clear.

In a large land mass it is possible for cultures to become effectively separated and display diversity reflecting the time depth of the occupation in the area. New Zealand is an archipelago which extends over 1600 km in length. This distance, in many parts of the prehistoric world, could cover several diverse cultures and some aspects of the cultural differences between New Zealand and Rarotonga or New Zealand and the Chatham Islands illustrate the degree of cultural separation which could have been reached within New Zealand had the available time depth been utilised. Yet it is clear that despite certain regional differences, these cannot be attributed to persistent and isolated regional traditions. This situation poses some interesting problems, because although archaeological explanations seem to concentrate upon explaining differences, processes which limit such diversity are as worthy of our attention as those which create it.

There are two broad alternative explanations possible for New Zealand's lack of wide diversity in A.D. 1800; that the low diversity does represent a short time depth, or that cultural isolation did not occur, to any marked degree, in any area of New Zealand. The first explanation requires a cultural replacement to have taken place relatively recently in New Zealand with a group originating from one confined area in the country coming physically to dominate all of it. While arguments have been developed (Duff 1956:7, Groube 1970: 153-155, Simmons 1971:95) for population movements from northern New Zealand these have not met wide acceptance, as the proposed movements are in part founded on Maori tradition and have little factual support from excavated evidence.

The second explanation; lack of isolation, takes New Zealand as a cultural whole. If culture is considered as commonly held ideas and models, then for

New Zealand to contain a more or less homogeneous culture those ideas and models must have been universally held and new ideas must have been transmitted through the whole region. That is, an information system effectively exchanged ideas on ways to behave. Effective communication of cultural information throughout New Zealand is, in fact, implicit in a lot of statements on Maori culture.

A third broad explanation might be that there was little cultural change through New Zealand prehistory. This is not supported by either internal archaeological evidence or external comparative evidence.

Some of the external differences are explicable by founder effect and an initial adaptation to a non-tropical environment, some internal regional differences (e.g. most of those described by Skinner 1926) are explicable through regional environmental differences, and some internal changes are ongoing adaptations to environmental modification and population pressure feedback. This does not, in my view, explain the totality of Maori culture. Cultures change as well because people are creative. The extent and nature of the response of Maori Society to the opportunities offered by European contact (Groube 1969:7-9) shows Maori Society of the 19th century as strongly creative, not merely adaptive.

In the rest of this paper I wish to review the processes operating in New Zealand at the time of European contact which were limiting diversity late in New Zealand prehistory, then review some archaeological evidence demonstrating effective interchange, and suggest some of the processes found ethnographically as potential models.

To some who perhaps do not know him, this may seem an unusual subject to raise in a volume offered to Ron Scarlett. Those more familiar with Ron will know of his eagerness to contribute to all facets of archaeology and will respect his dedication to issues of social change in our own society. My contacts with Ron over more than a decade have been a material contribution to my interests in archaeology.

Information Exchange—The Ethnographic Record

While a distinction was made above between homogeneity arising from replacement and that arising from interchange, the ethnographic record suggests that many of the population movements which took place were not commonly part of any process of replacement by newly dominant groups.

An important paper by Urlich (1972) has reviewed movements by North Island Maori groups in the period 1800 to 1840. She tabled some 46 movements in the period and estimated they involved about 20% of the population. The majority were clearly in response to stress resulting from warfare and their timing suggests many were occasioned by the inequitable introduction of guns into warfare. The majority are not conquests of new territory but rather groups in danger of complete defeat seeking a safer or more advantageous position. Relatively few were simple movements into a neighbouring territory, and many spanned a large part of the length of the North Island. This could be seen as an intensification of a prehistoric pattern. While warfare may at first sight appear to be likely to induce a more defensive attitude, restricting movements, the reverse seems to be the case. The result of this process is that inter-regional diversity must decrease, but, conversely, where incoming and resident groups have little contact diversity within a region could well be increased. Sullivan's review of ethnographic material from the Auckland Isthmus (1976) has shown that such movements there were reciprocal between distantly related kin groups, and groups allowed to reside in an area would be obliged to allow the hosts use of part of their own territory in the future.

On a more local scale there are clear accounts of exchanges between neighbouring groups which constitute occasions for information exchange. Territorial boundaries between neighbouring groups do not appear to have been closely drawn. Thus a number of groups living in a region could have recognised rights to exploit resources which were spatially distributed in a pattern which did not reflect their usual residences. This is not very surprising when considered in relation to the structure of a basic Maori social group, the hapu. Being a descent group which could recruit members from anyone who could trace a lineage from a person acknowledged to have been a member, or to a pivotal ancestor, it allowed considerable flexibility (Webster 1975:124, 144).

There was no requirement that a hapu be a territorial group or even that all its members be co-resident (ibid.:126) and there appears to be no ethnographic evidence to allow any dogmatism on these points. Some seem to have had geographically widely spread exploitation rights. The historical origins of such rights to land and resources are readily understood in terms of flexible recruitment into hapu. Sullivan's study (1976) again shows territorial overlap in the use of harbour resources in the Auckland Isthmus. Other examples are groups resident in the Waikato having rights for fishing on the West Coast (Roberton 1965:40) and a group from the Hauraki Gulf encountered in temporary encampment in Mercury Bay in November 1769 (Banks 1962:427-428). Clearly this pattern of resource use made for much more frequent inter-group contact and required certain mutual understandings. Indeed, in fishing, it seems clear that large netting operations were carried out by agglomerations of widely spread groups especially brought together just for that purpose. (Yate 1835:71, 85). The huge nets of the Bay of Islands for example, must have required the services of large numbers of people to be of use (Roux in Kennedy 1969:219, Duperrey 1971:92, Banks 1962:444). Similar co-operative ventures occurred in pursuing warfare (Roberton 1965:40) but only rarely (Vayda 1960:32-33).

Re-location of one party upon marriage is an effective mode of distributing cultural knowledge. It is very difficult to judge the extent of inter-group marriage from the ethnography. Certainly well-recorded instances of marriage between highly ranked people occurred but it must be doubted if choice of spouses in lower ranks was as often made from groups as widely spread. Marriage was most commonly within hapu (Biggs 1960). This form of direct information exchange cannot then be held to be important. However, the kinship ties which could almost always be traced between groups often stemmed from high rank inter-marriage and re-cementing these ties by social exchanges was commonly recorded. Likewise, there are many accounts of lavish feasts and gift-giving which seemed excessive to the canny European observers (Yate 1835:139, Polack 1838:75, 79).

The control on this distribution seems to have been concentrated in the higher class ranks of Maori society. The exchanges of food mostly seem to be in the form of hospitality lavished on visiting parties with presents of food restricted to portable choice items. Gifts were more commonly in goods and made to higher ranked people in opposite parties. Social visits and accompanying gift-giving were not just a local process. For instance, the Ngai Tahu group in the northern South Island maintained contact with the south of the South Island, to the West Coast of the South Island and to Hawkes Bay (Stack n.d.:192). Social exchanges, then, appear to be an important element in providing models leading to greater dispersion of new varieties of items of material culture. There is on record quite a complex set of rules governing gift reciprocity (Sahlins 1974:149-168), and Mair's (1972) review of the ethnographic records of the Southern Wairarapa contains several examples of the sorts of interaction listed above and illustrates well the reciprocity of gifts of food, other consumable items and artefacts which went on.

In contrast to the frequent records of these sorts of exchange, there seem to be very few which record barter between individuals. One can suggest that the presence of Europeans would re-focus any barter exchange which might otherwise have occurred between Maoris. Orchiston (1975) has suggested that the readiness with which the Marlborough Sounds Maoris entered into barter with the first Europeans bartering goods and food, was an adaptation from the situation they were often in, acting as middlemen in exchange of goods between the North and South Islands. (But it is worth pointing out that although many navigators praised the Maoris for their fair trading not a few suffered blatant acts of theft during barter).

In the light of Orchiston's suggestion, the trading between Maoris visiting the <u>Endeavour</u> in 1770 near Cape Palliser is interesting: "... on having presents made them immediately made presents to us in return (an instance we have not before met with in this Island)." (Banks 1962:465). This observation supports the importance of exchange to residents of the middle of the country. There are also some accounts of barter between individuals of different Maori groups. Tuki, the Maori abducted to Norfolk Island in 1793, was reported by King as saying that in times of peace they "... visit and traffic for Flax and Green Talc Stone of which they make Axes and Ornaments" (McNab 1914 Vol. 1:542). It would be dangerous to read too much into this. Nicholas describes barter he observed between individuals in 1814, when the goods being exchanged were flax mats for gannet feather hair dressing pieces, "... prepared exclusively in the Bay of Islands whence they are carried to the other districts and form a staple article of trade." (Nicholas1817 Vol. 1:398). The occasion for this barter was a visit on a European vessel to another district. Europeans were involved in the trade, and the Maoris were also bartering for pieces of iron; facts which militate against regarding this as a representative occurance. Some support however, is offered by the high frequency of occasions when early Europeans were offered cloaks in barter and the ease with which trade in prepared flax was developed by the first traders, suggesting these were already recognised items of barter.

In the historical records and in traditional histories it is possible to find accounts of the sorts of materials moving from one place to another but the social implication of these is obscure. Thus Taranaki seems to have been regarded as a source of fine cloaks (Skinner 1946:64), the west side of the Auckland peninsula produced Dentalium nanum shell used elsewhere (Hamilton 1899: 307), obsidian in core form was taken from Mayor Island to the north (Manning 1875:103), eels from Wairarapa to the Napier area (Mair 1972:210), slaves from Ureweras to Thames (bartered for muskets), (Best 1925:520), preserved birds from inland Hawkes Bay to Mahia in exchange for whale bone (Best 1925: 534), preserved kumara, "mats" and canoes were bartered between Cook Strait and areas further south in exchange for nephrite, white heron feathers, preserved birds, a scented vegetable oil and vegetable food product from Cordyline sp. (Shortland in McKay 1873:125), and from the west coast of the South Island nephrite (Heaphy in Taylor 1959:237) and a chewing pitch (Heaphy in Taylor 1959: 207) passed to other parts of New Zealand.

It cannot be held that these exchanges were solely on a barter basis and it is impossible to claim that barter was an important mechanism for the exchange of goods between Maori groups. In generating opportunities for cultural interchange and exchanging new models of material items other processes seem to have been more important.

In maintaining homogeneity of art forms in the male domain, travelling specialists appear to have been important. In 1820, Cruise observed a carver at work in the Bay of Islands and was told he had been brought there from Thames for the purpose (Cruise 1957:34), while Polack encountered a tattooist from Whakatiwai (in the Hauraki Gulf) whilst travelling between Hokianga and Kaipara. (Polack 1838: vol. 1:164). If such interchanges were common, skilled artisans must have been very familiar with the work of their fellow artists over a larger part of the country than solely the area in which they had relatives.

There is also some evidence that slaves with special skills were given a more respected place in society than their origin would normally allow. Thus the tattooist whom Earle came to know was well rewarded for his services in the Bay of Islands despite being a slave (Earle 1966:124-125), and Duperrey (1971:39) observed another slave in the Bay of Islands who had been given a wife and who acted as agent for his owner in dealings with Europeans. However, these cases might reflect the unusually large number of slaves owned by residents of the Bay of Islands in the 19th century. Historical accounts more commonly indicate that slaves, and other people of low rank, were usually allocated the load-bearing, food preparation and similar domestic tasks. If this was generally the case, and given that slaves were frequently captives from distant areas, then their allocation to domestic duties could have reduced their potential ability to introduce cultural diversity into the societies of their owners.

Turning now to information in a narrow sense, effective and rapid communication is recorded in the literature. On Cook's first voyage, Maoris at Thames knew of the Tahitian Tupaia, the information reaching them, no doubt from Mercury Bay, before the ship had travelled to Thames (Banks 1962:435). On the same voyage people near Cape Palliser asked for nails on first contact. Knowledge of these must have passed from the East Coast to Palliser in the period the ship was working around the North Island (Leach 1976:40; Cook 1951: 115). Both of these are instances of communication within tribal groups. It was not always effective beyond groups. For instance, the people at Queen

Charlotte Sound on Cook's second voyage had no knowledge of him despite his previous visit there.

An illustration which literally sets out an individual Maori's perception of the world, as it affected him, is the map of New Zealand drawn by Tuki while on Norfolk Island in the 18th century. (Milligan 1964) Figure 16.1 shows the North Island only, and the grid distortion shows the geographical interpretation advanced by Milligan. While Tuki's world was weighted to the territory in which he lived, this is certainly wider than the territory which he would need to know for subsistence. It could approximate to the area in which he had kin ties, or was involved in intergroup politics.

Summarising the ethnography, there seems to be good evidence that in the protohistoric period, cultural information was transmitted readily and widely in New Zealand and that differences, where they existed, could not have been maintained in ignorance of other ways of behaving or of other models to follow.

Prehistory

The sorts of data available from archaeology relate to the transmission of materials, particularly rock materials, through New Zealand and to the spread of new forms of behaviour. Neither can be viewed in isolation from the considerable diversity of human environments which New Zealand offers.

The similarity of the material culture of Archaic sites throughout New Zealand has been commonly cited (Golson 1959:44), Green considering that the low diversity was surprising given the differing economies which were adopted (1971:26). The rock shelter art of North Otago and South and North Canterbury also shows strong similarities (Trotter and McCulloch 1971:71-76, 1973). These, and the wide distribution of rock types (reviewed below), have lead to suggestions that the population was very mobile (Simmons 1969) and that the people were involved in long distance trading journeys (Keyes 1975:10). There is some evidence now that this was not the case. The people buried in the Wairarapa sites, particularly in the Washpool Midden Site (N168/22), differ in a number of respects from the people of Wairau Bar (S29/7). Their diet, as it has affected their teeth, appears to have differed; their general health appears to have differed also, and there are suggestions that while both groups were clearly Polynesian they belonged to separate Mendelian populations (Sutton 1974:67-70; 141; Houghton, 1975). In addition, the occupants of the Washpool sites appear to have lacked the adze-making skills necessary to work the metasomatised rocks as well as these were worked at Wairau Bar. Taken together these facts suggest that populations were to some extent localised by the time Wairau and the Washpool Site were occupied.

These suggestions of insularity can be contrasted with the wide-ranging sources of raw material used in the early Palliser Bay sites, the imported rock material being numerically dominant (Prickett 1975:35, Leach 1976). In fact wide movment of a variety of materials is common in the Archaic of New Zealand. Scallop shells used for ornaments were imported to Wairau Bar (Duff 1956:134) to the Washpool Midden (Leach 1976) and the Weka Pass Timpendean Shelter (S61/4 C_{14}, NZ 892 Shell A.D. 1514 $\pm$ 53, Trotter 1972b:46).

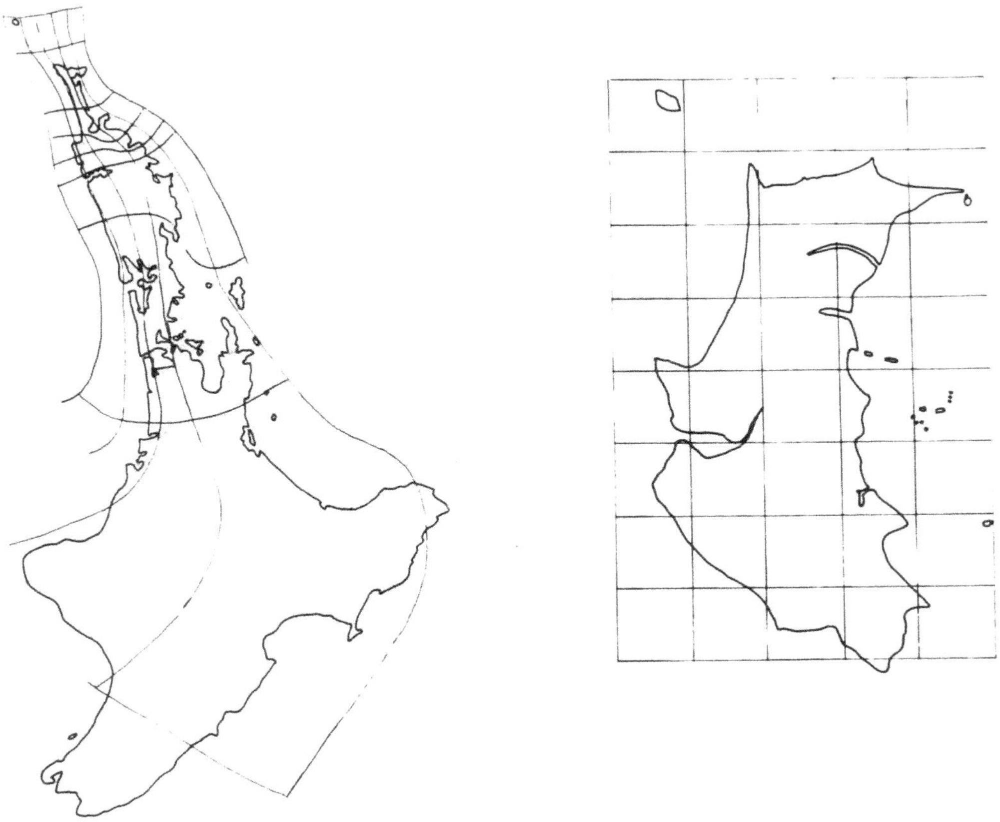

Fig. 16.1 Tuki's map and reconstructed grid distortion.

Shellfish were moved inland to a variety of sites in the South Island (Weka Pass, Pentland Downs (S61/20), Glen Gynk (S61/24) (Trotter and McCulloch 1973), and Gooseneck Bend Shelter (S117/8, Ambrose 1970:407) and in the north as well, to Whakamoenga (N94/7, Leahy 1976:45). Some of these shells were doubtless for use as tools. Moa bone was also moved from the South Island to Palliser Bay (Leach 1976) and to Paremata (N160/50, Scarlett 1974: 11). In the South Island moa bone was moved into Fiordland (ibid:9). Turning to rocks, basalt from Tahanga quarry on the Coromandel Peninsula is present in the earliest sites throughout the north of the North Island (Best 1975, Moore 1975, 1976) from Taranaki to the Eastern Bay of Plenty. Siliceous sinter occurring near Tahanga was moved to Houhora (N6/4) in the far north along with Tahanga basalt and Mayor Island obsidian (Best and Merchant 1976: 108).

Some rock materials of lower quality were not transported far. The greywacke used on Motutapu is only dominant locally (Davidson 1972) and a rhyolite with small spheroidal obsidian inclusions used as an abrader on the Coromandel (Crosby 1977) does not occur elsewhere.

Most importantly obsidian from a variety of northern North Island sources (Ward 1973) was used throughout New Zealand. The earliest North Island sites all appear to have obsidian and the nearest source is not always dominant. At Houhora (Best 1975) and Tokoroa (N75/1, Law 1973) obsidian from Mayor Island is dominant while other sources are closer. Judged on transmitted light characteristics it is unusual for only one obsidian source to be represented in the early sites (Green 1964). The only determinations available for an early, northern site are from the Sunde Site (N38/24) on Motutapu. Here obsidian from Layer 2 and 4 (C14 a.d. 1330 ± 60 and later, NZ 1899) is sourced to Huruiki in Northland and Mayor Island (Davidson 1972:6).

The Palliser Bay sites provide the best data we have on sources. Here obsidian from seven different sources including Mayor Island, Coromandel, Northland, Rotorua and Taupo appears in the lowest level of the Washpool Midden and the source range here, and at two other sites, does not increase with time, but if anything becomes restricted (Prickett 1975, Leach 1976). What is interesting in the present context, is that the proportion of the different sources shows little variation between the different Palliser Bay assemblages. If the obsidian was obtained by direct voyages between Palliser Bay and the sources it is remarkable that this consistent variety should ensue. The observed pattern is more consistent with a model which has occasional small supplies of obsidian arriving at Palliser through intermediaries. Filtering supplies through a chain (or chains) would limit sudden changes in proportion.

Obsidian movement extends to Southland in the Archaic (Lockerbie 1959: 83). It has been claimed to be present in the earliest sites but published evidence is wanting for this. Obsidian reached Waitaki River Mouth (S128/1) in core from at some stage in the occupation of the site (C_{14} SUA 61 moa bone a.d. 1350 ± 80). It seems clear from the variety of colours in obsidian from South Island Archaic sites that a variety of sources are represented in assemblages. (Millar 1971:163; Trotter 1970:473; Trotter 1972a:145), although Heaphy River (S7/1) is in contrast (Wilkes and Scarlett 1967:207 C_{14} NZ 509 shell a.d. 1400 ± 70).

The only rivals to obsidian, in their extent of distribution, are the metasomatised rocks of the Nelson mineral belt. (Keyes 1975; Walls 1974). The precise sources of these are difficult to identify. One source, distinctive in hand specimen, is the Ohana quarry on D'Urville Island. (S10/56, S10/90). This material occurs in adze form throughout New Zealand although it is rare in the northern North Island, and it is found in the earliest layer at the Washpool site (Prickett 1975:145). It seems unlikely that this rock, or the other Nelson metasomatised rocks, were transported outside the source area other than as finished adzes.

Nephrite appears in Archaic sites in Cook Strait and the South Island but there is only site-association evidence for it in the northern North Island at Wheritoa (N53/4 Crosby 1977). In the South Island its use extends to Southland (Lockerbie 1959:83). A similar material, bowenite, from Fiordland is found at Wairau Bar, Papatowai (S184/5), Tiwai Point (S181/16 C_{14} a.d. 1508 $\pm$ 53) (Coutts 1971:61), Kings Rock (S184/6) lower layer (Lockerbie 1940: 413), and at Shag Point (S146/5 C_{14} shell a.d. 1516 $\pm$ 50), (Trotter 1970:478). Some South Island rock materials have more restricted distributions, for instance Gawler Downs silicified tuff (Orchiston 1976). Silcrete from a variety of Otago sources is predominantly found in Otago sites. Simmons and Wright (1967) argue for some movements of it within Otago.

While this review has demonstrated good evidence for the movement of materials in the Archaic, little serious work seems to have been put into models to account for the distributions. The interrelation of Archaic communities is deserving of more attention. At present it appears that communities were localised but that they had sufficient contact with each other to recognise each others needs and meet these through exchange relationships. There are, however, some problems with the Archaic evidence in the North Island. The first is simply that the evidence is spatially discontinuous. Archaic sites are well attested from the Coromandel and the Hauraki Gulf. Though the survey data is sparse it seems likely sites can be attested on most of the east coast southwards, but there is very little evidence of similar sites from Taranaki north until Aupouri and there is a similar gap in the evidence from Doubtless Bay south to Great Barrier on the east coast. The distribution map for adzes given by Golson (1959:68) reflects well the discontinuous distribution of known sites.

The second problem is the apparent persistence of the Archaic in various areas. On Motutapu a number of Archaic adze attributes apparently continued in use for long after the 14th century ash eruption of Rangitoto and in sites with the usual early mixture of food and artefactual remains. (Scott 1970, Golson 1959). On the Coromandel a similar site was occupied in the 18th century (Opito, Moores Bach Site N40/16 C_{14} NZ 1992 a.d. 1770 $\pm$ 60, NZ 1993 a.d. 1760 $\pm$ 60), while the Hot Water Beach Site (N44/69) continuing to the 16th century, contains clear evidence of a large number of Archaic traits, again in a site with mixed food and artefactual deposits (Leahy 1974). It seems clear that Archaic portable artefacts persisted late in the Coromandel area as did an economy emphasising exploitation of marine birds and open sea, baited hook, and lure fishing. (However it should be noted that storage sites are dated to the early Archaic at Harataonga (Law 1975), Skippers Ridge (N40/7 Davidson 1975 C_{14} NZ 1740 a.d. 1143 $\pm$ 57) and Sarahs Gully Pa (N40/10,

Birks and Birks 1970, 1973)). At Tahanga Quarry some Archaic adze attributes apparently persisted to the end of prehistory, although probably only the attributes well adapted to the use of free flaking rock (cf. Best 1976).

A third problem is that, sites with a conventional Classic orientation appear disconcertingly early and in areas where the usual Archaic evidence is non-existent. Thus Otakanini (N37/37) appears by the 14th or 15th century as a very large defended site with food storage evidence and in a location offering little opportunity for open ocean exploitation or for exploitation of the range of birds found in Archaic sites. (Bellwood 1971:71, 1972, 1973). There are no earlier Archaic sites in the area. Likewise, Kauri Point (N53/5) is in close proximity to good horticultural soils, and has sophisticated fortifications dating from the 15th century. The 15th century garden on Moturua (N12/6, Peters 1975) in the Bay of Islands is also in an area where no case can be made for any substantial presence of Archaic portable artefacts, yet occupation evidence is available from this coast in Wellman's sediment profiles which lie under 14th century pumice deposits (1962). Aileen Fox's pioneer work in Hawkes Bay established that a defended site there extends back to the 16th century and it has earlier storage facilities. Again there is no suggestion of a close relationship to any occupation of the area using Archaic portable artefacts. (Fox 1974, 1976). On the Auckland Isthmus there is an early date from Mt Wellington (N42/4 C_{14} a.d. 1430 $\pm$ 40)—a fortified site, and Archaic adzes from there and from the Mangere area (Best 1975, Copsey 1974) in association with rich garden soils used relatively early (Sullivan 1975), fortification sites and estuarine resources. Lastly, the slight Archaic evidence on the North Island's west coast, at Kaupokonui, and Manukau and Raglan Harbours, does not recommend this region as a starting point of a local transition to later Maori adaptation to New Zealand. Rather, it suggests a pattern of unintensive and coastal Archaic use of the area followed by the more populous settlement of fully adapted Classic groups. A similar case could be made for the Waikato.

It seems, therefore, that we have established an extraordinary antiquity for non-Archaic behaviour, and certainly one which overlaps with much of our excavated Archaic evidence in the northern North Island. In addition, by the mid point of the North Island's prehistory there were areas where exploitation and artefact manufacturing behaviour still looked back to the initial adaptation to New Zealand. In other words, a high degree of cultural diversity remained. When and why did it decline ?

The Development of Cultural Uniformity

Groube, in an earlier review of New Zealand evidence wrote "it is very probable that some of the changes crucial to our understanding of the Classic Maori took place very rapidly and at an earlier date than suggested by South Island fish-hook evidence" (1969:10). This seems to be being reinforced by more recent evidence, which indicates that by the 15th century the basic pattern of a Classic economic adaptation had been established, that it was practised over a wide area in the North Island by people who built sophisticated fortifications, and had almost entirely replaced the tropical Polynesian portable artefacts. If this is correct, it must place a considerable time depth on the first reaching of this adaptation. The possibility of New Zealand's low diversity at contact arising from a recent spread of a late horizon is thereby contradicted.

There is archaeological evidence to demonstrate that the adoption of the different elements of late patterns of culture was not simultaneous. The most important evidence is from Kauri Point. The development there of the round-topped combs known ethnographically is demonstrably later than the first occupation of the adjacent knoll (Shawcross 1963, 1964). The development of this form from a frame of rectangular shape can be argued to parallel the development of curvilinear late Maori art from a more typically Oceanic art form. The development of the late comb form, at least, is demonstrably not contemporary with the first exploitation of microenvironments typical of the Classic phase— such as the harbour-side location of Kauri Point. This suggests that processes other than ecological adaptation may have been occurring, in particular, trait diffusion.

Some varieties of fishing gear appear to have achieved a rapid, late distribution throughout New Zealand. Crosby has characterised the late fishing gear in New Zealand as a development of localised secondary adaptations, but there are exceptions. Kahawai lures appear to have been distributed throughout the North Island. Evidence for any prehistoric occurrence of the shell inlay in archaeological sites appears to be lacking. It could then be regarded as a late spread of a new form. Similarly, multiple barbed and serrated points of two-piece fishhooks make a late appearance throughout New Zealand, although, in this case, prehistoric provenances are known. (Lockerbie 1940; Simmons 1967:34; McKinlay 1971:89). A similar case can be developed for one-piece hooks with internal shank and point barbs (Hjarno 1967:35). The more frequent use of barbs in later South Island Archaic sites is among the evidence referred to by Groube above. Trotter (1965) makes a strong case for a progressive adoption of barbs together with a shift to two-piece hooks in North Otago sites. This is also suggested in other Otago sites (Simmons 1967:48; Hjarno 1967: 38; Trotter 1970:477), and the adoption of barbs and two-piece hooks seems to have been occurring over a wide area of New Zealand (Millar 1971, Leach 1976). A case has been made for a late 18th century conquest of Otago by a group bearing Classic Maori Culture (Simmons 1967:55), but the artefact forms supposedly introduced are not much more numerous than those which appeared in Otago and elsewhere earlier. The argument is weakened by the dates from a palisade at the important Classic site of Murdering Beach (S164/16 C_{14} a.d. 1640, 1617, 1610; errors of the order of $\pm$ 50). It would seem more likely from this evidence that Otago was actively contributing to Classic Maori in the 18th century rather than simply having it introduced from further north.

The economic evidence suggests that the exchange of goods, in the late period was more restricted than it had been earlier. Marine resources reached the inland Waikato, (Shawcross 1968:22; Bellwood 1971:87; Peters 1971:137), and Taupo (Leahy 1976:45), and the occurrence of toheroa (*Paphies ventricosum*) at Skippers Ridge (Davidson 1975:23) is outside the modern range of this species. This evidence is less impressive than that for the earlier period.

Of the adze materials, Tahanga basalt and the Nelson metasomatised rocks have much more limited distributions. Best (1976) produces a convincing rationale for this and demonstrated what is perhaps a more typical late distribution pattern, that for gabbro from the Tangihua massifs in Northland (1976:60). Obsidian however continued to move throughout New Zealand. It is found in late South Island sites such as Little Papanui upper layers (S164/1, Simmons 1967:43, and Pariwhakatau (S55/7, Trotter 1975:150). The only source determinations available are from late prehistoric Southland. Mayor Island obsidian occurs at Garden Island, Chalky Inlet (Coutts No. GI1) and at Sandhill Point (Coutts Nos. SHP/1,2 and 4), while Taupo and Huruiki obsidian occur at the latter (Ward 1974:53, 56).

So far as the North Island is concerned, obsidian at Otakanini (not necessarily all late) was obtained from Great Barrier, Huruiki and Mayor Island. At Mangakaware (N65/35) in the Waikato, obsidian came from Mayor Island and Taupo while in the later site at Skippers Ridge (II, N40/73 C 14 > a.d. 1818, > a.d. 1737) obsidian came from Whitianga and Mayor Island. (Armitage et al 1972:418). On Motutapu obsidian has been sourced from the Station Bay sites, and it originates from Mayor Island, Great Barrier and Whitianga (Davidson 1972:7).

Nephrite continued to be used throughout New Zealand late in prehistory with considerable quantities occurring in East Coast South Island sites such as Murdering Beach, and Houhou-Pounamou (S76/7). Finds of nephrite artefacts in late contexts are occasionally made in the northern North Island (McKinlay 1971:89) but it must be emphasised that they are rare.

This brief survey of archaeological evidence suggests that areal diversity decreased late in New Zealand prehistory. There is some evidence for information exchange in trait diffusion and evidence of exchange distributing at least a few raw materials. To that extent the ethnographic evidence suggesting there was information interchange throughout New Zealand seems to be confirmed in the archaeology of the later sites. On the other hand, there is no convincing evidence of a wave-like spread of Classic culture; rather a continuous process of change shared between groups spread over most of New Zealand, and New Zealand's prehistory, is suggested.

Ethnographic Models

The specific modes of information interchange recognised in the ethnographic evidence are worth reviewing as models for application to archaeological evidence. Group movement is a long established model in prehistory but the ethnography places it in a different light: not one of strophic wave-like replacement by newly dominant groups but rather a continuing adjustment process affecting the disadvantaged as well, and involving individual movements which are decidedly 'un-wavelike', and incoherent in direction.

Modes of information interchange other than group movements seem unsuited to overt use in building models beyond recognising that information interchange occurred. But there are some specific instances where they may be worthy of testing. Slavery seems to be a case point. The development of the institution of slavery in New Zealand could be linked with the institutionalising of warfare. The greater spread of status within small groups may have lead to greater status separation of tasks. This is a linkage of some power in that it predicts the separation of food preparation from other industrial activities, a separation which is not characteristic in the Archaic. The ethnographic model of interior groups having coastal fishing rights finds application in the Waikato archaeological evidence and is of potential explanatory value for these areas, in suggesting specific modes of information interchange, for testing against other modes.

A Wider View

This subject leads repeatedly to comparisons outside New Zealand. It appears that intergroup contacts in prehistoric Australia, which ethnographically were perhaps less frequent than those of the Maori, have been sufficient to maintain a surprising degree of cultural (but not linguistic) uniformity over the entire continent (Mulvaney 1976). A contrast is found in Hutterer's discussion of Southeast Asia (1976), where he stresses the prehistoric and historic persistence of marked cultural differences between neighbouring groups despite their frequent contact in trading situations being a matter of historic record.

In the New Guinea Highlands quite marked diversity has persisted although there is, as well, a number of common elements which cannot have great time depth. Well-known movements of goods have occurred in the Highlands; moving salt, feathers, rock materials and sea shells, but these are often in contexts which minimise intergroup contact.

Elsewhere in Melanesia, there are well-known trading linkages and many of these social contacts extended well beyond barter, between groups covering the entire spectrum of the characteristic diversity of the area. This near-contradiction is a focus of current research (Terrell 1976). Contrasting this region with New Zealand, by applying Terrell's Solomon Island area-language relationship (1976:8) to half the land area of New Zealand's North Island (approximating to the area suitable for use by a horticultural people), results in a predicted 116 languages. Clearly evidence of some contact between neighbouring groups cannot generally be held to be sufficient to explain homogeneity, especially if the contact cannot be demonstrated to extend beyond bartering of goods.

Turning to island Oceania, Pawley and Green suggest that communities separated by more than 450 km of ocean, and which initially have the same language, will not be able to resist differentiation. (Pawley and Green 1973: 40). In Polynesia, linguistic diversification has been slight, with Samoa and Tonga each being outstanding examples. Yet these and others had, at least late in prehistory, rank systems which, at the top, gave authority over areas greater than single islands. In contrast, Fiji did not maintain a single language, but split into two broad groups (Pawley and Green 1973:46), but only after "all the Fijian languages underwent a long period of more or less unified development " (op. cit.). Even after this split a good deal of cultural unity was still maintained throughout Fiji.

The sorts of interchange which maintain cultural homogeneity in primitive societies are not obvious from a superficial review. It might seem that there is some vital difference between, on the one hand, Southeast Asia and Melanesia, and on the other, Australia and Polynesia. Until this is better understood, it would be as well to treat many of the sorts of interchange listed here for New Zealand as necessary to limit diversity but not a sufficient explanation of a lack of diversity.

The view of the Maori advanced here is that they were a stateless nation. Cook expected kings and was puzzled not to find them, but it is clear that the organisational infrastructure necessary to maintain the authority of a state was quite foreign to Maori Society. Observations of the King Movement in the 19th century underline this (Gorst 1864:245-285). What was the future direction of Maori Society when it was interrupted in the early 19th century? Movement towards a state or larger political units could only have been slow and development of stronger regional differences with eventual appearance of strong linguistic diversity seems more likely, using Fiji as a model. One can only suppose that the cultural momentum of Maori society would be weakened by such a development. But perhaps the way was open for the apolitical cultural unity to continue. If it was, it would seem that the second millennium of New Zealand prehistory might have been of unusual interest.

BIBLIOGRAPHY

Ambrose, W., 1970. Archaeology and rock drawings from the Waitaki Gorge, central South Island. Rec. Cant. Mus., 8:383-437.

Armitage, G. C., Reeves, R. D. and Bellwood, P., 1972. Source identification of archaeological obsidians in New Zealand. N.Z. Journal of Science 15:408-420.

Banks, J., (Ed.), 1962. Beaglehole, The Endeavour Journal of Joseph Banks. 2 Vols. Angus and Robertson, Sydney.

Bellwood, P., 1971. Fortifications and economy in prehistoric New Zealand. Proc. Prehist. Soc., 37:56-95.

Bellwood, P., 1972. Excavations at Otakanini pa, South Kaipara Head. J. Roy. Soc. N.Z., 2:259-291.

Bellwood, P. 1973. Radiocarbon dates for Otakanini. N.Z. Arch. Assn. Newsletter. 16:173-174.

Best, E., 1925. Tuhoe. Polynesian Society, Wellington.

Best, S., 1975. Adzes, Rocks and Men. Unpublished M.A. research essay. University of Auckland.

Best, S., 1976. Hard rock and the classic adze. N.Z. Arch. Assn. Newsletter 19:66-70.

Best, S. and Merchant, R. J., 1975. Siliceous sinter and the early Maori. N.Z. Arch. Assn. Newsletter. 19:106-109.

Biggs, B., 1960. Maori Marriage. Polynesian Society, Wellington.

Birks, L. and Birks, H., 1970. Radiocarbon dates for a pa site at Sarah's Gully, Coromandel Peninsula. N.Z. Arch. Assn. Newsletter. 13:63.

Birks, L. and Birks, H., 1973. Additional dates for Sarah's Gully pa site. N.Z. Arch. Assn. Newsletter. 16:73.

Cook, J., (Ed.), Reed, A. H. and A. W., 1951. Captain Cook in New Zealand. Reeds, Wellington.

Copsey, J., 1974. Surface collection of Maori adzes from a farm on Ihumatao Rd, Mangere. In: A. Sullivan(Ed.), Course Projects in N.Z. Prehistory, 1974. Working Papers in Anthropology, No. 33,University of Auckland.

Crosby, E., 1977. Wheritoa; a post-settlement dune midden on the Coromandel Peninsula. Oceanic Prehistory Records, No. 2. University of Auckland Archaeological Society.

Cruise, R. A., 1957. Journal of Ten Months Residence in New Zealand. Pegasus, Christchurch.

Davidson, J. M., 1972. Archaeological investigations on Motutapu Island, New Zealand. Rec. Auckland Inst. Mus., 9:1-14.

Davidson, J. M., 1974. Further identifications of sources of obsidian flakes from N38/37 on Motutapu Island, New Zealand. Rec. Auckland Inst. Mus., 11:11-12.

Davidson, J. M., 1975. The excavation of Skipper's Ridge (N40/7), Opito, Coromandel Peninsula, in 1959 and 1960. Rec. Auckland Inst. Mus., 12:1-42.

Duff, R., 1956. The Moa-Hunter Period of Maori Culture. Government Printer, Wellington.

Duperrey, L. I. (Ed.), Sharp, A., 1971. Duperrey's Visit to New Zealand in 1824. Turnbull Library, Wellington.

Earle, A., (Ed.), McCormick, E. H. 1966. Narrative of a Residence in New Zealand. Oxford.

Fox, A., 1974. Tiromoana pa, Te Awanga, Hawkes Bay. Interim report. N.Z. Arch. Assn. Newsletter, 17:163-170.

Fox, A., 1976. Tiromoana pa, N135/1, Te Awanga, Hawkes Bay, second interim report 1975. N.Z. Arch. Assn. Newsletter, 19:27-29.

Golson, J., 1959. Culture change in prehistoric New Zealand. in J. D. Freeman and W. R. Geddes, (Eds.), Anthropology in the South Seas. Avery, New Plymouth.

Gorst, J. E., 1864. The Maori King. MacMillan.

Green, R. C., 1964. Sources ages and exploitation of New Zealand obsidian. N.Z. Arch. Assn. Newsletter, 7:134-143.

Green, R. C., 1971. Moa-hunters, agriculture and changing analogies in New Zealand prehistory. N.Z. Arch. Assn. Newsletter, 15:16-39.

Groube, L. M., 1969. From archaic to classic Maori. Auckland Student Geographer, 6:1-11.

Groube, L. M., 1970. The origin and development of earthwork fortification in the Pacific. In: R. C. Green, and M. Kelly, (Eds.), Studies in Oceanic Culture History, Vol. 1, Pacific Anthropological Records, 11: 133-164.

Hamilton, A., 1899. Maori Art. Part 4, New Zealand Institute, Wellington.

Hjarnø, J., 1967. Maori fish-hooks in southern New Zealand. Rec. Otago Mus., Anthropology No. 3.

Houghton, P., 1975. The people of Wairau Bar. Rec. Canterbury Mus. 9:231-246.

Hutterer, K. L., 1976. An evolutionary approach to Southeast Asian Cultural sequence. Current Anthropology, 17(2):221-227.

Kennedy, J., 1969. Settlement in the Bay of Islands, 1772. Studies in Prehistoric Anthropology, No. 3. Otago University.

Keyes, I. W., 1975. The D'Urville Island-Nelson metasomatised rocks and their significance in New Zealand prehistory. Historical Review, 12: 1-17.

Law, R. G., 1973. Tokoroa moa-hunter site, N75/1. N.Z. Arch. Assn. Newsletter 16:150-164.

Law, R. G., 1975. C_{14} dates from Harataonga Bay, Great Barrier Island. N.Z. Arch. Assn. Newsletter, 18:48-52.

Leach, B. F., 1976. Prehistoric communities in Palliser Bay, New Zealand. Unpublished Ph.D. thesis, Otago University.

Leach, B. F., 1977. A rapid method of sourcing New Zealand lithic material using a low power XRF analyser. Oceanic Prehistory Records. No. 3. University of Auckland Archaeological Society.

Leahy, A., 1974. Excavations at Hot Water Beach (N44/69), Coromandel Peninsula. Rec. Auckland Inst. Mus., 11:23-76.

Leahy, A., 1976. Whakamoenga Cave, Taupo, N94/7. Rec. Auckland Inst. Mus., 13:29-75.

Lockerbie, L., 1940. Excavations at King's Rock, Otago, with a discussion of the fish-hook barb as an ancient feature of Polynesian culture. J. Polynes. Soc., 49:393-446.

Lockerbie, L., 1959. From moa-hunter to classic Maori in southern New Zealand. In: J. D. Freeman and W. R. Geddes (Eds.), Anthropology in the South Seas. Avery, New Plymouth.

McKay, A., 1873. A Compendium of Official Documents Relative to Native Affairs in the South Island. 2 Vols. Government Printer, Wellington.

McKinlay, J., 1971. Waioneke, 1968-69. N.Z. Arch. Assn. Newsletter, 14:86-91.

McNab, R., 1914. Historical Records of New Zealand. 2 Vols. Wellington.

Mair, G. M., 1972. The Protohistoric Period of Wairarapa Culture History. Unpublished M.A. Th sis, University of Otago.

Manning, G. E., 1875. Extract from a letter read before Wellington Philosophical Society 1875. Trans. N.Z. Inst., 8:102-103.

Millar, D. G. L., 1971. Excavation of an archaic site at Tahunanui, S20/2 Nelson. N.Z. Arch. Assn. Newsletter, 14:161-172.

Milligan, R. R. D., 1964. The Map Drawn by the Chief Tuki-Tahua in 1793. Privately published, Mangonui.

Moore, P., 1975. Preliminary investigation on the Tahanga basalt, Coromandel Peninsula. N.Z. Arch. Assn. Newsletter, 18:32-36.

Moore, P., 1976. The Tahanga basalt: an important stone resource in North Island prehistory. Rec. Auckland Inst. Mus., 13:77-93.

Mulvaney, D. J., 1976. The chain of connection. In: N. Peterson, (Ed.), Tribes and Boundaries in Australia. Australian Institute of Aboriginal Studies, Social Anthropology Series No. 10, Canberra.

Nicholas, J. L., 1817. Narrative of a Voyage to New Zealand. 2 Vols. James Black and Son, London.

Orchiston, D. W., 1975. Maori material culture change in early prehistoric New Zealand: the greenstone trade at Queen Charlotte Sound. The Artefact, 39:40-77.

Orchiston, D. W., 1976. Petrological studies in South Island New Zealand prehistory - 1. Maori use of Gawler Downs rhyolitic tuff. J. Roy. Soc. N.Z., 6:213-219.

Pawley, A. and Green, R. C., 1973. Dating the dispersal of the oceanic languages. Oceanic Linguistics. 12:1-67.

Peters, K., 1971. Excavations at Lake Mangakaware Site 1, N65/28. N.Z. Arch. Assn. Newsletter. 14:127-140.

Peters, K., 1975. Agricultural gardens on Moturua Island in the Bay of Islands. N.Z. Arch. Assn. Newsletter, 18:171-180.

Polach, J. S., 1838. New Zealand: Being a Narrative of Travels and Adventures During a Residence in That Country Between the Years 1831 and 1837. 2 Vols. Richard Bently, London.

Prickett, K. E., 1975. Prehistoric Exploitation and Knowledge of Geological Resources in the Lower Wairarapa. Unpublished M.A. thesis, University of Otago.

Roberton, J. B. W., 1965. Maori Settlement of the Waikato District. Te Awamutu Historical Society Bulletin No. 2.

Roe, N., 1969. An Archaeological Assemblage from Houhora, Northland. Unpublished M.A. thesis, University of Auckland.

Sahlins, M., 1974. Stone Age Economics. Tavistock Publications.

Scarlett, R. J., 1974. Moa and man in New Zealand. Notornis, 21:1-12.

Scott, S. D., 1970. Excavations at the "Sunde site", N38/24, Motutapu Island, New Zealand. Rec. Auckland Inst. Mus., 7:13-30.

Shawcross, W., 1963. Kauri Point Swamp. N.Z. Arch. Assn. Newsletter, 6:50-56.

Shawcross, W., 1964. An archaeological assemblage of Maori combs. J. Polynes. Soc. 73:382-398.

Shawcross, W., 1968. The Ngaroto site. N.Z. Arch. Assn. Newsletter, 11:2-29.

Simmons, D., 1967. Little Papanui and Otago prehistory. Rec. Otago Mus. Anthropology, No. 4.

Simmons, D., 1969. Economic change in New Zealand prehistory. J. Polynes. Soc., 78:3-34.

Simmons, D., 1971. Regional traditions and culture history. N.Z. Arch. Assn. Newsletter, 14:92-97.

Simmons, D. and Wright, J. B., 1967. Use of the polarising microscope for classifying quartzite artefacts from South Island sites. Trans. Roy. Soc. N.Z. General, 2, No. 4.

Skinner, H. D., 1926. Culture areas in New Zealand. J. Polynes. Soc., 30:71.78.

Skinner, W. H., 1946. Reminiscences of a Taranaki Surveyor. Avery, New Plymouth.

Stack, J. W., n.d. The Sacking of Kaiapohia (bound with W.T.L. Travers, the stirring times of Te Rauparaha) first published 1893. Whitcombe and Tombs, Christchurch.

Sullivan, A., 1975. Radiocarbon dates from Wiri (N42/24). N.Z. Arch. Assn. Newsletter, 18:206-207.

Sullivan, A., 1976. Paper delivered to N.Z.A.A. conference, Whakatane.

Sutton, D. G., 1974. Resurrection of the Prehistoric Dead. Unpublished M.A. thesis, University of Otago.

Taylor, N. M., 1959. Early Travellers in New Zealand. Oxford.

Terrell, J., 1976. Island biogeography and man in Melanesia. Archaeology and Physical Anthropolog in Oceania 11:1-17.

Trotter, M. M., 1965. The barbed fish-hook: its place in the Murihiku culture sequence. J. Polynes. Soc. 74:347;355.

Trotter, M. M., 1970. Excavations at Shag Point, North Otago. Rec. Canterbury Mus., 8:469-485.

Trotter, M. M., 1972a. A Moa-hunter site near the mouth of the Rakaia River, South Island. Rec. Canterbury Mus., 9:129-150.

Trotter, M. M., 1972b. Investigations of the Weka Pass shelter S61/4. N.Z. Arch. Assn. Newsletter, 15:42-50.

Trotter, M. M., 1975. Further investigation at Pari Whakatau, Southern Marlborough. N.Z. Arch. Assn. Newsletter, 18:145-151.

Trotter, M. M. and McCulloch, B., 1971. Prehistoric Rock Art of New Zealand. Reeds, Wellington.

Trotter, M. M. and McCulloch, B., 1973. Radiocarbon dates for South Island rock shelters. N.Z. Arch. Ass. Newsletter, 16:176-178.

Urlich, D. U., 1972. Migrations of the North Island Maoris 1800-1840: A Systems View of Migration. N.Z. Geographer, 28:(1):23-25.

Vayda, A. P., 1960. Maori Warfare. Polynesian Society, Wellington.

Walls, J., 1974. Argillite quarries of the Nelson mineral belt. N.Z. Arch. Assn. Newsletter, 17:37-43.

Ward, G. K., 1973. Obsidian source localities in the North Island of New Zealand. N.Z. Arch. As'sn. Newsletter, 16:85-103.

Ward, G. K., 1974. A paradigm for sourcing New Zealand archaeological obsidians. J. Roy. Soc. N.Z., 4:47;62.

Webster, S., 1975. Cognatic descent groups and the contemporary Maori: a preliminary reassessment. J. Polynes. Soc., 84:121-152.

Wellman, H. W., 1962. Holocene of the North Island of New Zealand: a coastal reconnaissance. Trans. Roy. Soc. N.Z., General 1, No. 5.

Wilkes, O. R. and Scarlett, R. J., 1967. Excavation of a Moa-hunter site at the mouth of the Heaphy River. Rec. Canterbury Mus. 8:177-208.

Yate, W., 1835. An Account of New Zealand. Seeley and Burnside, London.

MAORI GODSTICKS AND THEIR STYLISTIC AFFINITIES

Peter Gathercole

Maori godsticks have been discussed extensively in the ethnographic literature (Barrow 1959a, 1961, with earlier references), and it is not my intention to recapitulate that discussion here. This paper, offered to Ron Scarlett as a token of a friendship extending over more than twenty years, might perhaps be seen an unfashionable excursion into antiquarianism. But it attempts to be something more, namely a study of structure and form in Maori art, which endeavours to relate the morphology and style of the godstick to the wider panoply of Classic Maori carving. The first part of the discussion is historical; the second is speculative. I hope that the result is to Ron Scarlett's taste.

In his book _Te Ika a Maui_ the Rev. Richard Taylor said: "The Maori could scarcely be said to be idolators, although they certainly had idols, yet they were not generally worshipped, but only used by the priests as adjuncts to their karakias" (Taylor 1870:211). Some of these "adjuncts" were the so-called godsticks, which Taylor described as follows: "The _Wakapakoko_, or images, thus used, were little more than wooden pegs with a distorted figure of the human head carved on the top; these were about eighteen inches long, the other end was pointed so that they could be stuck in the ground". According to Barrow, who has described 27 examples (Barrow 1959a, 1961), there are two forms: one consists of a carved mask or human head, below which is a shaft, usually recessed to take a flaxen binding, and a peg base, generally pointed. The other form is similar, but with a complete human figure, not simply a head. The length is usually about 14 inches.

Godsticks have received considerable attention from ethnologists, largely, I imagine, because they are one of the few classes of artefact within the corpus of Maori carving about which there is good documentation. This comprises two long statements by Taylor and a letter from the Rev. John Aldred to Sir Walter Buller, dated 1892, enclosing a copy of a letter from the Rev. T. G. Hammond (Barrow 1959a:183-186; Barrow 1961:219-221).

Taylor discussed godsticks both in his manuscript _Journal_, now in the Turnbull Library, Wellington, in a section probably dated 1844, and in _Te Ika a Maui_ (1855, 1870). The relevant remarks in the _Journal_ are as follows:

"The natives of Wanganui had many Gods, and likewise images of them, the principal of these were Maru, Kahukura, Reua Korongomai.
"In the northern part of the Island I never met with any of these images nor am I aware that any of the Europeans have seen any prior to those which I obtained; for long before the gospel was generally embraced the sound of it

had reached these natives who learning from it the folly of worshipping senseless Idols literally cast away their Gods to the bats and moles concealing them in clefts of rocks or hollow trees, it was only when I enquired about them that they were induced to search for them and bring them to me. Their manner of worshipping these Idols was singular if not unique the head of God alone was represented which was carved on the top of a short stick and ornamented with feathers of the parrot, this was called the Pahau or beard the rest of the stick was bandaged round, when worshipped this image was stuck into the ground with a string attached to it which the worshipper held in his hand, who sitting down and leaning by a stone pillar (called a ngahu or praying stone about 2 feet high) gave the God a little jerk to arrest his attention and then repeated a prayer which being done he stuck a little stick in the ground near the image and then giving another jerk and uttering another prayer he placed another stick by the side of the former to remind the God how many prayers had been said, this was done as often as a prayer was uttered" (Taylor n.d. (1):542-543).

In Te Ika a Maui, Taylor's comments contain more detail on the ritual involved:
"The Wakapakoko, or images, thus used, were little more than wooden pegs with a distorted figure of the human head carved on the top; these were about eighteen inches long, the other end was pointed so that they could be stuck in the ground. In respect to idolatry, the Maori differed from almost every other heathen people, in not having images of their gods of gigantic size,... they seemed to have a more spiritual idea of their gods; the only exception, a large human figure over the chief entrance of the Ohine Mutu pa at Rotorua, the door being between the legs of the figure, but this was not an idol, merely a figure to strike terror in the enemy. These images were only thought to possess virtue or peculiar sanctity from the presence of the god they represented when dressed up for worship; at other times they were regarded only as bits of ordinary wood.

This dressing consisted in the first place of the pahau, or beard, which was made by a fringe of the bright red feathers of the kaka, parrot,—next of the peculiar cincture of sacred cord with which it was bound; this mystic bandage was not only tied on in a peculiar way by the priest, who uttered his most powerful spells all the time he was doing it, but also whilst he was twisting the cord itself, and lastly, painting the entire figure with the sacred kura; this completed the preparation for the reception of the god who was by these means constrained to come and take up his abode in it when invoked. But the presence of the atua was not supposed to be confined to these images, he more frequently took up his abode in the priest himself, who, when thus filled with his spirit, was so violently distorted in all his limbs, so frantic in his movements, foaming at his mouth as to appear filled with the god, and no longer the master of his own actions, his body was then entirely surrendered to him; every word he uttered whilst in that state was regarded as proceeding from the god, and not from himself; This was thought to be the case after uttering certain karakias, at other times he was only an ordinary person

When the chief wanted to carry any measure which he felt he had not sufficient influence to do by his own mana, or power, he usually called in the aid of the tohunga, who not only managed to be inspired when thus wanted, but to utter just what agreed with his chief's wish, which coming direct from the atua at once convinced the people.

To return to the subject of idols. The chief use to which they were devoted was to render sacred their plantations; the image was stuck in the ground and powerful spells uttered to make the men scared, that none of those employed should leave the work before it was finished, pass over, or even enter them, except for their cultivation. This was done at the tokanga, or planting; the ceremony was again repeated when the kumara began to grow, and again at the haukakenga, or in-gatering" (Taylor 1870:211-213).

In his book Taylor published two of his own drawings, one illustrating the ritual of a tohunga addressing a decorated godstick, to which he is attached by a cord, and another of four godsticks in a basket (cf. Barrow 1959a:185, 187). Two of the artefacts illustrated are in the University Museum of Archaeology and Anthropology, Cambridge, one is in the Pitt Rivers Museum, Oxford; and one is in the Oldman Collection in the National Museum, Wellington.

Taylor's remarks indicate that, prior to their conversion to Christianity, tohungas in the Wanganui area engaged in ritual whereby they became possessed by atua, which also possessed carved and appropriately decorated sticks. This transference from a profane to a sacred state would only occur if both tohunga and stick undertook a precise ritual, the purpose of which could be to assist the chief "to carry any measure which he felt he had not sufficient influence to do by his own mana, or power" and "to render sacred their plantations." The gods which could be invoked were not apparently limited in number or interest. The Wanganui Maoris "had many Gods, and likewise images of them, the principal of these were Maru and Kahukura, and Reua Korongomai." According to Buck (1950:461) Maru and Kahukura were important war gods of the Taranaki and Wanganui tribes, and Rongomai was a tribal god of the Ngati Tuwharetoa, his symbols being shooting stars and comets. The Pitt Rivers' specimen was identified by Taylor as being Rongomai (Barrow 1961:228), the Cambridge ones as Hukere (according to Buck "probably a local god in the Whanganui district" (Buck 1950:469)) and Kahukura (Barrow 1959a:188-9), while the Oldman specimen in Wellington has a label which says "mokotiti a New Zealand God from NZ" (Barrow 1959a:192). Other specimens attributed by Barrow to Taylor were apparently associated with Maru (Barrow 1959a:189,190,193).

Taylor's evidence is very convincing. He went to New Zealand as a missionary in 1839, aged 34, with prior experience at Parramatta, a Cambridge graduate with a strong interest in botany and geology. For a time he was at Waimate North, but in 1843 he was appointed to the Wanganui mission. He stayed in Wanganui for most of the remainder of his life, dying there in 1873. He made two visits to England; in 1855, when he saw the first edition of Te Ika a Maui through the press, and between 1867 and 1870, when he published The Past and Present of New Zealand. He was thus a notable observer and scholar. His Journal entry for 1844 can be regarded as important field evidence concerning the use of godsticks, and the account in Te Ika a Maui as a more considered and literary version of that evidence. The fact that he sought for and obtained some actual specimens from converts who had previously hidden them "in the clefts of rocks or hollow trees" indicates that Taylor had a strong interest in the sticks themselves, in the associated ritual, and its paraphernalia.

The information provided by Taylor is confirmed and to some extent amplified by information given by the Rev. John Aldred in a letter to Sir Walter Buller in 1892 (Barrow 1961:219-222). Aldred (1818-1894) was a Wesleyan missionary who possessed three godsticks, now in the Auckland Museum. Aldred said that these "came into my hands ... between the years 1845 and 1847. I obtained them through Mr. William Hough, who was at that time the Wesleyan catechist stationed at or near Waimate, ... between Wanganui and New Plymouth. The principal tribes then living in that district were the Ngatiawa and the Ngatiruanui, and the gods now in question probably belonged to the former. The information I received regarding them was given to me personally by Mr. Hough, and was confirmed by certain of the older Maoris from the same district who were known to me, and whose trustworthiness and knowledge of the mythology of the tribes named gave weight to their statements. The images were parted with by the Maoris when they embraced Christianity, and were said to be of great age... Among the Maoris the sole custody of the gods lay with the tohunga (priest or diviner), and they were seen by eyes other than his upon special occasions only .... When invoking a particular spirit the tohunga would hold the image of that god in his hand while he uttered his incantations" (Barrow 1961:219-220).

Aldred had forgotten the names of two of the three gods concerned, and enlisted the help of the Rev. T. G. Hammond (1846-1926), another Wesleyan missionary then at Patea, who was "an acknowledged authority" on Maori traditions. Hammond identified the three gods as Tangaroa (or Turanga), Maru, "the most important atua of these people" (some of whose images still survived at the time of writing), and Rongo, "as the descendants of the Turi always offered incantations to Rongo in order to secure good crops" (Barrow 1961:221). These identifications amplified Aldred's own recollections. The latter included the remark that "Te Maru is short and thick set, well corded, and coloured red. The cord that entwines his body was tapu (sacred), and within it was said to lie the secret of this deity's mighty power" (Barrow 1961:220).

It is, of course, probable that both Aldred and Hammond had read Taylor's Te Ika a Maui, and may have been influenced by his remarks on the sticks. But it is clear from the letters concerned that each was casting his mind back to his own earlier experiences: Aldred to 1845-1847, and Hammond to about twenty years later. They added new information. Aldred said the sticks were held in the hand during a tohunga's ritual, not stuck in the ground as Taylor reported. Hammond named another god, Tangaroa, in addition to those referred to by Taylor. We thus have evidence that godsticks were associated with at least six gods, at two or even three levels of importance in the Maori pantheon. Tangaroa, god of seafarers and fishermen, and Rongo, god of agriculture, were two of the six main "departmental" gods, to use Best's classificatory term. Maru, Kahukura, Rongomai and Hukere were recorded as tribal gods in the Wanganui—Taranaki areas[1]. Moreover, as Taylor said that a tohunga could be used by a chief to "utter just what agreed with his chief's wish" (Taylor 1870:212), it appears very likely that a number and variety of gods could be invoked in the operation of the ritual on different occasions. To judge from Taylor's rather cynical comment, it is possible that rituals were occasionally used to invoke "family and inferior gods", as Buck termed them (1950:462-464), sometimes, as he noted, by charlatans. Be that as it may, the godstick ritual was not only very systematic and controlled but also extremely flexible. Some sticks were clearly associated with particular gods, but no stylistic design was peculiar to any one of them. Indeed Barrow has maintained that at least 15 of his 27 sticks, whatever their god associations, were made in the Wanganui carving style. However ambiguous might have been the relationship between tohunga and atua, the material expression in carving style appears to have been a matter more of social tradition and aesthetics rather than specific ritual affiliation.

This leads me on to three other aspects of this body of evidence to which I wish to refer. Firstly, the styles represented are quite varied. Aside from the Wanganui specimens, one is in the Taranaki style; one in the East Coast—Bay of Plenty; at least two are indeterminate by the canons of Classic Maori carving, and five have features reminiscent of carvings, such as the Kaitaia lintel, which appear to pre-date Classic Maori art (Barrow 1959a:188-189, 191-192; 1961:222-226; cf. Mead 1975). Some of these five specimens may be in a style contemporary with the Archaic phase of New Zealand prehistory. One, from Kai-iwi, Wanganui (Barrow 1961:222-223), has a mask stylistically reminiscent of a recently discovered fragment from the western Bay of Plenty, which Simmons considers to be the head of a godstick, carved in a style similar to that known ethnologically in the Cook Islands (Simmons 1973). Two others, from Moa Flat and Wickliffe Bay, Otago, respectively (Barrow 1961:223-226), have features which invite comparison with ethnological material from Easter Island. The specimen from Wickliffe Bay was found in an undated archaeological deposit. It is similar in form to a possible godstick found by Dr. Peter Coutts in 1968 in a cave at Southport, Chalky Inlet, Fiordland (SP/5) in a prehistoric context (Coutts 1972)[2]. In other words, although the evidence is sparse and rather ambiguous, there is no reason to assume that godsticks were either confined to the Wanganui area or restricted in their use to the late prehistoric or protohistoric periods. Indeed, the existence of similar stick gods in Hawaii (Barrow 1961:217-218, with refs.) and the fact that the pattern of the flaxen binding on some Maori specimens is similar to that found "on the shaft of a god symbol in Mangaia where it is termed inaere and is highly valued" (Buck 1950:469), goes some way to suggest that godsticks, sui generis, had considerable antiquity in eastern Polynesia.

Secondly, four of the sticks discussed by Barrow, i.e. specimens from Wanganui, Taranaki, Waimate and Moa Flat, have double, so-called janus, heads, facing opposite ways (Barrow 1959a:191-192; 1961:218, 223-224). Another Wanganui-style specimen now in Ipswich Museum has a rudimentary janus design not on its head, but on its peg base, below the median recess (Barrow 1961:216-217). Of these janus examples, the Waimate specimen is the only one to have an association with a particular god (Tangaroa), and it is one of six godsticks in Barrow's corpus to have their bindings intact. The janus form was a well-established constituent of Maori carving. It is known, though rarely, on tiki (e.g. a specimen (D.36.901)in Otago Museum, Dunedin), and more commonly on taiaha and shell trumpets. The form is echoed in the way identical heads are sometimes found at the opposite ends of feather boxes, serving as handles. In eastern Polynesia the form is found, for example, in the Austral Islands on the handles of fans and fly-whisks, and in the Marquesas Islands on stone pounders, and on stone and bone tiki. Here also, therefore, is an indication that the Classic Maori godstick, in this case with janus head, embodied characteristics of some antiquity in eastern Polynesia.

The third point I wish to make refers to some of Skinner's remarks concerning the Cambridge specimens acquired by Taylor (Skinner 1922). More or less in passing he pointed out that the local god Maru was "presumably a deified ancestor, as it is recorded of him (Taylor 1870:138) that he was killed and eaten when on earth" (Skinner 1922:169). In other words, godsticks associated with Maru, at least, may well have occupied a similar position in the minds of Maoris as did the carvings of ancestors which graced so much of Maori architecture. Structurally, the Maru godsticks might have provided a bridge between the worlds of gods and of men. The dichotomy may have been partial rather than absolute.

Skinner went on to stress a point noted but not given prominence by other commentators. While remarking that sets of godsticks "are by no means common", he noted that "they appear usually to go in threes" and he listed four such sets: the three Aldred specimens in the Auckland Museum, the three from Taylor at Cambridge, three in the British Museum, and three in a private collection. In two of the four cases quoted, one stick of the three has a janus head (Skinner 1922:172). It seems to me very likely that the "ideal" set of godsticks comprised three specimens, one of which was a janus. It is worth asking why this arrangement should be so, but in order to do so, the question must be set in a wider frame of reference.

Sufficient has been said about the function and likely antiquity of godsticks for it to be accepted that the objects had conventionalised meanings as physical expressions of persistent patterns of ideas, even if their specific designs changed over time. These changes would have been particularly sensitive, I believe, in the treatment of the three major components of Maori carving, namely, the representation of the human figure (especially the head), the manaia motif, and associated curvilinear patterns. To put the point in another way, I wish to explore the question why certain godsticks, e.g. those examples from western Bay of Plenty (Simmons 1973) and Otago (Barrow 1961:223-226) have a simple (Archaic phase ?) decorative patterning, which others, notably in the Wanganui style, conform to designs we call Classic Maori. Logically, however flimsy our evidence to document it, a change of style must have occurred. This change must also have reflected a larger process of social transformation. The very hegemony of Classic Maori carving revealed to Cook and his colleagues in the 1770s, demonstrating a close interrelationship of material and esoteric life, indicates that this was so.

In his paper "Aspects of Symbolism and Composition in Maori Art", Jackson devoted much attention to the symbolism of lintel (pare) carving

(Jackson 1972:40-60). Lintels, of course, bring together human figures, the manaia, and curvilinear designs, as carved motifs which have to be considered in an interrelated way. But, as Jackson showed, the significance of these three elements did not rest there. It permeated the whole composition of the lintel at numerous levels. Jackson demonstrated that a trinary principle operated in the composition of the lintel. For example: "At this point it is possible to account for the 3-fingered hand in Maori art on the basis of a logical consistency which it shows with other aspects of composition. Sufficient has been said to make it clear that the number three is central to the principles of composition underlying pare design. I have noted how the three elements are brought into unity through powerful and rhythmic deployment. The three fingers unified in the single upraised arm might be considered to be consistent with the overall concept of the design. Its meaning is to be found in the total positional sense, not in isolation. Taken as an isolated referential symbol it would defy attempts to elucidate it " (Jackson 1972:48). Later in the paper, when discussing "The pare and Maori social life", Jackson argued that the lintel was an example of condensation symbolism for the dynamic structure of Maori society. "The process described here is logically and symbolically connected with the processes of fission and fusion by which lineage segments in Maori society (or even individuals) merged together to become a single identity—the tribe—or fragmented to live and act as partial units. The pare compositions permit both a statement of independence for the half-groups (as it were) and a statement of interdependence concerning the way in which these half-groups combined with one another to make up a whole, a wider unity, a tribe. The necessities and actualities of tribal segmentation and fragmentation are admitted. But the principles underlying this fission and fusion of tribal elements or individuals are condensed into a single powerful symbolic composition in the pare" (Jackson 1972:57).

This argument provides a lead for a deeper understanding of the meaning of godsticks and their relation to the development of Maori carving. Jackson's trinary principle can be put alongside Skinner's comment that the sticks "appear usually to go in threes." Moreover, if one accepts the notion of the "ideal" set of sticks comprising two sticks with single heads and one with a janus head, then the trinary symbolism is taken further. One can suggest that the three sticks correspond to the structual arrangement of figures on the lintel, with a central figure and two manaia at the margin. But one could go further and suggest that the janus headed godstick is itself structurally related to the lintel in that it incapsulates the lintel design. If the design on the janus is unrolled so that it is seen not in the round, but in one plane, the lintel design appears. One of the human heads is now in the centre of the design, flanked by curvilinear patterns, with two emergent manaia, one at each side, which are "created" out of the single human head from the opposite side of the janus, this head having been split down its centre. To carry out this unrolling extends the argument made by Archey (e.g. 1933) and others, on morphological grounds, that the manaia is the representation not of a bird, but of a human head divided into two parts.

This argument suggests that due to the basic trinary structure of the symbolism of Maori carving (and so of Maori life), godsticks could have been organised in trinary sets. Such an arrangement conformed to an ideological "ideal". But why was the design of the stick itself transformed from its Archaic-like form to a Classic Maori form, to revert to a question I posed earlier? I suggest that it might be possible to answer this by considering the form and possible origin of the manaia design, a design, incidentally, which has always appeared to be the least Polynesian of motifs, and has caused most debate as to the origins of the Classic Maori style. What I shall suggest is a paradigm of artistic transition, in which a carving tradition, epitomised by "earlier" godsticks, is transformed into something recognisably similar in form and function, but is more flexible in design and is geared to new social needs and aspirations (just as the Kaitaia ridge cresting is to later Classic Maori lintels).

Firstly, what does the word manaia mean? Mc Ewen has said: "Williams' Dictionary of the Maori Language gives the following meanings for manaia: a grotesque beaked figure sometimes introduced into carving; ornamental work, a lizard; the sea-horse; a raft; and as an adjective fastidious. It is interesting that in Samoa the word (with the causative prefix) fa'amanaia means to decorate or embellish. In Niue the cognate word fakamanaia means the same. As the main use of the manaia is to embellish the principal figures, it seems very likely that the name simply means "embellishment" or decoration" (McEwen 1966:410a).

These meanings, with the exception of "raft", are consistent with one another. "Raft" might be regarded either as an aberrant meaning or one which is linked by some sort of metaphorical association with the others. Information available about Maori rafts indicates that they were flimsy, expendable things, usually made of reeds bound into a symmetrical shape with an identical bow and stern, the shape made inevitable by the method of production (Best 1925:136-140). So one of the meanings of manaia might be "like a raft" or "something symmetrical".

The manaia design, however, is not itself symmetrical. Therefore, rather than looking for symmetry in the manaia itself, we might look for examples of carving where the manaia is used specifically to bring symmetry to a larger design. The clearest example where this occurs is its use at each end of the lintel, balancing in its overall design the central human figure and associated curvilinear patterns. Lintels were much esteemed objects, and exhibit such control of workmanship in the balance achieved between these three elements in their design, that they probably reflect a well-established carving tradition. One of the functions of the manaia, therefore, could have been to complete a design on an artefact which had an important meaning; for example, in the case of the lintel, to neutralise the possibly harmful effects of the tapu of persons who passed under it. So the lintel and its attendant design must have possessed its own sacred quality. But why was this particular design, incorporating the manaia, the one that was conventionally employed? Did it also have its own sanctity, independent of its associations with the lintel?

Let us go back to the Maori raft for a moment. Like the lintel, the raft is symmetrical in shape. In addition, however, because its terminals are identical, it looks the same from either end, and it is symmetrical from any viewpoint. This is not true of the lintel, which must be viewed only in one plane. If manaia can mean "raft", "like a raft" or something symmetrical", it may also mean "something that looks the same from each end"—which a lintel does not. One might suggest, however, that the sacred quality of the lintel design is due, in part at least, to the fact that it is also associated with the design of objects which are themselves sacred, symmetrical, and look the same from each end, for example janus-headed godsticks, the sacred quality of which is so well documented. This form of godstick has two of the design elements found on the lintel: the human figure and curvilinear ornament. Moreover, when the stick is viewed from the side, the head is remarkably reminiscent of the manaia, and may, in fact, be its prototype. As we have already seen, when unrolled, the janus is transformed into the lintel. In its new position, therefore, the manaia is indeed like a raft. On the one hand, it can be incorporated in the representation of a janus head, which "looks the same from each end." On the other hand, it can make the design on a lintel "become symmetrical." It is created as an independent element when an all-round design is re-expressed in one plane.

This hypothesis may reflect nothing more than a jigsaw game with shapes which, because they were part of the stylistic formulae of Classic Maori carving, inevitably possessed numerous technical and symbolic interrelationships in common. However, there may be more to the hypothesis than this. It seems likely that Maori carving incorporated visual games[3]. Generally speaking, Maori designs were not just formal alignments of discrete parts. The parts were dynamically related, expressing relationships between the human world and the spirit world of gods and ancestors (Mead 1975:177). The division of a phenomenon into two opposing yet complementary segments, as I have suggested for the genesis of the manaia, was a well-known formula in Maori epistemology. We have seen a good example in the sacred/profane complementary dichotomy found in godstick ritual. The principle of artistic transformation, where a design was transferred from one medium or locale to another, was also known. A dramatic example was the practice by non-literate Maoris in the early 19th Century to sign a document by drawing on it their own moko, quite unprompted. It can be suggested, therefore, that the transformation of sacred godstick designs into lintel designs can be seen as an analogue of, if not the occasion for, the emergence of the manaia. One ideal aesthetic expression was transformed into another, by expressing the same stylistic formula in a new way. Innovation was a novel statement of tradition.

Can we suggest when this transformation process might have occurred? Mead has noted that work by Roberton (1965) on tribal traditions, and by Groube (1970) and Simmons (1971) on the relationship between tradition and archaeology, provides a "growing body of evidence to strengthen the hypothesis that the great population movements associated with the Hekenga or Great Migration were really internal movements like those of the Awa people" (Mead 1975:182). In other words, the Hawaiki of tradition lay not in eastern Polynesia, as Smith and Best had thought, but in North Auckland, and the migration took place southwards to the Bay of Plenty, Taranaki and Hawkes Bay. Mead has suggested that these movements were associated in some way with a change in ideology and social organisation, when canoe captains replaced time-honoured ancestors as eponymous reference points in the reckoning of genealogies (Mead 1975:202). If these hypotheses can indeed be substantiated, then the momentous social and ideological changes involved could have been expressed in carving. They could have provided the catalyst for the coming together of hitherto discrete design elements to form the patterns we label as distinctively Classic Maori, incorporating the transformation process suggested above. It may have been the social upheavals implicit in the Hekenga which changed Maori carving from being predominantly an art form oriented to the dialogue between man and his gods, of which the godstick rituals were an important surviving component, into one which emphasised social solidarity and kinship with the ancestors, as epitomised in the canoe traditions.

One might recall in conclusion Barrow's remarks in 1959 when introducing his discussion on free-standing Maori images. He noted that representations of ancestors "are abundant especially in relation to architecture whilst the manufactured symbols of the Maori gods are comparatively rare. This situation reflects the shift from the essentially religious symbolism of central Polynesian carving to the more secular and social function of the New Zealand forms" (Barrow 1959b:111). This discussion on godsticks and their stylistic affinities underscores Barrow's comments.

NOTES

1. The appellation "mokotiti" on the Oldman specimen in the National Museum, Wellington, could be a variant of "mokotiki", meaning "decorated head."

2. Now in Otago Museum (L75.6). The stick came from the sole cultural layer (layer 2), 60 cm thick. Coutts accepts two C14 dates for this layer: 615±61B.P. and 706±30B.P., but rejects a third of 980±80 B.P. as being too early (information from G. S. Park).

3. Mrs Jean Smith has pointed out to me in correspondence that, similarly, Maori mythology subsumed verbal games.

BIBLIOGRAPHY

Archey, G., 1933. "Evolution of Certain Maori Carving Patterns", Journal of the Polynesian Society, 42:171-190.

Barrow, T., 1959a. Maori Godsticks collected by the Rev. Richard Taylor, Dominion Museum Records in Ethnology, 1(5).

Barrow, T. T., 1959b. "Free-standing Maori Images", in Freeman, J. D. and Geddes, W. R. (eds.), Anthropology in the South Seas, pp. 111-120. New Plymouth, Avery.

Barrow, T., 1961. Maori Godsticks in Various Collections, Dominion Museum Records in Ethnology, 1(6).

Best, E., 1925. The Maori Canoe, Dominion Museum Bulletin, 7. Wellington, Government Printer.

Buck, Sir Peter, 1950. The Coming of the Maori (2nd ed.). Wellington, Whitcombe and Tombs.

Coutts, P., 1972. The Emergence of the Foveaux Strait Maori from Prehistory. Unpublished Ph.D. thesis, University of Otago, Dunedin.

Groube, L. M., 1970. "The origin and development of Earthwork Fortifications in the Pacific", in Green, R. C. and Kelly, M. (eds.), Studies in Oceanic Culture History, Vol. 1, Pacific Anthropological Records, 11, pp. 133-164. Honolulu, Bernice P. Bishop Museum.

Jackson, M., 1972. "Aspects of Symbolism and Composition in Maori Art", Bijdragen tot de Taal-, land- en Volkenkunde, 128:33-80.

McEwen, J. M., 1966. "Maori Art" in McLintock (ed.), An Encyclopaedia of New Zealand, Vol. 2, pp. 408-429. Wellington, Government Printer.

Mead, S. M., 1975. "The origins of Maori Art: Polynesian or Chinese?", Oceania, 45:173-211.

Roberton, J. B. W., 1965. "A consecutive account of the traditional history of the Whakatane District", Whakatane and District Historical Society, Memoir 4, Whakatane.

Simmons, D. R., 1971. "Regional traditions and culture history", New Zealand Archaeological Association Newsletter, 14:92-97.

Simmons, D. R., 1973. "'Godstick' Head from Katikati", Records of the Auckland Institute and Museum, 10:65-67.

Skinner, H. D., 1922. "Maori 'God-sticks' (Whakapakoko Rakau)", New Zealand Journal of Science and Technology, 5:168-172.

Taylor, R., n.d. Journal, Vol. 1. (Manuscript in the Turnbull Library, Wellington).

Taylor, R., 1870. Te Ika a Maui, (2nd ed.), London, Macintosh, and Wanganui, Jones.

www.ingramcontent.com/pod-product-compliance
Ingram Content Group UK Ltd.
Pitfield, Milton Keynes, MK11 3LW, UK
UKHW061845041026
15591UKWH00015B/922

9 780860 540625